GEOGRAPHY AND ENLIGHTENMENT

GEOGRAPHY AND

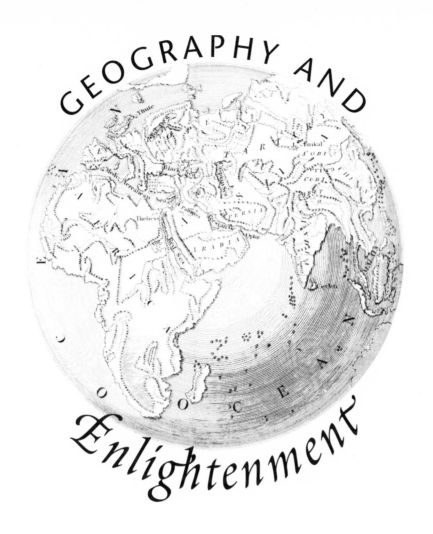

Enlightenment

EDITED BY
David N. Livingstone and Charles W. J. Withers

THE UNIVERSITY OF CHICAGO PRESS
CHICAGO AND LONDON

DAVID N. LIVINGSTONE is Professor of Geography and Intellectual History at the Queen's University of Belfast and a Fellow of the British Academy. He is the author of *Nathaniel Southgate Shaler and the Culture of American Science* (1987), *Darwin's Forgotten Defenders* (1987), *The Preadamite Theory* (1992), and *The Geographical Tradition* (1992). He is currently working on a historical geography of science. CHARLES W. J. WITHERS is Professor of Geography at the University of Edinburgh. He is the author of *Gaelic in Scotland 1698–1981* (1984), *Gaelic Scotland: The Transformation of a Culture Region* (1988), and *Urban Highlanders: Highland-Lowland Migration and Urban Gaelic Culture, 1700–1900* (1998). His recent work in the history of geography includes discussion of encyclopedism and geographical knowledge, and geography and the public sphere. He is currently writing a book on geography, science, and national identity in Scotland since 1550.

The University of Chicago Press, Chicago 60637
The University of Chicago Press, Ltd., London
© 1999 by The University of Chicago
All rights reserved. Published 1999

08 07 06 05 04 03 02 01 00 99   1 2 3 4 5
ISBN: 0-226-48720-2 (cloth)
ISBN: 0-226-48721-0 (paper)

Library of Congress Cataloging-in-Publication Data

Geography and enlightenment / edited by David N. Livingstone and
    Charles W. J. Withers.
        p.   cm.
    Includes bibliographical references and index.
    ISBN 0-226-48720-2 (alk. paper). — ISBN 0-226-48721-0 (pbk. :
alk. paper)
        1. Geography—History—18th century.   2. Enlightenment.
    I. Livingstone, David N., 1953– .   II. Withers, Charles W. J.
    G70.G4432   1999
    910'.9—dc21   98-52359
                                                              CIP

Contents

Preface and Acknowledgments

The papers collected in this volume derive from the conference "Geography and Enlightenment," held in the Department of Geography in the University of Edinburgh in July 1996, in association with the University of Edinburgh's Institute of Advanced Studies in the Humanities, and the Queen's University of Belfast. The conference examined, from interdisciplinary perspectives, the ways in which geographical knowledge was implicated in the Enlightenment. The result was a rich exchange between geographers, historians of science, historians, and social anthropologists; the papers presented here thus draw upon recent work in intellectual history generally and more particularly in the historical sociology of science and the history of geographical thought and practice. Taken together, these perspectives have underscored the situated nature of scientific knowledge, and so it is understandable that consideration is given in what follows not only to the geography *of* (the) Enlightenment, and to the ways in which geographical knowledge was practiced *in* the Enlightenment, but also to how questions of a geographical nature fitted into and were actively engaged with those matters of rationality and human nature routinely held to characterize 'the' Enlightenment.

A great many people have helped in seeing this project through to publication. The editors and contributors owe a great debt to the anonymous readers who offered detailed and insightful comments on earlier drafts and to Derek Gregory whose thorough reading and many suggestions have helped improve the work immeasurably. Individual contributors have, by and large, acknowledged the work of fellow scholars in their chapters but it is a pleasure to here record our thanks (in no particular order other than the alphabetically conventional) to Kay Anderson, Wolfgang Boeker, Marie-Noëlle Bourguet,

Dan Brown, David Cobb, Catherine Delano Smith, Robin Donkin, Michael John Gorman, Andrew Grout, Frances Herbert, Christian Jacob, Nuala Johnson, Bruno Latour, Robert Mayhew, Johannes Mohm, Mark Noll, Miles Ogborn, Hester Parr, Alessandro Scafi, Simon Schaffer, Ruediger Schreyer, David Sibley, Michael Wintroub, and Paul Wood.

For permission to reproduce maps and illustrations in their care, we would like to thank the Syndics of Cambridge University Library for the figures used in chapter 5 and the Cambridge Philosophical Society for those in chapter 7. Figures 6.1 and 6.2 are included with the permission of the Osher Map Library and of the Harvard Map Collection, and we acknowledge the latter for the illustrations in chapter 8. Figure 3.1 is reproduced here with the permission of the Royal Geographical Society (with the Institute of British Geographers), and figure 4.1 with the permission of the Queen's University of Belfast. Figures 2.1 and 2.2 are reproduced by courtesy of the British Library. Figures 13.1 and 13.2 were reproduced with the assistance of the library of the University of Glasgow.

Several academic bodies were generous in their support of the conference on which this collection is based: we gratefully acknowledge the British Academy for its contribution towards the travel expenses of Paul Carter, Matthew Edney, Anne Godlewska, Dorinda Outram, and Nicolaas Rupke; the Royal Society of Edinburgh; The Council of the City of Edinburgh; The Royal Geographical Society (with the Institute of British Geographers); the Royal Scottish Geographical Society; the University of Edinburgh Inter-Disciplinary Symposium Fund: Lothian and Edinburgh Enterprise; and, for funding which enabled postgraduates to attend, the History and Philosophy of Geography Research Group and the Historical Geography Research Group of the Royal Geographical Society (with the Institute of British Geographers). Thanks are due, too, to Gosia Sozanska, who was a splendid conference assistant.

We owe a great deal to Penny Kaiserlian of the University of Chicago Press for her support and faith in the proposal. Roy Porter graciously accepted an invitation to undertake an Afterword, and we thank him for 'closing' the collection in this way. Finally, we thank all our fellow authors for their scholarship and for their patience.

Chapter One
Introduction: on Geography and Enlightenment

CHARLES W. J. WITHERS AND DAVID N. LIVINGSTONE

1.1. Reconsidering Geography and the Enlightenment

The origins of modern geography have frequently been traced to the period of the Enlightenment. The monumental writings of figures like Alexander von Humboldt and Carl Ritter and the scientific expeditions of men like James Cook, Louis-Antoine de Bougainville, and Joseph Banks were as diagnostic of modern geography as they were of an Enlightenment *mentalité*. Indeed, Stoddart has argued that modern geography is a constitutively European science whose decisive moment of formation, he suggests, was Cook's entry into the Pacific in 1769. The work of the naturalists and illustrators who accompanied Cook displayed three features which Stoddart takes to be of strategic importance for the formation of geography as a modern empirical science; realism in description, systematic classification in collection, and the comparative method in explanation. For Stoddart, it was the extension of these scientific methods of observation, classification, and comparison to peoples and to nature that made geography possible (Stoddart 1986).

Other existing work on geography in the age of Enlightenment has very largely conflated 'the enlightenment' with 'the (long) eighteenth century' and disengaged geographical knowledge from the social and intellectual contexts of its time. For one commentator upon geography's history, "geography [in the eighteenth century] remained almost entirely in the hands of textbook writers who produced compilations of regional descriptions and showed little concern with the theory of the subject or its relation to significant issues in the scientific thought of the time" (Bowen 1981, 124–25). Certainly, the age of Enlightenment was marked by the production of multivolume universal geographies (Downes 1971). Many, however, were produced for different audiences, and only recently has it been possible to chart the publishing histories of these various geographical grammars and to see them in the *longue durée* and not simply as products of an Enlightenment urge to order and classify (Sitwell 1993). In Bowen's reading of the late eighteenth century, it was only "as increasing challenges came to the dominance of the Newtonian form of scientific empiricism, that geography again became the focus of active intellectual concern" (Bowen 1981, 125), a focus that was first apparent in Kant's natural science enquiries and, notably, in his 1757 *Outline and Prospectus for a Course of Lectures in Physical Geography*. Kant certainly saw in geography a means to unify learning, and in this sense, it is possible to claim that his understanding of geography was closely related to concerns to expound rational empiricism as the basis to knowledge (Richards 1974). But it is another thing to claim, as does Bowen, that Kant's concept of nature as a dynamic system was the bedrock out of which Humboldt's synthesizing enquiries of the nineteenth century were hewn, and that—to change the image—the real roots of modern geography lie here exposed.

Indeed, there are now good grounds for suspecting that such a genealogy for geography fails to take with sufficient seriousness the much deeper historical roots of the discipline and the complexities of the Enlightenment. More recent work on the nature of geography in the age of Enlightenment has shown it to be an altogether more complicated and contested affair (Livingstone 1992). In broad terms, geography's perennial fascination with the far away, with mapping the world, with exhibiting and classifying knowledge, and with the imposition of European ways of thinking on global realms are all recognizably Enlightenment preoccupations (Gregory 1994; Withers 1995). This is apparent in mathematical conceptions of natural order; in natural theological perspectives that interpreted the world in teleological terms; in geography as a visualizing practice; in the geographical natural histories of Buffon, Linnaeus, Zimmerman, and others; and in the voyages and travels of explorers (Livingstone 1992, 102–138). Work on the teaching of geography within university curricula and in the public sphere during the Enlightenment suggests that while there was some emphasis placed upon description of the world and upon geography's practical utility, contemporary recognition of the subject's 'usefulness' varied with context: from its use in military campaigning (Staum 1987), to mining and survey (Dorflinger 1976), to polite sociability (Withers 1999), and to that knowledge of trade winds for commercial purposes that Thomas

Reid, the "Common-sense" philosopher of Enlightenment Scotland, included in his geography lectures in 1752 (Withers 1996b).

It is clear, too, that both the enlightenment idea of human progress, and the tenets of that stadial or stage-by-stage theory on which the idea was founded, were intrinsically geographical issues, both in the sites of their conception and in relation to the places used as 'models' within their normative framework (Heffernan 1994). The intellectual sphere which geography has encompassed has been examined in texts such as the *Encyclopédie* (Withers 1993, 1996a). Other work suggests that the terms 'geography' and 'geographer' may not have been constant in the period. For England, for example, Mayhew has shown how those terms connoted very particular precepts and practices in the encyclopedic thinking of Samuel Johnson (Mayhew 1997, 1998), and others have demonstrated how particular notions of rational geography were central to Edward Gibbon's view of history and historical change (Abbatista 1997).

Yet there is still much to know about why and how geography took the shape it did in the Enlightenment. And what is true of contemporary research on 'Enlightenment' and on 'Geography,' of course, is that recent work has found both terms exceedingly slippery. Writings on the history of both have disclosed a pronounced historiographical shift away from the search for essential disciplinary or defining characteristics towards a recognition of the centrality of process and of context. For historians of geography anyway, geography is now understood more as a set of discursive situated practices and much less readily as a formal and unified academic discipline (Driver 1992, 1994, 1996). There is now widespread recognition that essentialist questions about the nature of geography are inappropriate without attention to context—to the particular meanings the subject has had in different times and places—and that there have been different sorts of geographical knowledge practiced and promoted at various times in various settings. Quite what sort of knowledge received the authority and the audiences it did, and why, has much to do with the values placed upon it, the intentions of its practitioners, and connections with other disciplines (Livingstone 1984, 1992, 1994, 1995). In the same way, some Enlightenment historians now consider it more helpful to look at the Enlightenment "as a series of debates, which necessarily took different shapes and forms in particular national and cultural contexts" (Outram 1995, 3). But that, too, is a relatively new historiographical departure (Bender 1992).

At the same time, scholars of the Enlightenment, only too ready to debate essential characteristics, such as the 'what,' and even the 'why,' have been less confident about the 'where' of Enlightenment. The existing corpus of work on this theme has tended to assume that the geography of Enlightenment necessarily and neatly followed the contours of national boundaries. Yet Enlightenment, we would want to argue, was also about the *movement* of ideas across borders and over time; it was about encounters with *particular places* such as Tahiti, however much it was also about the assumed universal nature of human society (Porter 1990c). It was about ideas that took shape in what historians of science term certain *analytic sites,* such as courts, museums, laboratories, botanic gardens, and in the *salons* of polite culture as well as in the lecture halls, coffee houses, and excursions of the public sphere. Far from simply being understood as a European movement with particular national expression, it becomes possible to conceive of the Enlightenment as being sited, produced, debated, and contested in *local spaces and circumstances* as well as being apparent at national levels. In this way, thinking of *the* geography of *the* Enlightenment may not be as helpful as conceiving of geographies of enlightenment in which geography is understood to embrace the sites and the practices in which enlightenment, as a process, a set of ideas, was produced, debated, and consumed.

We should not be misunderstood here: this is not to see the spaces of Enlightenment knowledge as just matters of geographical location. In fact, no one would deny that an understanding of the Enlightenment may be enhanced by reference to location, to geography's rhetoric and metaphors of central place, cultural hearth, diffusion, map, territory, networks, and hierarchies and to other spaces of human knowledge. But there are, we would claim, further insights to be gained in paying attention to the geography *of* Enlightenment ideas and in the local articulations of particular Enlightenment *mentalités.* There are also insights to be gained by widening and deepening the focus of our enquiries to include further questions about the place of Geography as a discipline and a form of knowledge in the Enlightenment. How, for example, were geographical practices like exploring, mapping, and naming drawn upon by Enlightenment theorists and philosophers as part of their attempts to understand the world? If, as we would wish to suggest in what follows, we might conceive of Enlightenment as a series of processes and problems, rather than as a definable period, and with local expressions rooted in particular contexts, what did contemporaries understand by the practical and intellectual processes of discovery, traveling and mapping, and how exactly did Enlightenment thinkers draw upon these geographical crafts?

CHARLES W. J. WITHERS AND DAVID N. LIVINGSTONE

In one sense, of course, geography as metaphor is central to the Enlightenment: the description by Diderot and d'Alembert of the *Encyclopédie* as a "map of the world" and of Burke's reference to Enlightenment stadial theories as "the Great Map of Mankind" come to mind. But in another and more concrete sense, geographical knowledge generated in the Enlightenment was about the processes involved in, and following from, conceptual and empirical encounters with the world: mapping new lands and understanding foreign 'others'; describing difference at home and overseas; using such newly mapped and measured data to pose new questions about the human place in nature and about the nature of human society; engaging with epistemological dilemmas posed by knowledge claims about entities distant from knowledge producers and conventional social authorities. And just as we can trace shifts in the history of geography towards such discursive questions, so it is also possible to discern a similar drift among historians of the Enlightenment itself.

1.2. What Is Enlightenment?

The question "What is Enlightenment?" posed by the educationalist Johann Friedrich Zollner in the *Berlinische Monatsschrift* in December 1783, prompted several responses. For Moses Mendelssohn, 'enlightenment' was to be understood as "related to theoretical matters: to (objective) rational knowledge and to (subjective) facility in rational reflection about matters of human life." For Immanuel Kant, the public use of reason and its challenge to established authority meant that Enlightenment consisted in "mankind's exit from its self-incurred immaturity." For Reinhold, it meant "the making of rational men out of men who are capable of rationality" (Nisbet 1992; Schmidt 1996, 1–44, 53–77; Staum 1985).

Despite the seeming agreement over the centrality of reason and its use in improving the human condition, definitions of Enlightenment have for a long time been characterized by difference and diversity. In seeking to come to terms with just what 'Enlightenment' might name, Shea and Huff (1995, 8) summarize the definitional options as "the rise of individual autonomy over traditional community, the rise of secularizing reason over inherited authority, the disengagement of nature from a supernatural worldview, the rise of methodical and institutionalized criticism, the rise of science as both technique and worldview, the rise of historical consciousness and the practice of historical method, and the establishment of bourgeois institutions and a democratic ethos in public life." In part, too, the lack of a clear referent for the term stems from the ways in which the grounds of the debate have shifted: from initial concerns with censorship regulations and how much a citizenry could or should be 'enlightened,' to tensions between enlightenment and faith, and, by the end of the eighteenth century, to the relationship between enlightenment and political authority: "Thus by the close of the eighteenth century, answering the question 'what is enlightenment' meant exploring the relationship between public discussion, religious faith, and political authority" (Schmidt 1996, 2). The lack of agreement over the question is also the result of differing interpretation by later writers. Understanding Enlightenment as an historical, and, as we hope to show, a geographical phenomenon—either as a formal entity, signaled by the definite article, or as the more process-oriented and debate-centered 'enlightenment'—is a matter of historiographical interpretation.

Although it is only in the last 50 years, according to Schmidt (1996, 15) "that the question of enlightenment has been reopened in earnest," it is already possible to identify a number of interpretative strands. Ernst Cassirer's *The Philosophy of the Enlightenment* (1951), first published in 1932, is a key text in those interpretations of *the* Enlightenment as a relatively homogenous affair with sharp temporal definition. Like Smith's *The Enlightenment* (1962) and Isaiah Berlin's 1956 *The Age of Enlightenment,* Cassirer (1964) fastens upon the power of reason to change human society and of science to demystify understanding of the world during an epoch delimited by the lifespans of Liebniz (1646–1716) and Kant (1724–1804). Broadly similar is Peter Gay's interpretation, which, while rotating around the dates of great thinkers, elucidates three distinct periods: a first phase defined by Voltaire (1694–1778) and Montesquieu (1689–1755), succeeded by Diderot (1713–1784) and d'Alembert (1714–1780), and culminating in the "late" Enlightenment of Kant (Gay 1966–1969). Temporal periodization is also to the fore in Norman Hampson's *The Enlightenment* (1968), which distinguished an "early" (1715–1740) and "late" (1740–1789) Enlightenment, and Leslie Crocker's *The Age of Enlightenment* (1969), which located it between the 1685 Revocation of the Edict of Nantes and the 1793 guillotining of Louis XVI (Hulme and Jordanova 1990, 2; see also Macary 1977).

Since the 1970s, historians have focused more on the socially embedded nature of Enlightenment ideas, and upon their dissemination and reception. The monumental enterprise of Diderot and d'Alembert's *Encyclopédie,* published in seventeen volumes between

1751 and 1772, for example, has been seen as a crucial *machine de guerre* in the Enlightenment "war" against "irrationality," the power of the Church, and conventional social authorities (Kafker 1981; Kafker and Kafker 1988; Yolton 1991, 5, 145–50). At the same time, the work's publishing history and its differing social and geographical receptions have been examined to ascertain the diverse ways in which Enlightenment ideas were domesticated (Chartier 1988; Darnton 1979, 1984). Their differential impact on particular sciences and political theories (Hankins 1985) has further underscored the diversity of debates and prompted recognition that the debates themselves confirm a fundamental diversity within unity (Outram 1995; Porter 1990c).

It is tempting to reduce the historiographical interpretations of the Enlightenment to the twin dualisms of date and type: the focus of writers in the 1960s on a unified and datable movement of great thinkers, and that of the 1970s and later on the social use and meaning of Enlightenment ideas (Darnton 1971). Such a simple reading should be avoided, however, for several reasons. For one thing, some nineteenth-century thinkers—like Hegel and Nietzsche—devoted attention, albeit limited, to the nature of the Enlightenment (Schmidt 1996, 20–27). More important, the interpretation offered by historians in the 1960s, though dominant, was never universally espoused. Already by the beginning of this century, for example, Gustave Lanson had argued for a broader understanding of Enlightenment, with 'origins' earlier, in the early seventeenth and even in the sixteenth centuries, and in English philosophers such as Francis Bacon, Isaac Newton, and John Locke rather than in French thinkers. Lanson, echoing the earlier diagnoses of Paul Hazard, effectively widened the debate by pushing back the intellectual "origins" of Enlightenment, by emphasizing the fundamentally "restless" character of its thought forms, by stressing the nationally diverse European nature, and by treating it as a way of thinking, with roots in Cartesian philosophy (Hulme and Jordanova 1990, 2). These recent interpretations are indicative of a more searching set of questions about 'origins,' about the longer-run challenge to authority, and about the impact of ideas, not just the ideas themselves. Such claims invite geographical scrutiny of the sort disclosed in Yolton's (1991, 2–3) observations on the impact of the "electrifying enlargement of geographical and celestial space," on the expansion of the European mind and its different renditions in England, which was home to "the first bearers of the torch," and in France, where already by the mid-seventeenth century, authority "was beginning a process of weakening itself" (see also Jacob 1977, 1981).

Alongside these fundamentally historical readings is a range of interpretations taking theoretical inspiration from the work of such figures as Max Horkheimer and Theodor Adorno, Jurgen Habermas, and Michel Foucault, all of whom unite in sensing contradictions at the heart of Enlightenment thinking. In their *Dialectic of Enlightenment* (written in 1944, published first in German in 1947 and in English in 1972), Horkheimer and Adorno (1972, 3), recalling that "The Enlightenment had always aimed at liberating men from fear," ask why is it that "the fully enlightened earth radiates disaster?" Composed as Europe emerged from the ravages of world war, their critique aimed to expose the Enlightenment's dark side, its alter ego, and thereby to call into question the assumption that the "clear light" of reason precipitated liberation from myth and superstition. But since there was no common agreement over what constituted rationality, the power of reason was expressed as the power of some people over other people. As Horkheimer and Adorno noted (1972, xiii), the Enlightenment already contained "the seeds of the repression apparent everywhere today." That this indeed was so, largely stemmed from what they termed the "administered life," that rational management of human society, knowledge and nature so often found at the heart of the modern condition. In short, if Enlightenment was about reason and the potential for human freedoms, it was no less about restriction and domination: if knowledge was a means to challenge authority, so, too, was it the power of social authorities using claims to reason to regiment, restrain, control.

That these contradictions lie at the very core of Enlightenment can be discerned in discussions about the relationship between the Enlightenment and the French Revolution. Just as the terrors of 1789 marked for some historians the end of Enlightenment, the Second World War constituted (to Horkheimer and Adorno) the final horrendous nadir of a misplaced Enlightenment confidence (Schmidt 1996, 21). What is advertised here is less the 'failure' of Enlightenment rationality than its unforeseen, even its unimaginable consequences. And it is precisely because of this paradoxical capacity to liberate and oppress, to ennoble and brutalize that the Enlightenment's complicity with domination attracted the sustained interrogation of Michel Foucault. In his *Birth of the Clinic* (1973) and *Discipline and Punish* (1977) most especially, Foucault showed how the *ideas* of Enlightenment were implicated in repressive *practices*. Order and reason meant surveillance and control. Reexamining Kant's diagnostic 1784 essay, Foucault stressed the incompleteness of the Enlightenment—an incompleteness demanding renewed consideration of how its central ideas connect with the present—and conceived of Enlightenment not "simply as a general process affecting all humanity" or "only as an obligation prescribed to individuals," but "as a political problem" (Foucault 1984, 37; 1970; 1986; Rousseau 1973).

Habermas similarly takes Kant's essay as a key text for understanding Enlightenment and likewise sees the need to address its connections with the present. Habermas differs, however, from Foucault and from Horkheimer and Adorno not just in his judgments about the Enlightenment's dark side, but also in the weight he places on its incompleteness and still unrealized emancipatory power. He differs, too, in his treatment of what he called the "public sphere," that public realm of ideas and cultural materials which, for Horkheimer and Adorno, was part of the restrictive use of knowledge by the powerful in society. For Habermas, the public sphere with its emphasis upon the production by the middle classes of the apparatus for informed public opinion—debating societies, public lectures, coffeehouse conversazione—could yet allow Enlightenment ideas to be a means to political freedom (Habermas 1972, 1994; Wood 1994).

Last, it is possible to see in these debates, and even in some of the writings of earlier historians, a recognition of that very diversity that is now taken as characteristic of the Enlightenment. Consider, for example, Gay's 1969 use of the term "the geography of hope." Although amounting to little more than gesture, there are grounds for suspecting that he had in mind not just the fact that Enlightenment ideas—of moral progress, of reason, of the utility of scientific advance—are placed in time and space, namely that they have a history *and* a geography, but also that Enlightenment writers like Condorcet, Turgot, and Rousseau knew well that such ideas had unevenly distributed consequences and contradictions: "Reasoning, itself a good, may produce error as easily as it produces truth: in turn error, itself an evil, may become an unwitting tool of progress" (Gay 1969, 116). Porter seems to sense just such an interpretation when he claims that "More recently, historians have attempted to advance Gay's programme of a 'social theory of ideas' beyond Gay's own horizons. They have focused on the fluid and fruitful interplay between wider groupings of thinkers and their ebbings and flowings; they have emphasised how the Enlightenment was a broad collective endeavour, not just the work of a few giants; and they have characterised the intimate interconnections between material circumstance, people's lifestyles, and their thought" (Porter 1990c, 44–45). This is apparent in earlier work by Porter himself on science and provincial culture in Enlightenment England (Porter 1980), on the sexual underworlds of the Enlightenment (Rousseau and Porter 1987), medical knowledge and the Enlightenment (Bynum 1980; Cunningham and French 1990; Porter 1992, 1996; Wellman 1992), exoticism and the Enlightenment (Outram 1995, 63–79; Porter 1990b), and on the anthropology and literary histories of the Enlightenment (Moravia 1980; Rousseau 1991; Wokler 1996).

It is clear from even these preliminary reflections that the links between geography and enlightenment are many and diverse. The following sections consider two distinct but related ways by which we take further consideration of geography and enlightenment.

1.3. The Geography of Enlightenment

Towards the close of a 1971 lecture "Why was the Enlightenment?", Gay reflected upon "yet another distinctive characteristic of the Enlightenment: the variety of its shapes across the Western World." "Why, then" [he asked], "did the Enlightenment take the particular form it did in each country?" and noted, simply, "Each country had the Enlightenment it deserved" (Gay 1972, 70). While it is difficult to know precisely what to make of this claim, two observations suggest themselves. First, Gay implicitly recognized the importance of local variety within Enlightenment thought. To him the divergence of the *philosophes'* views are a consequence of the way "they adjusted their aspirations to their estimate of local possibilities," though he remained careful not to accord cognitive primacy to the social world as *the* basis to those particular national forms: "the relation of ideas to their environment is not quite so predictable as all that" (Gay 1972, 70–71). And yet, while eschewing socially determinist or ecologically reductionist readings of the nature of national Enlightenments, he did presume that it was at the scale of the nation that the geography of the Enlightenment was to be understood. This perspective has been dominant in work on this theme.

Alongside these affirmations are more recent revisionist accounts of the Enlightenment as an inherently European phenomenon. In some writing of the 1970s and in later work, this idea has been revised in two main ways. Thus, attention has been paid to renditions

CHARLES W. J. WITHERS AND DAVID N. LIVINGSTONE

of the Enlightenment in national contexts beyond Europe, notably in the Americas (Aldridge 1971; Leventhal 1976; May 1976). In addition, more detailed attention has been paid to the national geographies of Enlightenment within Europe. Venturi, for example, has shown how Enlightenment ideas were widely debated on the 'periphery' of Enlightenment Europe in Greece, Italy, Poland, Hungary, and Russia, not just within the French 'hearth' (Venturi 1971, 1989). Porter and Teich's 1981 edited collection, *The Enlightenment in National Context,* for example, is a key work in this regard, recognizing as it does that traditional accounts were insufficiently appreciative of the Enlightenment's "geographical, social, and political *location* as a cultural movement" [original emphasis]. Accordingly, that volume sought "to grasp the meaning of Enlightenment in thirteen national contexts" (Porter and Teich 1981, vii) by considering England, Scotland, France, the Netherlands, Switzerland, Italy, Germany, Austria, Bohemia, Sweden, Russia, and America. The specific claims of individual authors need not be recounted here, but attention can be called to the earthing of Enlightenment ideas in particular social, economic, and demographic circumstances. National differences in the reception of those ideas and in the institutions—political, educational, popular, and public—that helped shape encounters are recorded, including what Frängsmyr, talking of the Enlightenment in Sweden, considered "Resistance to the Enlightenment" (Frängsmyr 1981, 166–70).

While *Enlightenment in National Context* thus represents an important way marker, it is far from the last word on the geography of Enlightenment. For one thing, the scale of analysis obscures variation *within* the nation in the production and reception of Enlightenment ideas, to say nothing of the fact that the idea of the nation itself was uncertain in this period anyway. Indeed, many of the essays exhibit a tendency to essentialize national Enlightenments, either in terms of defining characteristics or in regard to the lives of great thinkers (see also Wade 1971). The Enlightenment in Italy, for example, is marked by four "national characteristics": by internal geographical identities so distinct and different that "Few thought about Italy"; by the hegemony of the Pope; an emphasis upon practical utility, and by, "simply, the genius of Italians" (Chadwick 1981, 90). The idea of Italy itself was remarkably fluid, a point which makes it difficult to talk of an *Italian* Enlightenment. For Pole, Enlightenment had both fixed geographical expression and certain chronology: the Enlightenment in America came to an end with the lives of Presidents John Adams and Thomas Jefferson (Pole 1981, 214).

The question of geography is understood here, where it is clearly articulated at all, either as a matter of location, or one of scale, chiefly that of the nation. However much these accounts understand the Enlightenment in terms of ideas and their varied reception, it is still portrayed as an essentially European and North American experience. But it has become increasingly clear of late that the European Enlightenment was substantially moulded through engagement with other parts of the world, notably with the Americas as a whole and with the Pacific (Hulme 1986; Pagden 1993; Porter 1990a; Smith 1985). On one level, colonialism provided what Nicholas Dirks has called "a theatre for the Enlightenment project" (Dirks 1992, 6); but more than that, the Enlightenment *itself* was also powerfully shaped by contact with the exotic 'other' (Fox, Porter, and Wokler 1995; Outram 1995). Contact with New Worlds demanded reliance not upon geography as national or disciplinary space but as a set of particular ideas and practices shaped by the peoples and the environments encountered. This is not to reject the idea of there being national geographies of Enlightenment. Enquiries into the nature of Enlightenment in England, for example, have cast important light upon the nature of English society in the eighteenth century (Clark 1982, 1985; Porter 1981), as well as upon the critical role of seventeenth-century English philosophers in the 'origins' of Enlightenment thought (Yolton 1991). Rather, it is to consider relationships between places, to recall that Enlightenment ideas are themselves sited and have a social and a gendered history (Darnton 1971; Tomaselli 1985), and to attend to the contested meanings of geography and enlightenment. Robertson (1997), for example, stresses both the international nature of the Enlightenment and the local places of its making in examining connections between Scotland and Naples.

Such considerations about the global, the local and the *connected* nature of Enlightenment—that it was not just a national phenomenon—reinforce Hulme and Jordanova's suspicion that "The Enlightenment's self consciousness was to some extent a geographical consciousness based on the distinctiveness of the part of the world that came to be called 'Europe'" (Hulme and Jordanova 1990, 7). They also invite more detailed scrutiny of precisely how that "distinctiveness" was made, not just as a matter of consciousness, or of the imagined geographies of the South Seas, or through the manifold ways in which the New World of the Pacific "enlarged the sphere of contemplation" (Marshall and Williams 1982, 258–298), but as processes of situated practical enquiry—mapping (Godlewska 1995), and also in traveling, naming, displaying, and representing (Stafford 1984, 1994)—at a variety of scales. Like Jacob in her discussion of the eighteenth-century public sphere, we want to use geography "to employ a comparative approach and search for a broader vision of the Enlightenment, one that draws upon local settings to enrich a Euro-american perspective," while, at the same time, considering how ideas of enlightenment and of geography, and perhaps retaining even reinforcing national identities, "also transcended national boundaries, gender divisions, and the influence of local settings" (Jacob 1994, 112). Having thus reflected on the diverse ways in which it may be possible to think of Enlightenment themes having their own geographies, and mindful of earlier remarks on changing trends in the history of geography, it is appropriate now to turn to the character of geographical knowledge during the Enlightenment era.

1.4. Geographical Knowledge and Enlightenment

Geographers tracing the 'origins' of modern knowledge (including geography) to the Enlightenment have perhaps been too ready to see the Enlightenment as a singular, rather monolithic, enterprise sometimes designated as "The Enlightenment Project." For David Harvey, for example, "That project amounted to an extraordinary intellectual effort on the part of the Enlightenment thinkers to develop objective science, universal morality and law, and autonomous art according to their inner logic. . . . the development of rational forms of social organization and rational modes of thought promised liberation from the irrationalities of myth, religion, superstition, release from the arbitrary use of power as well as from the dark side of our own human natures. Only through such a project could the universal, eternal, and the immutable qualities of all of humanity be revealed" (Harvey 1989, 12). For others, "modern social theory still bears the marks of its Enlightenment origins, and its claims to know continue to respond to Kant's attempts to instill reason as the undisputed arbiter in all spheres of social life" (Gregory 1994, 12).

We would not necessarily dispute these views. But we do want to move beyond claims to a presumed unity and to consider more directly the complex and negotiated ways in which geographical knowledge was an active part of Enlightenment knowledge. Hulme and Jordanova acknowledge something of the complexity of geographical knowledge in the age of Enlightenment by their recognition that the geographical discovery of the New World was both a material *and* a metaphorical enterprise. First and foremost, as a depiction of the nature of scientific advance, discovery evoked "images of marching into new territories, taming what one found there, and giving a coherent account of fresh terrain" (Hulme and Jordanova 1990, 5). Further, that characteristic Enlightenment trope—the idea (and ideal) of 'progress'—was rooted in a fundamentally geographical comparison between the institutions of civilized Europe and those living in the state of Nature. Geographical discovery had brought all the world into view, in all its different stages of 'development.' This is what Edmund Burke meant when he wrote "But now the Great Map of Mankind is unroll'd at once; and there is no state or Gradation of barbarism, and no mode of refinement which we have not at the same instant under our View. The very different Civility of Europe and of China; the barbarism of Persia and Abyssinia. The erratick manners of Tartary, and of Arabia. The Savage State of North America, and of New Zealand" (Burke 1791, in Marshall and Williams 1982). This spectacle of the geography of the world, in which the different stages of human progress could be seen—what Gregory describes as "the world as exhibition" (Gregory 1994, 15–69)—was a theoretical matter promoted by European thinkers in ways which absolutized time and space by folding two distinctions into one another: that between "the West" and "the non-West," and one between "History" and what Wolf called "the people without History" (Wolf 1982).

Hegel was to affirm these distinctive global geometries in his *Lectures on the Philosophy of History,* first delivered in 1822–23, in arguing that it was History which made real the absolute spaces of world Geography: "The history of the world travels from east to west for Europe is the absolute end of history, just as Asia is the beginning" (Hegel 1956, 103). As others have noted, what was "beyond Europe" was also "before Europe" (Gregory 1998; McGrane 1989, 94). These views routinely found expression in maps. James Wyld's 1815 "Chart of the World Shewing the Religion, Population and Civilization of Each Country" was based upon a five-point scale, for example, from "I, comprehending the savage" to "V, the most civilised nations as England, France &c." Clark's 1822 "Chart of the World" claimed to exhibit "the Prevailing Religion, the Form of Government, State of Civilization, & the Population of each Country," and did so through a similar five-point scale from "Savage" to "Enlightened." Wyld claimed his map to be an objective representation of "the State of the Globe," what we might now call a world political geography.

As Fabian has put it, such "geopolitics" was importantly based on "chronopolitics" (Fabian 1983, 26). Both maps articulate the views of an educated European authority at a particular stage in the imperial and empirical relationship of Europe with the rest of the world and remind us of the ways in which Enlightenment and geographical knowledge were connected with notions of empire (Bell, Butlin, and Heffernan 1995; Godlewska and Smith 1994; Smith 1994).

Recognizing that, so to speak, geographical knowledge was everywhere, but not everywhere the same during the Enlightenment is to raise methodological and conceptual questions not just about the nature of geographical knowledge but about the geographies of knowledge. Our thinking about such matters has drawn upon a range of recent work in the historical sociology of science and in the history of geography which has paid attention to this question of the spaces of knowledge in a variety of ways. In their survey of work on what they call "the place of knowledge," Ophir and Shapin (1991) insist that it is no longer appropriate to consider science as a universalist and essentialist enterprise, and in so doing, they raise questions about the situatedness of knowledge, and, particularly, its *localist* nature: "What if knowledge in general has an irremediably local dimension? What if it possesses its shape, meaning, reference, and domain of application by virtue of the physical, social, and cultural circumstances in which it is made, and in which it is used?" (Ophir and Shapin 1991, 4). This emphasis on the local nature of knowledge (see also Shapin 1998) draws upon a more widespread recognition of the contextualized character of scientific practice but extends beyond this to consider the way spatial metaphors inform a variety of disciplinary perspectives.

This is what we would see as the geographies or setting dependencies of conceptions of truth, order, warrant, reason, and so on. This notion may be further expanded by exploring the relationships *between* local situated geographies of knowledge. Thus, local sites for the conduct of knowing may be seen as productive of certain meanings. They may also be intimately related to other local but more distant sites through practices of circulation and processes of negotiation designed to warrant the credibility of knowledge, the credibility of certain claims to knowledge. As has been noted, "A division in the map of knowledge flows from placement in physical and social space: on the one side, immediate experience; on the other, reliance on authority and trust. Within knowledge-making sites, epistemological and disciplinary distinctions are related to spatial arrangements that differentiate degrees of visibility, directness of access to objects of research, facility of movement between workplaces, and density of interaction among persons occupying different positions at different locations" (Ophir and Shapin 1991, 9). This is not just to identify the localist nature of scientific knowledge and of ideas. Neither is it simply to recognize that knowledge is made and sustained through situated practical activity. It is to claim that, if we can talk of a practically constituted social history of truth (or, at least, of what passes for truth) (Shapin 1994), then we may be able to talk of social geographies of both warranted assertibility in general, and of science in particular, in ways sensitive to the context-dependent nature of meaning *and* to the negotiated transfer and movement of ideas between sites. Of course, this raises questions *both* about the mechanisms through which knowledge is attained and through which certain forms of knowing become dominant, *and* about the uses put to certain knowledges. As Foucault and others have shown, knowledge is a means to power, a power inscribed in space and in time in certain institutionalized settings.

Our attention to the geographies of knowledge also reflects wider interest in the study of science as a social construction and a more recent "recovery of geographical discourse" within the humanities and social sciences (Livingstone 1994, 1995; Thrift, Driver, and Livingstone 1995). In one way, this is apparent in those, like Foucault and Edward Said (1978, 1993), who have emphasized the geographically contingent nature of power and knowledge, and in others like Paul Carter (1987), who stress the spatiality of history. In another, it is evident in attention to the local in anthropological and sociological enquiry (Geertz 1973, 1983; Giddens 1990, 1991), and in the interest of philosophers in the settings, the situatedness, of standards of rationality. MacIntyre has argued, for example, that universal moral principles such as rights and welfare are only rendered practically articulate in specific settings: practical rationality is, then, always located reasoning (MacIntyre 1988, 363). For Wolterstorff, too, what counts as rational varies depending on the context: "Rationality is always situated rationality" (Wolterstorff 1983, 154). And Donna Haraway speaks of "positioned rationality" where she argues that truth is never the view from nowhere, but is always socially grounded, the result of a positioning in social space (Haraway 1991).

It is also possible to discern the preliminary traces of what we would call the geographies of scientific knowledge in three further contexts: those works that discuss the regionalization of scientific

CHARLES W. J. WITHERS AND DAVID N. LIVINGSTONE

style (for example, Crombie 1994; Glick 1974, 1987; Nicholson 1989); in work which has demonstrated how particular political topographies condition the conduct and reception of scientific endeavor (for example, Inkster and Morrell 1983; Morrell and Thackray 1981; Morus, Shaffer, and Secord 1992); and, most especially, in that range of work which has discussed the social spaces and local practices of scientific sites: in the laboratory (Hillier and Penn 1991; Latour 1987; Latour and Woolgar 1979; Lynch 1991), the lecture hall (Bourdieu 1990), the library (Chartier 1994), the pub (Secord 1994), the museum (Findlen 1994; Withers 1995), royal courts (Moran 1991), and the public sphere (Moravia 1980; Wood 1994). Again, this is not to see the geographies of Enlightenment as just local knowledge. Latour's scrutiny of centers of calculation has shown how scientific practices locally articulated depended crucially on the utilization of information gained from afar and on the ways in which hitherto unfamiliar events and peoples were represented, portrayed, and classified not just in physical sites, such as museums, botanic gardens, and laboratories, but in different texts—in maps, paintings, even novels. For Latour, "these centres of calculation construct manageable local representational spaces . . . through which scientists move and inside which worlds can be mobilised" (Latour 1987, 228).

For the geographer as for the natural historian, the botanist, the landscape painter, and the geologist, Enlightenment knowledge was never, then, simply a local or domestic affair, but was bound up with continental- or global-scale problems to do with discerning order in 'the field' that was the New World (Bourguet 1997). Western science in the Pacific treated that space, for example, at one and the same time as a laboratory for testing theories and as "a vast classroom for educating the European mind" (MacLeod and Rehbock 1988, 1). Yet, as several of our authors show, the meaning of the Pacific world was actually constructed through *particular practices* of enquiry focused on *local sites* from which certain forms of knowledge were made real in given ways by particular people according to what was there and then understood.

Insights of precisely this kind encourage us to attend to the intrinsic spatiality in other Enlightenment discourses. Considerable attention has been paid, for example, to the history and scientific form of natural history in the Enlightenment (for overviews, see Jardine, Secord, and Spary 1996; Miller and Reill 1996; see also Browne 1983; Larsen 1986; Sloan 1996; Withers 1995). Traveling, mapping, naming, and putting in place were all practices shared by natural historians and geographers. Knowledge thus accumulated was displayed and classified according to taxonomic schemes like the binomial nomenclature of Linnaeus (Frängsmyr, Heilbron, and Rider 1990; Pratt 1992). The origins of such universal 'natural' knowledges thus turn out to be rooted in local encounters with particular parts of the world. Enlightenment conceptions of environmentalism and of the fragile plenitude of nature were likewise a product of encounters with particular places, not least beyond the European terrain (Grove 1994). And not surprisingly, the ordering of Nature and the ordering of Empire through geographical natural history were reciprocally related discourses throughout the Age of Enlightenment (Gascoigne 1994, 1996; Miller 1996; Schaffer 1996). John Gascoigne (1994), for example, uses his enquiry into Joseph Banks's conduct of natural history as a form of imperial knowledge to reveal how Enlightenment had several 'local' geographies: the localities at which Banks undertook his work in the South Seas; those sites in England which were, variously, concerned with the production, reception, and exhibition of knowledge about the South Seas, as well as what we might consider that varying topography of knowledge among those elite groups within England for whom enlightenment through natural knowledge was synonymous with self-improvement. As for Banks himself, his reputation as "the Enlightened Cultivator of Natural Knowledge" as one contemporary termed him (Gasgoigne 1994, 32), was secured through his combining certain sorts of geographical enquiry with the more local classificatory spaces of his natural history cabinets, and in the networks of correspondence from his London center of calculation (Miller 1996). Similarly, medical enlightenment in eighteenth-century England, for example, might be understood not just in terms of the plural values of Enlightenment knowledge or the discoveries of particular people in diverse countries (Porter 1982, 53) but in terms of its local shaping in particular institutions and in an environment where what it was to be 'ill' or 'mad' was contested and, routinely, geographically expressed (Philo 1995).

So, too, can the Enlightenment's encounter with the 'exotic' and the 'erotic' be seen as rooted in local worlds of restraint and release, themselves embedded in more widespread notions of reason and order (Outram 1995; Porter 1990b; Pratt 1992; Rousseau 1994). And that central vision of the Enlightenment, the ambition of creating a science of man, can also be seen to be a matter of geography (Fox et al. 1996; Wood 1996). What Enlightenment travelers encountered

CHARLES W. J. WITHERS AND DAVID N. LIVINGSTONE

were certainly new peoples, typified variously as 'savages' or 'primitives.' But, in important ways, geography in this sense provoked an encounter not just between travelers and 'others,' but among travelers themselves. Travel across space prompted reflection not just on the human condition over time, but upon the physical and moral distinctions between the 'natives' of these new worlds and the 'civilized' societies that had discovered them (Carrithers 1996; Wokler 1996).

The acquisition of geographical knowledge about faraway peoples and places also raised crucial epistemological questions of testimony and trust. How reliable was any particular witness in recounting what had been seen and touched and heard? In order that the claims of scientific travelers and exploring geographers could secure the status of knowledge, such persons had to exemplify perceptual and moral integrity: it was, after all, such testimony that bridged the gap between 'here' and 'there.' But how were the reports of scientific travelers and explorers to be trusted, and how were their claims to be authenticated? In facts, the marks of travel and of bodily deprivation —their wounds, scars, and so on—were frequently taken as legitimating epistemic insignia of the rigors of "doing geography" in order to secure genuine observation or accurate measurement. The reliability of knowledge claims rested upon the status of the person, and as others have shown, the production of scientific knowledge is, and always has been, intimately connected with believing the reports of the right witnesses because of who they are, not alone because of what they tell us (Shapin 1988, 1994).

1.5. On Geography and Enlightenment

In his discussion "What is Enlightenment?", Foucault noted that

> We must never forget that the Enlightenment is an event, or set of events and complex historical processes, that it is located at a certain point in the development of European societies. As such, it includes elements of social transformation, types of political institution, forms of knowledge, projects of rationalization of knowledge and practices, technological mutations that are very difficult to sum up in a word, even if many of these phenomena remain important today. (Foucault 1984, 43)

Geography and Enlightenment is informed by a similarly strong but nuanced conceptualization both of Enlightenment and of geography. Like Foucault, we take Enlightenment to be neither a definitive period nor an essential movement delimited by the lives of great thinkers but rather as discursive processes of situated critical reasoning, processes to do with how knowledge was made, how people drew upon certain parts of the world in framing stadial theories, or came to base their notions of reason and order upon certain places. It is also understood as a means by which we may explore the situated nature of Enlightenment knowledge: the sites in which certain sorts of reason were proposed and/or opposed, the particular moments and local places of encounter between 'savage native' and 'enlightened' traveler, the ways in which the world was brought to view through map, text, geographical grammar, expedition, and so on. The process of enlightenment is not only to be understood as complex *historical* process. We would suggest that these 'events' and 'social transformations,' 'forms of knowledge,' and 'projects of rationalization' were also complex *geographically.* They were, after all, placed, located in space as well as in time. Indeed, geography was itself a particular form of such situated knowledge. At the same time, we want to consider the potential for thinking about geographies of enlightenment knowledge as information was produced in and about certain places, for example, as it traveled by word of mouth, in manuscript or in printed form, and as it was trusted and acted upon by others elsewhere.

In trying to make sense of this complexity, we have structured the book in four sections: beginnings, mappings, travelings, placings. Individual introductions to each section summarize the arguments of the several chapters in that section and connect them to questions of more general concern. The emphasis upon process is deliberate. The distinction between the sections is not, however, absolute since several authors discuss more than one, if not all, of the themes we have been elaborating in the course of their particular focus. After all, if it is now more appropriate to conceive of geography and of Enlightenment as located in time and in space, as processes that were negotiated, resisted, contradicted, is it not also more fruitful to move away from the search for one essentialist beginning to *the* Enlightenment and talk rather of the *beginnings* of Enlightenment? This may be a matter of history. But it is also, crucially, one of geography since to allow for situated and particular beginnings is also to concede that the meaning of Enlightenment was different in different places. In the same way, the use of the term "mappings" is intended to embrace in a variety of ways ideas of bringing to order, together with that more precise sense of practical process closely bound up with the nature of geographical knowledge. It is also understood as a metaphorical project through which the political state

CHARLES W. J. WITHERS AND DAVID N. LIVINGSTONE

of the world and of individual disciplines and practices were themselves placed on that larger "map of human knowledge" that the Enlightenment sought to delineate. Mapping is, thus, central to what we would see as the discursive complexity of geography as a form of knowledge and a means of practical Enlightenment rationalization.

By "travelings," we mean that sense in which Enlightenment knowledge was a product of encounter with the non-European world, and at the same time a process that was heavily dependent upon the credibility that Enlightenment audiences placed upon the texts produced. Traveling meant getting to the New World that was the Pacific, or the west coast of America, or the inner reaches of the Amazonian river system. More important, it should be understood both as the means by which Enlightenment writers encountered these new geographies and the ways their ideas about such places were brought to the attention of others. If we are to talk of 'sites' of Enlightenment in the ways discussed above, to what extent may it also be possible to chart the movement of Enlightenment ideas between sites, or to represent what have hitherto been understood as 'national' Enlightenments as, perhaps, the result of ideas being differently received? "Placings" picks up this theme in its emphasis upon the situated character of Enlightenment knowledge.

Geography and Enlightenment cannot pretend to have dealt in full with all the issues raised above. But in bringing together these essays, and by drawing upon and extending existing work in the history of geography and of Enlightenment, this volume does hope to take further what is known of geography's relationship with Enlightenment, as well as highlighting new possibilities for further work. These essays, like the Enlightenment itself, offer the possibility of questioning intellectual and disciplinary boundaries. Ideas are sited and their reception charted. If only in outline form, these papers suggest ways in which new maps of knowledge may be drawn of the geographies of Enlightenment and, indeed, of Enlightenment itself.

References

Abbatista, Guido. 1997. "Establishing 'the Order of Time and Place': 'Rational Geography,' French erudition and the emplacement of history in Gibbon's mind," in David Womersley, ed., *Edward Gibbon: Bicentenary Essays: Studies on Voltaire and the Eighteenth Century* 355: 45–72.

Aldridge, A. Owen, ed. 1971. *The Ibero-American Enlightenment*. Urbana, IL: University of Illinois Press.

Bell, Morag, Robin Butlin, and Michael Heffernan, eds. 1995. *Geography and Imperialism 1820–1940*. Manchester: Manchester University Press.

Bender, John. 1992. "A New History of the Enlightenment?" *Eighteenth-Century Life* 16: 1–20.

Berlin, Isaiah. 1956. *The Age of Enlightenment.* New York: New American Library.

Bourdieu, Pierre. 1992. *In Other Words.* Cambridge: Polity Press.

Bourguet, Marie-Noëlle. 1997. "The Explorer," in Michel Vouvelle, ed., translated by Lydia Cochrane, *Enlightenment Portraits.* Chicago: University of Chicago Press, 257–315.

Bowen, Margarita. 1981. *Empiricism and Geographical Thought: From Francis Bacon to Alexander von Humboldt.* Cambridge: Cambridge University Press.

Browne, Janet. 1983. *The Secular Ark: Studies in the History of Biogeography.* New Haven: Yale University Press.

Bynum, William. 1980. "Health, Disease, and Medical Care," in George S. Rousseau and Roy Porter, eds., *The Ferment of Knowledge.* Cambridge: Cambridge University Press, 211–254.

Carrithers, David. 1996. "The Enlightenment Science of Society," in Fox, Porter, and Wokler, eds., 232–270.

Carter, Paul. 1987. *The Road to Botany Bay.* Cambridge: Cambridge University Press.

Cassirer, Ernst. 1951. *The Philosophy of the Enlightenment.* Princeton, NJ: Princeton University Press.

Chadwick, Owen. 1981. "The Italian Enlightenment," in Porter and Teich, eds., 90–105.

Chartier, Roger. 1988. *Cultural History: Between Practices and Representation.* Cambridge: Polity Press.

———. 1994. *The Order of Books.* Cambridge: Cambridge University Press.

Clark, Jonathan. 1982. *The Dynamics of Change: The Crises of the 1760s and English Party Systems.* Cambridge: Cambridge University Press.

———. 1985. *English Society, 1688–1832: Ideology, Social Structure and Political Practice during the Ancien Regime.* Cambridge: Cambridge University Press.

Crocker, Leslie, ed. 1969. *The Age of Enlightenment.* London: J. M. Dent.

Crombie, Alistair. 1994. *Styles of Scientific Thinking in the European Tradition.* 3 volumes. London: Duckworth.

Cunningham, Andrew, and Roger French, eds. 1990. *The Medical Enlightenment of the Eighteenth Century.* Cambridge: Cambridge University Press.

Darnton, Robert. 1971. "In Search of the Enlightenment: Recent Attempts to Create a Social History of Ideas." *Journal of Modern History* 63: 113–132.

———. 1979. *The Business of Enlightenment: A Publishing History of the Encyclopédie, 1775–1800.* Cambridge, MA: Harvard University Press.

———. 1984. *The Great Cat Massacre, and Other Episodes in French Cultural History.* London: Allen Lane.

Dirks, Nicholas, ed. 1992. *Colonialism and Culture.* Ann Arbor: University of Michigan Press.

Dorflinger, Johannes. 1976. *Die Geographie in der "Encyclopédie" ein Wissenschaftsgeschichliche Studie.* Wien: Müller.

Downes, Alan. 1971. "The Bibliographic Dinosaurs of Georgian Geography (1714–1830)." *Geographical Journal* 137: 379–387.

Driver, Felix. 1992. "Geography's Empire: Histories of Geographical Knowledge." *Environment and Planning D: Society and Space* 10: 23–40.
———. 1994. "New Perspectives on the History and Philosophy of Geography." *Progress in Human Geography* 18: 92–100.
———. 1996. "Histories of the Present? The History and Philosophy of Geography, Part III." *Progress in Human Geography* 20: 100–109.
Fabian, Johannes. 1983. *Time and the Other: How Anthropology Makes its Object.* New York: Columbia University Press.
Findlen, Paula. 1994. *Possessing Nature: Museums, Collecting, and Scientific Culture in Early Modern Italy.* Berkeley, CA: University of California Press.
Foucault, Michel. 1970. *The Order of Things.* New York: Vintage Books.
———. 1973. *The Birth of the Clinic.* New York: Vintage Books.
———. 1977. *Discipline and Punish.* New York: Vintage Books.
———. 1984. "What is Enlightenment?", in Paul Rabinow, ed., *The Foucault Reader.* London: Penguin, 45–56.
———. 1986. "Kant on Enlightenment and Revolution." *Economy and Society* XV: 88–95.
Fox, Christopher, Roy Porter, and Robert Wokler, eds. 1996. *Inventing Human Science: Eighteenth-Century Domains.* Berkeley, CA: University of California Press.
Frängsmyr, Tore. 1981. "The Enlightenment in Sweden," in Porter and Teich, eds., 164–175.
Frängsmyr, Tore, Johann Heilbron, and Richard Rider, eds. 1990. *The Quantifying Spirit in the Eighteenth Century.* Berkeley, CA: University of California Press.
Gascoigne, John. 1994. *Joseph Banks and the English Enlightenment: Useful Knowledge and Polite Culture.* Cambridge: Cambridge University Press.
———. 1996. "The Ordering of Nature and the Ordering of Empire: A Commentary," in Miller and Reill, eds., 107–116.
Gay, Peter. 1966–1969. *The Enlightenment: An Interpretation.* 2 volumes. New York: Knopf.
———. 1972. "Why Was the Enlightenment?" in Peter Gay, ed., *Eighteenth-Century Studies Presented to Arthur M. Wilson.* Hanover, NH: University Press of New England, 59–72.
Geertz, Clifford. 1973. "Thick Description: An Interpretive Theory of Culture." *The Interpretation of Culture: Selected Essays.* New York: Basic Books, 3–30.
———. 1983. *Local Knowledge: Further Essays in Interpretive Anthropology.* New York: Basic Books.
Giddens, Anthony. 1990. *The Consequences of Modernity.* Cambridge: Polity Press.
———. 1991. *Modernity and Self-Identity: Self and Society in the Late Modern Age.* Cambridge: Polity Press.
Glick, Thomas, ed. 1974. *The Comparative Reception of Darwinism.* Austin, TX: University of Texas Press.
———. 1987. *The Comparative Reception of Relativity.* Dordrecht: Reidel.
Godlewska, Anne. 1995. "Map, Text, and Image. The Mentality of Enlightened Conquerors: A New Look at the *Description de l'Egypte.*" *Transactions of the Institute of British Geographers* 20: 5–28.

Godlewska, Anne, and Neil Smith, eds. 1994. *Geography and Empire*. Oxford: Blackwell.

Gregory, Derek. 1994. *Geographical Imaginations*. Oxford: Blackwell.

———. 1998. "Power, Knowledge and Geography." *Geographische Zeitschrift*. 86: 70–93.

Grove, Richard. 1994. *Green Imperialism: Colonial Expansion, Tropical Island Edens and the Origins of Modern Environmentalism, 1600–1860*. Cambridge: Cambridge University Press.

Habermas, Jurgen. 1972. *The Structural Transformation of the Public Sphere*. Cambridge, MA: MIT Press.

———. 1994. "Taking Aim at the Heart of the Present: On Foucault's Lecture on Kant's *What Is Enlightenment?*", in Michael Kelly, ed., *Critique and Power: Recasting the Foucault-Habermas Debate*. Cambridge, MA: MIT Press.

Hampson, Norman. 1968. *The Enlightenment*. London: Penguin.

Hankins, Thomas. 1985. *Science and the Enlightenment*. Cambridge: Cambridge University Press.

Haraway, Donna. 1991. *Simians, Cyborgs and Women: The Reinvention of Nature*. London and New York: Routledge.

Harvey, David. 1989. *The Condition of Postmodernity*. Oxford: Blackwell.

Hazard, Paul. 1935. *La Crise de la Conscience Europeene 1680–1715*. Paris: Boivin.

Heffernan, Michael. 1994. "On Geography and Progress: Turgot's *Plan d'un Ouvrage sur la Geographie Politique* (1751) and the Origins of Modern Progressive Thought." *Political Geography* 13: 328–343.

Hegel, Georg. 1956. *The Philosophy of History*. New York: Dover.

Hillier, Bruce, and Andrew Penn. 1991. "Visible Colleges: Structure and Randomness in the Place of Discovery." *Science in Context* 4: 23–49.

Horkheimer, Max, and Theodor Adorno. 1972. *Dialectic of Enlightenment*. New York: Herder and Herder.

Hulme, Peter. 1986. *Colonial Encounters: Europe and the Native Caribbean 1492–1797*. London: Routledge.

Hulme, Peter, and Ludmilla Jordanova, eds. 1990. *The Enlightenment and Its Shadows*. London: Routledge.

Inkster, Ian, and John Morrell, eds. 1983. *Metropolis and Province: Science in British Culture, 1780–1850*. Philadelphia: University of Pennsylvania Press.

Jacob, Margaret. 1977. "Newtonianism and the Origins of the Enlightenment." *Eighteenth-Century Studies* 11: 1–25.

———. 1981. *The Radical Enlightenment: Pantheists, Freemasons and Republicans*. London: Allen and Unwin.

———. 1994. "The Mental Landscape of the Public Sphere: A European Perspective." *Eighteenth-Century Studies* 28: 95–113.

Jardine, Nicholas, James Secord, and Emma Spary, eds. 1996. *Cultures of Natural History*. Cambridge: Cambridge University Press.

Kafker, Frank, ed. 1981. "Notable Encyclopedias of the Seventeenth and Eighteenth Centuries: Nine Predecessors of the *Encyclopédie*." *Studies on Voltaire and the Eighteenth Century* 194.

Kafker, Frank, and Serena Kafker. 1988. "The Encyclopedists as Individuals: A Biographic Dictionary of the Authors of the *Encyclopédie*." *Studies on Voltaire and the Eighteenth Century* 257.

Larson, James. 1986. "Not Without a Plan: Geography and Natural History in the Late Eighteenth Century." *Journal of the History of Biology* 19: 447–488.

Latour, Bruno. 1987. *Science in Action: How to Follow Scientists and Engineers through Society.* Milton Keynes: Open University Press.

Latour, Bruno, and Stephen Woolgar. 1979. *Laboratory Life.* Beverly Hills, CA: Sage.

Leventhal, Herbert. 1976. *In the Shadow of the Enlightenment: Occultism and Renaissance in Eighteenth-Century America.* New York: New York University Press.

Livingstone, David. 1984. "The History of Science and the History of Geography: Interactions and Implications." *History of Science* 22: 271–302.

———. 1992. *The Geographical Tradition: Episodes in the History of a Contested Enterprise.* Oxford: Blackwell.

———. 1994. "Science and Religion: Foreword to the Historical Geography of an Encounter." *Journal of Historical Geography* 20: 367–383.

———. 1995. "The Spaces of Knowledge: Contributions Towards an Historical Geography of Science." *Environment and Planning D: Society and Space* 13: 5–34.

Lynch, Michael. 1991. "Laboratory Space and the Technological Complex: An Investigation of Topical Contextures." *Science in Context* 4: 51–78.

Macary, Jean, ed. 1977. *Essays on the Age of Enlightenment in Honor of Ira O. Wade.* Geneva: Droz.

MacIntyre, Alastair. 1988. *Whose Justice? Which Rationality?* Notre Dame, IN: University of Indiana Press.

MacLeod, Roy, and Philip Rehbock. 1988. *Nature in Its Greatest Extent: Western Science in the Pacific.* Honolulu: University of Hawaii Press.

Marshall, Peter, and Glyndwyr Williams. 1982. *The Great Map of Mankind. British Perceptions of the World in the Age of Enlightenment.* London: J. M. Dent.

May, Henry. 1976. *The Enlightenment in America.* New York: Oxford University Press.

Mayhew, Robert. 1997. *Geography and Literature in Historical Context: Samuel Johnson and Eighteenth-Century English Conceptions of Geography.* University of Oxford School of Geography, Research Papers, 154.

———. 1998. "Was William Shakespeare an Eighteenth-Century Geographer? Constructing Histories of Geographical Knowledge." *Transactions of the Institute of British Geographers* 23: 21–38.

McGrane, Bernard. 1989. *Beyond Anthropology: Society and the Other.* New York: Columbia University Press.

Miller, David Philip. 1996. "Joseph Banks, Empire, and 'Centers of Calculation' in Late Hanoverian London," in Miller and Reill, eds., 21–37.

Miller, David Philip, and Peter Hanns Reill, eds. 1996. *Visions of Empire. Voyages, Botany and Representations of Nature.* Cambridge: Cambridge University Press.

Moran, Bruce, ed. 1991. *Patronage and Institutions: Science, Technology and Medicine at the European Court 1500–1750.* Woodbridge: Boydell Press.

Moravia, Sergio. 1980. "The Enlightenment and the Sciences of Man." *History of Science* 18: 247–268.

Morrell, John, and Arnold Thackray. 1981. *Gentlemen of Science: Early Years of the British Association for the Advancement of Science*. Philadelphia: University of Pennsylvania Press.

Morus, Iwan, Simon Schaffer, and James Secord. 1992. "Scientific London" in Fox, ed., *London—World City 1800–1840*. New Haven, CT: Yale University Press.

Nicholson, Malcolm. 1989. "National Styles, Divergent Classifications: A Comparative Case Study from the History of French and American Plant Ecology." *Knowledge and Society: Studies in the Sociology of Science Past and Present* 8: 139–186.

Nisbet, Harold. 1992. "Was Ist Aufklarung? The concept of enlightenment in eighteenth-century Germany." *Journal of European Studies* 12: 77–95.

Ophir, Adir, and Steven Shapin. 1991. "The Place of Knowledge: A Methodological Survey." *Science in Context* 4: 3–21.

Outram, Dorinda. 1995. *The Enlightenment*. Cambridge: Cambridge University Press.

Pagden, Anthony. 1993. *European Encounters with the New World: From Renaissance to Romanticism*. New Haven and London: Yale University Press.

Philo, Chris. 1995. "Journey to Asylum: A Medico-Geographical Idea in Historical Context." *Journal of Historical Geography* 21: 148–168.

Pole, John. 1981. "Enlightenment and the Politics of American Nature," in Porter and Teich, eds. 192–214.

Porter, Roy. 1980. "Science, Provincial Culture and Public Opinion in Enlightenment England." *British Journal for Eighteenth-Century Studies* 3: 20–46.

———. 1981. "The Enlightenment in England," in Porter and Teich, eds. 1–18.

———. 1982. "Was there a Medical Enlightenment in Eighteenth-Century England?" *British Journal for Eighteenth-Century Studies* 5: 49–63.

———. 1990a. "The Terraqueous Globe," in George Rousseau and Roy Porter, eds., *The Ferment of Knowledge*. Cambridge: Cambridge University Press, 285–324.

———. 1990b. "The Exotic as Erotic: Captain Cook at Tahiti," in George Rousseau and Roy Porter, eds., *Exoticism in the Enlightenment*. Manchester: Manchester University Press, 117–144.

———. 1990c. *The Enlightenment*. London: Macmillan.

———. 1992. *Doctor of Society: Thomas Beddoes and the Sick Trade in Late Eighteenth Century England*. London: Routledge.

———. 1996. "Medical Science and Human Science in the Enlightenment," in Fox, Porter, and Wokler, eds., 53–87.

Porter, Roy, and Mikulas Teich, eds. 1981. *The Enlightenment in National Context*. Cambridge: Cambridge University Press.

Pratt, Mary Louise. 1992. *Imperial Eyes: Travel Writing and Transculturation*. London: Routledge.

Richards, Paul. 1974. "Kant's Geography and Mental Maps." *Transactions of the Institute of British Geographers* 61: 1–16.

Robertson, John. 1997. "The Enlightenment above National Context: Political Economy in Eighteenth-Century Scotland and Naples." *Historical Journal* 40: 667–697.

Rousseau, George. 1973. "Whose Enlightenment? Not man's: The case of Michel Foucault." *Eighteenth-Century Studies* 6: 238–256.

———. 1991. *Perilous Enlightenment: Pre- and Post-modern Discourses—Sexual, Historical.* Manchester: Manchester University Press.

Rousseau, George, and Roy Porter, eds. 1987. *Sexual Underworlds of the Enlightenment.* Manchester: Manchester University Press.

Said, Edward. 1978. *Orientalism.* London: Penguin.

———. 1993. *Culture and Imperialism.* London: Penguin.

Schaffer, Simon. 1996. "Visions of Empire: Afterword," in Miller and Reill, eds., 335–352.

Schmidt, James, ed. 1996. *What Is Enlightenment? Eighteenth-Century Answers and Twentieth-Century Questions.* Berkeley, CA: University of California Press.

Secord, Anne. 1994. "Science in the Pub: Artisan Botanists in Early Nineteenth-Century Lancashire." *History of Science* 32: 269–315.

Shapin, Steven. 1988. "The House of Experiment in Seventeenth-Century England." *Isis* 79: 373–404.

———. 1994. *A Social History of Truth.* Chicago: University of Chicago Press.

———. 1998. "Placing the View from Nowhere: Historical and Sociological Problems in the Location of Science." *Transactions of the Institute of British Geographers* 23: 5–12.

Shea, William, and Peter Huff. 1995. "Introduction" in William Shea and Peter Huff, eds., *Knowledge and Belief in America: Enlightenment Traditions and Modern Religious Thought.* New York: Cambridge University Press, 1–14.

Sitwell, Francis, 1993. *Four Centuries of Special Geography.* Vancouver: University of British Columbia Press.

Sloan, Philip. 1996. "The Gaze of Natural History," in Fox, Porter, and Wokler, eds., 112–151.

Smith, Bernard. 1985. *European Vision and the South Pacific.* Sydney: Harper and Row.

Smith, Neil. 1994. "Geography, Empire and Social Theory." *Progress in Human Geography* 18: 491–500.

Smith, Preserved. 1962. *The Enlightenment, 1687–1776.* New York: Knopf.

Stafford, Barbara. 1984. *Voyage into Substance: Art, Science, Nature and the Illustrated Travel Account, 1760–1840.* Cambridge, MA: MIT Press.

———. 1994. *Artful Science: Enlightenment Entertainment and the Eclipse of Visual Education.* Cambridge, MA: MIT Press.

Staum, Martin. 1985. "The Enlightenment Transformed: The Institute Prize Contests." *Eighteenth-Century Studies* 19: 153–179.

———. 1987. "Human Geography in the French Institute: New Discipline or Missed Opportunity?" *Journal of the History of the Behavioural Sciences* 23: 332–340.

Stoddart, David. 1986. *On Geography.* Oxford: Blackwell.

Thrift, Nigel, Felix Driver, and David Livingstone. 1995. "Editorial: The Geography of Truth." *Environment and Planning D: Society and Space* 13: 1–3.

Tomaselli, Sylvana. 1985. "The Enlightenment Debate on Women." *History Workshop Journal* 20: 101–124.

Venturi, Franco. 1971. *Utopia and Reform in the Enlightenment.* Cambridge: Cambridge University Press.

———. 1989. *The End of the Old Regime in Europe 1768–1776: The First Crisis*. Princeton, NJ: Princeton University Press.

Wade, Ira. 1971. *The Intellectual Origins of the French Enlightenment*. Princeton, NJ: Princeton University Press.

Wellman, Kathleen. 1992. *La Mettrie: Medicine, Philosophy and Enlightenment*. Durham, NC: Duke University Press.

Withers, Charles. 1993. "Geography in its Time: Geography and Historical Geography in Diderot and d'Alembert's *Encyclopédie.*" *Journal of Historical Geography* 19: 255–264.

———. 1995. "Geography, Natural History and the Eighteenth-Century Enlightenment: Putting the World in Place." *History Workshop Journal* 39: 136–163.

———. 1996a. "Encyclopaedism, Modernism, and the Classification of Geographical knowledge." *Transactions of the Institute of British Geographers* 21: 363–398.

———. 1996b. "Notes toward a Historical Geography of Geography in Early Modern Scotland." *Scotlands* 3: 111–124.

———. 1999. "Towards a History of Geography in the Public Sphere." *History of Science.* 34: 45–78.

Wokler, Robert. 1996. "Anthropology and Conjectural History in the Enlightenment," in Fox, Porter, and Wokler, eds., 31–52.

Wolf, Eric. 1982. *Europe and the People without History*. Berkeley: University of California Press.

Wolterstorff, Nicholas. 1983. "Can Belief in God be Rational?" in Anne Plantinga and Nicholas Wolsterstorff, *Faith and Rationality: Reason and Belief in God*. Notre Dame, IN: University of Notre Dame Press, 47–63.

Wood, Paul. 1994. "Science, the Universities and the Public Sphere in Eighteenth-Century Scotland." *History of Universities,* XIII: 99–135.

———. 1996. "The Science of Man," in Jardine, Secord, and Spary, eds., 197–210.

Yolton, John, ed. 1991. *The Blackwell Companion to the Enlightenment*. Oxford: Blackwell.

Beginnings

The chapters in this section explore several of the intellectual and geographical contexts in which Enlightenment ideas about reason, scriptural authority, and the power of geographical knowledge have a beginning. This is not to advance a common claim for the origin of *the* Enlightenment or even to suppose that there is such a thing. There is, however, a shared sympathy among the authors here with those who see the Enlightenment as a broad intellectual movement with 'origins' in the later seventeenth century if not earlier and apparently more widely than in the minds of 'great thinkers' alone. More particularly, the chapters demonstrate the complex ways in which contemporaries challenged scriptural and ancient authorities from the later 1600s and discuss the ways in which geography differently figured both as the sites of Enlightenment ideas and of the challenges to them, and as the means by which knowledge about the terraqueous globe was brought to light.

For Denis Cosgrove, there are close connections between global illumination and enlightenment as a process of shedding Western intellectual light upon the world in the work of the late seventeenth-century cosmographers Vincenzo Coronelli and Athanasius Kircher. Both men envisioned their expanding known world through related notions of emblematic geography and global (particularly Catholic) enlightenment by way of maps and texts and globes, and, for Kircher, in what Cosgrove calls "an empire of letters and printed images." In such ways, suggests Cosgrove, we may trace the origins of Enlightenment emphases upon visual images of the new worlds to longer running forms of artistic representation within Europe, as, at the same time, we can see other beginnings to Enlightenment in the connections between particular moral and imperial visions, royal authority, and the networks of courtly knowledge and patronage in places such as Versailles and Rome.

Enlightenment thus understood as a process of critical intellectual enquiry about the world and a material and spiritual imperative for the future also demanded knowledge of the Earth's past. Charles Withers shows how answers to the question "Where was terrestrial Paradise?" focused upon the geography of scriptural authority and upon the ways in which geographical enquiry was differently used to cast light upon the intrinsically enlightenment questions of human origins and the origin of language. If we may understand Enlightenment as a process of critical reasoning centering upon the tensions between the 'truths' of the scriptures and what were held to be those 'truths' derived from the empirical rational sciences in which geography should be included, then the search for Paradise, for Christians anyway, is of central importance to Enlightenment beginnings as the search for *the* site of all beginnings. As Withers argues, it was, moreover, a question that drew differently upon different geographies in search of answers. In this sense, then, the beginnings of Enlightenment may also be found in the different *ways* in which questions of reason were understood, by whom they were so understood, and more through encounters with particular places than with the world as a whole.

Situating beginnings to so complex a thing as the Enlightenment thus depends in some measure upon our being able to consider the possibility of different beginnings. It depends, too, upon the importance of local and specific originating contexts, intellectual and geographical, in which, for example, questions about rational religion, moral philosophy and the human condition past and present were grounded. These are the concerns of David Livingstone's chapter. For Livingstone, geographical enquiry in the early modern period was crucial to the beginnings and later development of Enlightenment scepticism, not just because of the ways in which different parts of the globe were used to test extant authority, but also because what contemporaries understood as geographical knowledge led to profound scepticism in matters about the age of the Earth and the account of human origins. For Livingstone, the speculations of the mid-seventeenth-century French thinker Isaac de la Peyrère about the polygenetic origins of humans may be seen, at one and the same time, as an individual exercise in biblical exegesis, and as symbolic of that undermining of scriptural authority in which new geographical information played an important part. Elsewhere, and over a century later, matters of racial geography were differently used to support other moral and political visions of Enlightenment. As Livingstone shows in discussing the late eighteenth-century Americans Samuel Stanhope Smith and Jedidiah Morse, ideas about climate and race—the substance of Peyrère's work and so crucial to contemporary enlightenment theories about comparative human development—could be drawn upon in their work to legitimate ideas of Enlightenment based not upon biblical scepticism but upon Christian moderatism and, for Morse, even liberal republicanism.

Taken together, then, these chapters suggest the Enlightenment to have had different 'beginnings' in different places and in different questions, not in single given years, individual works, or even in the rise of secular reason over scriptural authority. Such different beginnings, it is argued, depend greatly upon geography, understood both as a matter of particular location and, as the chapters later in the book show, as particular practices of enquiry.

Chapter Two

Global Illumination and Enlightenment in the Geographies of Vincenzo Coronelli and Athanasius Kircher

DENIS COSGROVE

2.1. Introduction

In his 1566 book of *imprese,* the Venetian Girolamo Ruscelli used images of the globe as the key to emblematic devices he designed for both Francis II of France and Philip II of Spain. These rival monarchs, rulers of the two most powerful territorial states then disputing imperial hegemony over Europe, the oceans, and newly discovered coasts, had equal claim to an icon inherited from Charles V's proclamation of Christian empire and his global motto: *plus ultra* (Pagden 1995). Ruscelli was a member of the Venetian *Accademia della Fama,* publisher of Ariosto's geographical poem *Orlando Furioso,* Italian translator of Ptolemy's *Geography,* and author of a treatise on the manufacture of globes. His interest in matters global thus extended well beyond the simply rhetorical. The signifying globes of his imperial *imprese* are scientific constructions, mathematical graticules, that can act simultaneously as spatial ordering devices, accurately recording the territorial extent of sovereign power, and as symbols of monarchy's Apollonian perspective, seeing and illuminating an inferior, subject world (see also Part 2, "Mappings"). They signify the profound attachment of cosmographic science in Europe's early modern culture to an imperial mission of spreading the light of Christian empire across the surface of the earthly globe.

In what follows I explore this metaphor of light and the related idea of illumination, showing how permeable are the conventional historical and conceptual limits of 'the Enlightenment.' When Hulme and Jordanova (1990, 3) claim that "light was a central metaphor for knowledge long before the Enlightenment," they identify only one aspect of a much richer set of associations for light and illumination. Emblematically, light also had long signified both the redeeming grace of Christ and the power and authority of monarchy. The sun's symbolism had been significantly elaborated by sixteenth-century Neoplatonists and Copernicans in connection with ideas of metaphysical, natural, and social harmony. Both the Counter-Reformation Church and secular monarchs of absolutist states appropriated the symbolic power associated with the sun, so that the "light of the world" might apply equally to the Creator Himself, His Son, His Son's Church, or a crowned monarch, whose authority was divinely derived. Spreading light across the world was a Christian imperative, tragically blocked in the East by schism and the darkness of Islam, sadly fractured within the European borders of Christendom by doctrinal strife, but triumphantly following the course of the sun, west across the globe's oceans, into newfound lands. For Renaissance cosmographers such as Ruscelli, the focus was on the Americas; by the mid-seventeenth-century Europe's positioning within a *global* space was much clearer. As Paula Findlen (1996, 81) points out: "whilst sixteenth-century nationalists faced the dilemma of incorporating the artifacts of the Americas into their cosmographies, the seventeenth-century Jesuits attempted to develop a moral religious and philosophical framework that connected *all* the different regions of the globe." Examining the geographical works of two seventeenth-century ordained Catholic scholars, the Jesuit Athanasius Kircher and the Franciscan Vincenzo Maria Coronelli, allows us to make linkages, through the metaphor of illumination, between the metaphysics of late Renaissance cosmography and the rational empirics of Enlightenment exploration. I shall compare the projects of these two scholars, generally discussed in separate literatures, and their related visions of emblematic geography and global enlightenment, the Minorite and the Jesuit, in the context of a Baroque language of light and vision, paying attention to the rhetoric of Catholic empire as a missionary project in which Paris was rivaling Rome as the center from which the light of Faith was to be projected across the globe.

2.2. Vincenzo Maria Coronelli and Cosmography

Over a century after Ruscelli's emblem was published, another Venetian pressed the signifying potential of the geographical globe recognized by Ruscelli to unparalleled scientific and rhetorical extremes in order to flatter the imperial pretensions of France, in the person of its Sun King, Louis XIV. Two vast spheres, each 3.9 meters in diameter and raised on elaborately designed stands, were conceived and constructed by Vincenzo Maria Coronelli (1650–1718) as central elements within the global geographical iconography of the Palace of Versailles and presented to Louis in 1683. They stretch Ruscelli's emblematic theme to its logical conclusion. The printed emblem with its esoteric interplay of text and image here takes the form of illuminated, mathematically correct, scale models of the celestial and terrestrial globes, their textual devices inscribed within the borders of elaborate cartouches. The globes' dual claim, to empirical accuracy as illustrations of the state of nature and to moral authority as expressions of divinely ordained sovereignty, sustains their emblematic status. The dedicatory inscription on Coronelli's celestial globe proclaims it a picture of the sky

> In which all the stars of the firmament and the planets, are placed in the very locations where they were to be found at the birth of this glorious monarch, in order to conserve for all eternity a fixed image of that hour and disposition under which France received the greatest gift that the heavens had ever offered to the earth.[1]

The heavens themselves inscribe the auguring moment of Louis' birth, as John Milton's "Hymn on the morning of Christ's nativity" claimed they had done at Bethlehem. The accompanying terrestrial globe is equally explicit in its moral claims to be an emblem of Louis's character and purpose. Its image of the world known to Europeans at the time of its manufacture

> Renders continual homage to his glory and heroic virtues, showing the countries where a thousand great actions have been executed both personally and through his command, to the astonishment of the nations, which he could have subordinated to his empire had not his moderation arrested the progress of his conquests and prescribed the limits of his valour. . . . (quoted in Jacob 1992, 449, n. 112; DEC translation)

Such use of the terrestrial globe as an imperial symbol, and of the celestial orb as an emblem of divine approval and inspiration for an Apollonian ruler, reaches back in European cultural history to the first self-proclaimed Roman Emperor, Augustus. Indeed, in 1702 the French Petite Academie referred explicitly to the Augustan practice of picturing the location of the constellations on the obverse of coins and medals showing the emperor's profile. And Coronelli speaks in his written cosmographic text, the *Atlante Veneto,* of the armillary sphere, the absent third presence within the Versailles globe project, as an image of the *primum mobile,* whose invisible surface represents "the unity of all creation" (Coronelli 1690, I, 2). But the emblematic globes designed by Ruscelli, Coronelli, and others in the years between the later sixteenth and early eighteenth centuries are more than merely humanist references to Classical practice. Their makers are careful to mark the progress of modern discovery, negotiating the awkward path between declaring universal and metaphysical verities and accurately recording celestial and terrestrial spaces that were under more or less continuous revision from observation and navigation. They refer directly to contemporary intellectual and political practices, drawing structurally and iconographically upon the findings of theoretical cosmography and geography and empirical astronomical and geographical discovery, while illustrating and articulating moral absolutes of authority, virtue, and rule.

The globes made for Louis XIV were never in fact erected in the Palace of Versailles. They remained for twenty years in the Paris workshop where they were manufactured; in 1704 they were transported at the direction of the king to the Château at Marly (appropriately in terms of elemental emblems, Marly was where the hydrological machines which controlled the waterworks at Versailles were also housed). Square pavilions were erected especially for displaying the celestial and terrestrial globes, decorated with astronomical figures (the principal systems, eclipses, lunar phases) and geographical maps, respectively, complete with surrounding galleries so that the spheres could be viewed from above (Jacob n.d.). The globes remained at Marly until 1715, when they removed to the Royal Library at the Vieux Louvre (now the Bibliothèque Nationale), remaining there for two centuries before being stored, still unseen, in the Orangery at Versailles. They have thence emerged only once, for brief exhibition at the Centre Pompidou in 1980 (Pelletier 1993).

In a recent paper on these, by far the largest, Baroque globes, Monique Pelletier has suggested that the transient character of the knowledge represented on their surface, especially in the case of the terrestrial globe, designed at a moment when French geographical exploration and cartographic representation were entering their period of most rapid progressive change, rendered them an expensive and elaborate anachronism, unworthy of the expense demanded by their erection. Yet she also notes their appropriateness to the iconographic scheme of Versailles itself, which was global in conception, in both celestial and terrestrial senses. Referring to the contemporary French construction of a North American empire spearheaded by Cavelier de la Salle's exploration of the Mississippi, which provided vital source materials for Coronelli's representation of the continental interior, she states, "Louis XIV's colonial projects went beyond the limits of simple mercantilism: colonial conquest and propagation of the true faith cannot be disassociated from one another"

DENIS COSGROVE

(Pelletier 1993, 51). Thus Coronelli's work, although rigorously non-speculative in its cartography—in this sense a model anticipation of secular, Enlightenment geography—cannot be disconnected from another tradition of geographical study, concerned with illuminating the globe speculatively and actually with the light of Christian faith, mirrored from its cosmic source through the prince's body and projected thence across terrestrial space. In this tradition, enlightenment radiates from heaven to be reflected to the corners of the earth by the mirror of Catholic doctrine. Revocation of the Edict of Nantes in 1685, two years after Coronelli completed his globes, saw France assume the role of secular arm for Rome's missionary project of global conversion. As a Minorite friar whose geographical successes played a role in his rise to Vicar of his Order in Rome, Coronelli could not remain unaffected by this context (Miscellanea Francescana 1951; Wallis 1969). It is a context which will be illuminated much more clearly when we consider the role of Athanasius Kircher, whose influence in Rome remained powerful in the years of Coronelli's studentship there, and whose Society of Jesus played the central part in the missionary side of Louis XIV's imperial project.

Born in Venice, Coronelli studied for his holy orders at Rome and Padua, designed his globes in Paris, but worked for most of his life in the cartographic workshop he established at the Frari in Venice, achieving in the course of a long career both sacred and secular recognition. He rose to become briefly Vicar General of his Order, probably through the patronage of the French Cardinal Estrées at Rome, who, together with his brother, Admiral of France, had commissioned the Marly globes. Coronelli also enjoyed the titles of "Cosmographer of the Venetian Republic" and Reader at the city's university. Reviving the Venetian tradition of acting as scholarly clearing house and publication center for European geographical discovery, Coronelli established in 1680 what has been termed Europe's first Geographical Society, the *Accademia degli Argonauti,* under whose imprint his enormous output of geographical texts, globes and maps was published. This short-lived academy was in part Coronelli's means of acquiring cartographic materials and geographical information from a membership widely spread across Europe, and it included both scholars, such as Louis XIV's astronomer Jean-Dominique Cassini, his chief mechanician and engineer, the Abbé S. de Vauban, and political figures such as the King of Poland (Coronelli 1693; Mattelart 1996, 3–9). But more important, the Argonauti also offered a sophisticated means for financing Coronelli's publications. Each of the academy's 261 members, organized into local colleges at centers such as Rome, Milan, Florence, and Bologna in Italy and Paris and Moscow beyond, was committed to purchase six engraved plates sent out every month from Coronelli's Venice workshop. Of the eleven hundred such engravings eventually produced, only some two hundred and fifty were maps. The rest included topographic and hydrological diagrams, illustrations of naval architecture and ship designs from various parts of the world, emblematic devices, and portraits. The scale and influence of Coronelli's enterprise is indicated by the record of thirty Russian scholars sent to Venice by Peter the Great in 1697–98 to study cosmography, in anticipation of a visit (eventually cancelled) of the Tsar himself, as part of Peter's project of introducing Russia to Western European science, culture, and technology.

Coronelli published his lectures in cosmography attended by the visiting Russians as a two-volume *Corso Geografico Universale* (Coronelli 1692). The text summarizes much of the friar's vast multivolume cosmographic and geographic work, notably his *Atlante Veneto* (1690–97), many of whose illustrations are recycled in the *Corso,* and later in his *Biblioteca Universale,* an alphabetical encyclopedia of universal knowledge, initially projected to run to forty volumes and completed in manuscript up to the letter M, of which only seven volumes (A–C) were actually published, for lack of funds.[2] Even more condensed was the *Epitome Cosmografica* (1693), a single-volume summary of cosmographic knowledge produced under the inscription of the Argonauti. Between 1685 and 1701, when Coronelli left Venice for three, ultimately unsuccessful, years in Rome administering the Minorites, the global projects generated from his Frari workshop returned Venice to the central position it had enjoyed as Europe's clearing house for geographical publication in the days of Giacomo Gastaldi, Gianbattista Ramusio, and Girolamo Ruscelli (Bevilacqua 1984). By the end of Coronelli's life he had published nearly ninety works out of a projected total of 129, designed nine distinct globe types, and generated over seven thousand illustrations.

It is difficult to make coherent sense of Coronelli's vast geographical output. It is at once universal in conception and yet in execution theoretically limited and unoriginal, even parochial, often driven as much by the exigencies of immediate, local events as by any coherent philosophy or consistent method. His republication of texts and reuse of cartographic materials reveal an ambitious scholar but one pressed by both financial exigency and the need to amend, correct,

DENIS COSGROVE

and catch up with a fast-changing scientific picture. They reflect perhaps also the difficulties, despite his network of contacts, of a scholar working individually in Venice, a city which by Coronelli's time was distinctly marginal to the major currents of European thought and affairs.

The *Atlante Veneto* proclaims itself "a geographical, historical, sacred, profane, and political description of the empires, kingdoms, provinces, and states of the universe, their divisions and borders, with the addition of all newly discovered lands, and expanded by many newly published geographical maps" (Coronelli 1690, title page). Its first volume is the closest textual equivalent to the Paris globes in scope and content. It opens with a chronological listing of previous geographical authorities, plates which illustrate the great globes made for Louis XIV, the nomenclature of cosmography, summaries of the competing world systems (Ptolemeic, Copernican, and Tychonian), and 'ancient' and 'modern' theories of the age and material constitution of the earth. As with the detailed legends covering its terrestrial globe, there is little attempt in the *Atlante* to offer critical commentary on the authorities and views that Coronelli records. As an author he seems keen to avoid contention, more concerned to be comprehensive and contemporary in coverage. Cosmographical knowledge is presented as a hierarchy of representation, proceeding from the divine *Fiat,* through "the order of Creation" to the scale of individual regions and cities *(iconografia),* palaces *(scenografia),* and rivers *(potomografia).* The scale of Coronelli's cosmographic conception is graphically captured in his engraved *Idea dell'Universo,* whose concentric spheres expand from the lowest house of Hell at the very center of the earth, through Purgatory and Limbo immediately below the terrestrial surface, to the four elements, the seven heavens, the fixed stars (for whose numbers Coronelli evaluates competing claims), and finally the Primum Mobile. Smaller spheres illustrate the flux and reflux of the tides; relations between planets, zodiacal spirits, and the metals; eclipses, and the planetary horoscopes under which the continents and countries of Europe are positioned.

L'Idea dell'Universo broadly illustrates pre-Newtonian cosmography: a conventional Aristotelian synthesis of natural philosophy elaborated by the astronomical work of moderns from Copernicus through Galileo, Brahe, Kepler, and Scheiner. Coronelli's accompanying text also follows that cosmographic narrative which originated in medieval encyclopedias and had been elaborated by sixteenth-century European writers such as Martin Waldseemüller, Peter Apian, Sebastian Munster, and André Thevet, contrasting the classical *oikoumene* with the modern globe and describing systematically the pattern and contents of its continents, oceans, and rivers. Coronelli is by no means unconscious of the strains placed on the cosmographic model by recent geographical discoveries, scientific observations, and theoretical debates. He summarizes, for example, opposing views on planetary motion, eclipses, and earthquakes and discusses the problem of fixing the longitude. But these matters generally appear as interruptions rather than opportunities for argument in a work governed principally by taxonomic demands and empirical inclusiveness rather than any strong desire to engage in debate over such contentious matters as the Galilean controversy or the veracity of scriptural views about the physical creation. Coronelli's attachment to representing global unity seems directly connected to his skills as an image maker rather than as a writer or critic.

Coronelli's taxonomic endeavors and empirical inclusiveness are insufficient to hold the work together. Even in the first, and most coherent, volume, the maps and illustrations seem randomly chosen and often disconnected from the text, their form and subject matter driven by extra-academic concerns such as the financial exigencies of the Argonauti Academy, which required a fixed, regular output of images from Coronelli's workshop. Succeeding volumes of the *Atlante* lose coherence even further. Volumes 2 and 3 constitute an *Isolario,* the last expression of the Venetian tradition of geographical representation that had originated in the city's own insular topography and in the Republic's mercantile and navigational focus on the Adriatic and Aegean archipelagos. Choosing this anachronistic mode of representing the globe as islands of ever increasing size was also a response by Coronelli to the specific historical context of Venice's recovery of Morea (Peloponnese) from the Ottomans in 1687, which was seen optimistically by the Republic to augur a reversal of its secular decline. The mood in Venice was exploited by an ambitious and entrepreneurial Coronelli in dedicating his work to the Venetian Senate. Thus the second part of the island book opens with the visual conceit of Neptune presenting the emblem of the Argonauti to the victorious Venetian Doge Vandramin, signifying that universal empire is offered to Coronelli's native city in the form of global geographical knowledge. Volumes 4 and 5 constitute the *Corso Geografico Universale,* and the following two volumes a

Teatro della Città, illustrating the cities of the world and modeled on Braun and Hogenberg's chorographic project of a century earlier. The remaining four parts of the *Atlante* cover various subjects and documents, including the *Libro dei Globi* (1697) in which Coronelli reproduces his designs for the Marly globes, reduced in scale, as gores (Coronelli 1697). Global conception and local contingency clash directly throughout the *Atlante,* ultimately undermining its claims to encyclopedic cosmographic coverage and scholarly coherence.

The vast project represented by the *Atlante* was underwritten financially, if not intellectually, by Coronelli's *Accademia degli Argonauti.* Itself typically overcomplex, the Academy supposedly had branches across Europe. But there is little evidence of regular meetings, scientific investigation, or correspondence between members; rather, the group provided a secure list of subscribers to finance Coronelli's Venetian workshop production. Given its courtly membership, it is impossible to distinguish in the motives of the Argonauti academicians fashionable desires to possess finely wrought cartographic productions from a master who had worked for Louis XIV from scholarly or diplomatic intentions to secure the most recent cartographic knowledge from a recognized authority. The emblem designed by Coronelli for *Accademia degli Argonauti* (fig. 2.1) graphically captures the imperial scope of the unitary cosmographic ambition which sustained his projects and the syncretic nature of his learning. The emblem marries classical myth to modern empiricism and is resolutely secular. A trumpeting Fame unfurls a banner with the same motto of geographical discovery that Ruscelli had adopted from Charles V's device, *Plus Ultra,* while geographical and navigational instruments decorate the image as the triumphs on a Roman imperial arch. At its center a ship surmounts the globe's graticule, while Hercules' club and bearskin cast below suggest that the Moderns have superceded the greatest of the classical heroes, passing beyond the Pillars named for him and reaching to the ends of the earth.[3] Like previous cosmographers, including his Venetian predecessors Gianbattista Ramusio and Giacomo Gastaldi, Coronelli sought to bring the ends of the earth within the scope of a single illustrated text, but like them and like Sebastian Münster, their German contemporary, his text represents "a momentary and desperately ambitious effort to gather up in grand synthesis . . . the admirable variety of the world" (Lestringant 1994, 129). In some respects the failure of the *Atlante Veneto* appears to illustrate the progress over the succeeding century of a "crisis of cosmography," identified by Frank Lestringant in work of the sixteenth-century French cosmographer, André Thevet. In Lestringant's opinion, this crisis was religious, methodological, and epistemological:

> From the religious point of view, the cosmographer who raised himself to the level of the Creator in order to attain the latter's eternal and ubiquitist [sic] knowledge was guilty of pride, even blasphemy: he pretended to correct Scripture in the name of his sovereign, unlimited experience. At the level of method, he sinned by incoherence, confusing scales of representation and imagining that autopsy (or seeing for oneself) could guarantee the truth of a synthetic, and necessarily secondary, vision. Finally, from the epistemological point of view cosmography, which supposes a monumental compilation under the controlled authority of a single individual, was soon transcended by more supple and open forms of geographical knowledge. (Lestringant 1994, 130)

Coronelli's texts justify these claims of methodological and epistemological incoherence, in his case exacerbated by the financial arrangements that governed their production. They always multiplied conceptually, outrunning their author's capacity to contain them, either intellectually or financially. Yet, correct as his critique of the cosmographic project may be, Lestringant is mistaken in two respects. First, this "momentary effort" endured well beyond the late sixteenth-century decades of Thevet's life and was still being attempted at the very threshold of the Newtonian Enlightenment. Second, Lestringant may place too great an emphasis on the textual aspects of the cosmographic project and pay insufficient attention to the power of images themselves and the nature of their relations with text. Within cosmography, graphic images, and specifically cartographic-emblematic images, were always more than merely illustrations of a more significant narrative. Images had their own logic; their distinct claims to truth lay in their capacity to represent phenomena to the eye and in the graphic rules of mathematics and perspective which controlled their construction, "to recreate the world according to inherent principles of its own intelligibility" (Cavaillé 1991, 49). This was true above all of phenomena from 'new worlds': worlds apart in space through either distance (America or the Indies) or scale (visible only through microscope or telescope) (Alpers 1983), or their metaphysical qualities, and thus which could be seen but not necessarily spoken. Images brought such unnamed phenomena to light, giving meaning and coherence to a world which often defeated the synthetic capacities of text. Of these phenomena, perhaps none lent itself more powerfully to such a discourse of images than the sun itself, the source of light itself, at once

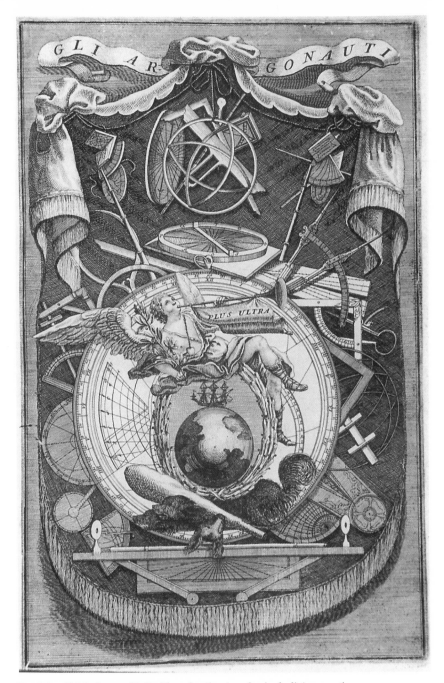

Fig. 2.1.   V. M. Coronelli: Emblem for the *Accademia degli Argonauti*

visible and yet blinding, and thus able to be observed only indirectly through the traces it burned through the lens onto paper (Panese 1997).

Light played a key role in the creation of truthful images, as did *enlightenment:* the intellectual and moral status of the observer, a point also made in Outram's discussion of travelers. When connected to words and text, the power of images could be very considerably enhanced. This was the fundamental principle of the *emblem,* or device, that was such a characteristic feature of early modern European culture (Cosgrove 1990; Ginzburg 1986; Gombrich 1948; Moseley 1989; Watson 1993). Although emblems were popular among both Protestants and Catholics, visual images attracted less suspicion within the powerfully iconographic Tridentine Catholic tradition than within a more frequently iconoclastic Protestant culture. Coronelli was first and foremost a creator of visual images, before he was a writer. His reputation rested during his life, as today, principally upon the globes he made so relatively early in his career for a Catholic sovereign not troubled in the least by those charges of pride and blasphemy Lestringant levels at cosmography. Coronelli's globes have a significance quite distinct from the texts of the *Atlante Veneto* in which they were described and reproduced as printed gores and which ultimately do fail the tests of epistemological or methodological coherence. It is therefore the globes, and the spaces of knowledge they assert, and into which they were inserted, that the significance of Coronelli's cosmographic project should be considered.

2.3. Globes and Enlightenment

The Versailles globes were commissioned from Coronelli by Cardinal César d'Estrées in 1680. Historians of cartography regard them as the greatest and most elaborate expressions of a tradition of Baroque globe-making that opened with Mercator's mid-sixteenth-century spheres and continued through the work of the Dutch manufacturers Hondius and W. J. Blaeu. Such globes were designed for public display in the palaces of European princes, evolving from the cosmographic rooms and *studioli* of the early Renaissance palace, for example, Ambrogio Lorenzetti's revolving planispheric disk designed for the Palazzo Pubblico at Siena in the 1340s or Egnazio Danti's mid-sixteenth-century work for the Medici in Florence and Rome. Globes had become during the sixteenth century the focal point of iconographic programs which incorporated architecture, mechanics, and decoration, systematically organizing the spatialities of a part of the palace as a microcosm of the universe. By the later seventeenth century, cosmographic images (heliocentric at least in their rhetorical spatialities), globes, and maps, were being elaborated everywhere in Europe as the pretensions of its monarchs, merchants, and philosophers competed to appropriate the symbolic authority of the globe. Adam Oleario, for example, produced three-meter diameter globes for Duke Frederick of Holstein, while Gerardo Weigel constructed a ten-meter armillary sphere, his *Pancosmo,* at Nuremburg in 1699. This powered mechanical wonder actively demonstrated universal time and space and illustrated the earth's physical and political geography. Held jointly by Hercules and Athene, the earth's internal fires were dramatized by erupting volcanoes, their smoke blown across the surface by breezes of air fed from the device (Stevenson 1921, 78). Manifestly, the primary conceit of such mechanical inventions was the prince's universal authority and power. However, the degree of public access to the spaces where these wonders could actually be witnessed varied, and thus the balance between their place as rhetorical expressions of power and legitimators of strategic scientific knowledge was by no means stable.

Coronelli's globes were initially destined to be placed in the Petite Orangerie at Versailles, a public space designed to flatter and celebrate royal authority, rather than for strategic deliberation or policy formulation. The minister charged with overseeing Le Vau's building plans and works at Versailles was Jean-Baptiste Colbert, also Louis's comptroller-general of finances and principal advisor on French overseas trade and commerce. It was Colbert who encouraged Louis's interest in Cassini's astronomical and geographical work, and who presented the King with model naval spectaculars at the Petite Venise. Colbert too was responsible for founding the Petite Academie, whose brief was the pursuit of esoteric knowledge in order to frame emblems and medals, those characteristic craftings of image and text designed simultaneously to illuminate knowledge for the cognoscenti and to obscure it from the ignorant and unworthy. The Petite Academie recommended that the decorative organization of the palace at Versailles be determined according to "relations among the influences and qualities attributed by mathematicians to the seven planets" (Pelletier 1982, 75), influences illustrated by Coronelli's image of the human microcosm in his *Idea dell'Universo.* The formal plan for the appartments at Versailles, into which the French court moved in 1682, was to be based on a heliocentric planetary system within which the various subsidiary rooms revolved around those of the Roi-Soleil himself, and Le Brun's decorative scheme for

Versailles owes much to the cosmogonic description provided by Ovid in the opening pages of the *Metamorphoses,* a basic text for iconographers and cosmographers alike (Schama 1995).

Coronelli's globes thus constituted a cosmographic emblem, a visible commentary on the symbolic spatialities of Louis's person, palace, court, and empire. Rather than circulating around his kingdom in the manner of the traditional sovereign, Louis gathered French space and society into the circuit of Versailles itself. A more fitting emblem of this spatial order than Coronelli's great globe at the heart of the palace could hardly be imagined. The rituals of Versailles revolved around the Apollonian conceit whereby the spreading light of kingly authority illuminated political and intellectual darkness, while the sovereign vision unified and controlled global space. This conceit, common to Baroque princes but taken furthest in France, had been anticipated at Louis's birth on 5 September 1638, the date commemorated by the arrangement of the heavens on Coronelli's celestial globe. The Italian utopianist Tomasso Campanella, heliocentrist and inheritor of the hermetic tradition of cosmography, had cast a horoscope for Louis, France's Apollo and the epitome of Catholic absolutism. The text of Campanella's theocratic *City of the Sun* (Campanella 1981) seems to anticipate the project for Versailles. The imaginary city is "divided into seven large circular areas named after the seven planets, and the way from each circle to the next is along four roads and through four gates which face the cardinal points of the compass." The city is dominated by a temple whose altar is decorated by celestial and terrestrial globes from which metaphysicians, furnished with astrolabes and telescopes, study the heavens, while geographic and ethnographic information is furnished by explorers and ambassadors (Pelletier 1982, 1993). Campanella, who had lectured in Paris just before his death in 1639, envisioned the supreme authority in his utopian city as a priest/philosopher/king, enlightened and enlightening in a similar manner to the philosopher-scientists of Francis Bacon's *New Atlantis.*

At Versailles the beams of Apollonian illumination radiated out from the palace into nature, moving along the axes of André Le Nôtre's designed landscape park. The principal path along which the royal gaze was directed moved from east to west as the line of the sun's progress, ending in the great fountain of Apollo where the gilded sun god emerges from the waters. The waterworks at Versailles, fed from Marly, represent the rivers of France while the largest of its canals was filled with a selection of the world's ships similar to those illustrated on the oceans of Coronelli's globe and later the subject of an entire volume of the *Atlante Veneto* (Schama 1995, 338–44). Versailles was a vast geographical as much as an astronomic conception:

> The four parts of the world were omnipresent, in garden statuary, in the pictures decorating the Ambassadors' staircase, and those of the royal appartments, which also represented the planets. The interest accorded by the King to the knowledge of the world was underlined especially in the rooms of Mercury, protector of the arts and sciences. In the King's rooms, one saw Ptolemy in conversation with the scientists in his Library, and Alexander receiving the animals of the whole earth, permitting Aristotle to write his natural history. . . . (Pelletier 1993, 47)

Within this context Coronelli's globes achieved a clear iconographic logic, acting as perfect emblems of the moral cosmos as viewed from Versailles:

> Symbols destined for the palace of Apollo, where the terrestrial globe would bear witness to a reality too distant for a king whose attention was focussed on the immediate eastern and northern frontiers, while the celestial globe is a 'relief figure of the Prince's nativity': the prince charged by Campanella to realise the project of religious unity and universal theocracy described in *The City of the Sun*. (Pelletier 1982, 88)

Within Versailles' scheme, spatial and iconographic—more than narrative—coherence counted. Observational accuracy was more critical than theoretical rigor. The emphasis was firmly on sight and vision. From the Hall of Mirrors (where at one stage Coronelli proposed locating his globes), the kingly eye could revolve from the ceiling decorated with Le Brun's image of Apollo crossing the Rhine and overawing the narrowly commercial Dutch, to majesty's own infinite reflection in the endless mirrors, and out through the gallery windows to pursue the line of the *grand allée* towards the western horizon. Coronelli's globes extended that gaze, empirically and symbolically, over terrestrial and celestial space towards infinity. They were a significant contribution to that "dialectic of reduction and expansion" which Christian Jacob has claimed as one of the keys to the Baroque imagination, allowing "us to rank together the astrolabe, globe, map and clock, whose transparency exhibits immutable mechanics" (Jacob 1992, 406). Coronelli's terrestrial globe has a bewildering excess of detail, both iconic and textual. Its illustrations include an inventory of ships from different marine cultures and of marine creatures filling oceanic spaces, and nearly one hundred painted scenes and emblems, including allegorical personifications of the continents, portraits of explorers and geographers, and images of the globe's faunal, vegetable, and ethnographic variety. Six hundred textual cartouches record both historical and modern geographical knowledge. These texts are not readily legible to the casual viewer; the scale of the work alone rendered them de facto esoteric knowledge. Like the contemporary published emblem, the globe's contents simultaneously illuminate and obscure the nature of the world, rendering it visible only to those intended to see it and specifically to the eyes of the king whose global authority is thus implied. Louis's courtier, François le Large, in a 1710 letter to Coronelli acknowledges this fact, stating that he has commissioned a special pair of lenses specifically for Louis to view the globes. Moreover, he has had their texts transcribed into two private notebooks which also contain detailed commentaries on the iconographic and allegorical significance of Coronelli's images and texts.[4] Le Large thus treats the globes precisely as emblems, whose knowledge is at once hidden and arcane, yet illuminating:

> Le Large, relying no doubt upon Coronelli's own archive, bore witness to the complexity of the Venetian cosmographer's project: a rich iconography appears in a collection of mnemonic images, in which bibliographic references, recent discoveries, theories, natural phenomena, technical processes, narratives of exploration, stories of conquest and wars are visually encoded. Thus, the globe illuminates as a network of memory places and mnemotechnical signs which refer back analogically to texts and descriptions. These analogical references are controlled by a strict graphic economy which recalls the logic of emblems: nothing is left to chance; objects, pictorial composition, detailing are all significant. (Jacob n.d., 11)

The king, whose Apollonian nativity is recorded on the celestial globe, and whose terrestrial position is recorded in the design of Versailles at the center of a divinely favored France, may encompass the earthly sphere with a royal vision, assisted by his lenses and Le Large's text, and thus access the arcana of global geographical knowledge.

A number of the inscriptions on Louis's globes later appear in the text of the *Atlante Veneto,* part of Coronelli's endless recycling of his work. They may be read as sober, factual statements of geographical science and discovery which seek to remain up to date with the most recent reports. Much has been made, for example, of Coronelli's inclusion of the details of the Frenchman Cavelier de la Salle's exploration of Louisiana, descending the Mississippi during the same years in which the Venetian cartographer was working on his terrestrial globe in Paris. Pelletier illustrates Coronelli's use of the la Salle expedition's manuscript maps to make a claim for the latter's familiarity in 1683 with the most recent knowledge and his access to the most accurate sources. Such concerns could not but render Coronelli's image of the globe a victim to changing knowledge in an age of rapid navigational progress that quickly rendered any such geographical image anachronistic. But cartographic and geographic empiricism cannot be divorced from the emblematic and iconographic purposes of the globes, nor their surface details and decoration from the globes' construction and use as elements of material culture within the play of representations constituted by Versailles. Viewed from this perspective, the globes, which all observers recognize as constituting Coronelli's principal claim to contemporary fame and which predate his great textual projects of the later 1680s and 1690s, reveal a coherent conception in the work of the youthful Coronelli. Such coherence challenges Lestringant's conclusions that cosmography was already a failed enterprise in the late sixteenth century. The textual project of the *Atlante Veneto* might lack theoretical integrity, the globes when understood emblematically and contextually do not. The continuing vitality of the cosmographic project is indicated not only by the overall conception of the globes and their iconography, it is echoed in Coronelli's concern with the same speculative geographical questions that had occupied his cosmographic predecessors more than a century earlier.

Some of these speculative concerns gain prominence from the actual spaces they occupy on the globes' surfaces. An example is the River Nile, whose subequatorial sources and annual flood constituted Europe's longest-standing geographical mystery. They are the subject of an elaborate cartouche and long dissertation on Coronelli's terrestrial globe, in which the Nile's numen is represented iconographically as a bearded river god occupying the reedy marshes of Lake Zaire in equatorial Africa. His accompanying nymphs support column, pyramid, and obelisk, symbols not only of ancient Egypt but of the earliest sources of esoteric knowledge, which Coronelli attributes to Hermes Trismegistus (some seventy-five years after the hermetic texts had been shown conclusively to derive from early Christian Neoplatonism). A nearby cartouche shows a contemporary geographer lifting a veil to reveal a second source of the Nile, connected via an underground channel to the River Niger at Lake Zarlan (Chad). The texts summarize the historical evolution of European knowledge of this mystery and indicate Coronelli's belief in the river's origin 12° south of the equinoctial line and passage through 40 degrees of latitude. Elsewhere, Coronelli describes oceanic tides as another great mystery of nature, subject of continued speculation among natural philosophers. He relates tidal movement to lunar influence, not because of Newtonian gravitational attraction, but through the moon's planetary sovereignty over cold things. While such commentaries suggest a more serious involvement with theoretical geographical discourse than initially appears from Coronelli's taxonomic work, they are conventional, drawing on long-standing cosmographic nostrums rather than the findings of contemporary natural science. Pursuing the sources of his theoretical observations both within his own texts and Le Large's extended commentary signals the continued fertility of ideas of metaphysical enlightenment which had fed the cosmographic project since the later fifteenth century and which found their greatest Baroque spokesman in Athanasius Kircher.

2.4 Athanasius Kircher: Geographical Enlightenment through Signs and Wonders

Athanasius Kircher (1602–1680) was Baroque Rome's most prominent natural philosopher and cosmographer, architect of monumental written works in which he sought to illustrate and explain physical and metaphysical worlds, from the earth's core to the highest heavenly regions (Findlen 1996; Godwin 1979; Rowland 1995; Schama 1995, 300–302; Stafford 1994). Kircher's writings play across the Ptolemeic scales of cosmography, geography (including a description of China), and chorography (in regional studies of Latium and the Alban lakes). Much of the material for his publications derived from the global correspondence he maintained through his Order (Gorman 1998) and from the vast collection of objects he gathered at the Jesuit College in Rome, imported from the furthest reaches of historic time and geographic space. Kircher's display of natural and ethnographic wonders was one of the spectacles of Baroque Rome, a genuine *Wunderkammer,* matched in scope and content only by a handful of private collections in other Italian cities and by the collections of fledgling scientific societies of London and Paris. Without a visit to Kircher's collection, no seriously intellectual—or even sight-seeing—visit to Rome was complete (Findlen 1991; 1996, 70). Kircher's life project was to conjure a syncretic understanding of creation's logic from the diverse materials that made up his collections, by mapping archetypes and correspondences between the natural world and diverse forms of knowledge. In this he was continuing and developing the Renaissance cosmographic tradition which originated in the scholarly *studiolo* and the princely cabinet of curiosities. André Thevet himself, for example, had possessed a collection of "'singularities': coins from the Greek and Latin peninsulas; feather ornaments from Mexico or Rio de Janeiro; coats, olas and arrows that testified to the reality of the incredible Patagonians; hippopotamus skins, toucan beaks, stuffed parrots and caimans" (Lestringant 1994, 126). These were the same curiosities illustrated on Coronelli's globes and described in Le Large's commentary. If they sought unity in diversity through images and texts illustrated on the external surface of manufactured globes, Kircher achieved it in the interior spaces of the Jesuit College at Rome and in the empire of letters and printed images. He coordinated it according to the principles of rational belief that had inspired the initial conception of the Society of Jesus and the synthetic epistemology that had governed the Jesuit curriculum since its foundation by Ignatius Loyola in 1540. As Findlen (1996, 46) points out, the Jesuit College and Kircher, its Professor of Mathematics from 1638, "lay at the heart of what made seventeenth-century Rome both *urbs* and *orbis,* a city and a world in which power and privilege, wisdom and curiosity, were mutually exercised."

Athanasius Kircher was among the sources upon which Coronelli drew for interpreting cosmographic and geographic arcana. Occasionally the globe maker directly acknowledged his intellectual debts to Kircher, more frequently they have to be inferred from the texts and cartouches. A direct reference appears in Coronelli's explanation of the tide race and whirlpool in the Straits of Messina (fig. 2.2), referred to by their classical names of Scylla and Charybdis. Kircher had devoted a long discussion to this ancient question in his *Mundus Subterraneus* of 1665, claiming that underground tunnels connected the straits to the open seas of the Malta Channel via caverns deep below Mount Etna. Superheated seawater caused the plume of steam from the volcano and the rush of water into the whirlpool on the western side of the strait.

It was from "Jesus," the *name* of God the Son, the Christianized Apollo, charged by the Father with bringing the light of redemption to a fallen material world, that Ignatius's Society of Jesus took its own inscription, mystically rendered as IHS, with the cross superscript. Ignatius's phrase, *ignem veni mittere in terram* ("I came to bring light into the world") became a motto for both the Jesuits' active missionary task across the terrestrial spaces opened up by European

Fig. 2.2.   V. M. Coronelli: Explanation of the tidal flow and whirlpool at the Straits of Messina (Scylla and Charybdis) based on Kircher's *Mundus Subterraneus* 1665, 1678 (*Atlante Veneto,* vol. 1, 1690)

Renaissance and Baroque discovery, and for the Society's intellectual task across metaphysical space, of comprehending and representing the cosmos opened up to the modern era by the instruments of visual observation deployed by scholars such as Christopher Scheiner and Giambattista Riccioli. Among the inventions in Kircher's own museum was a *Catoptrical Theatre,* an arrangement of mirrors which multiplied the images of its contents to infinity. And, as Kircher himself points out, his museum itself—like Louis XIV's Hall of Mirrors at Versailles—was intended to extend as far as global imperium itself: "one pope and many lands was best depicted as a hall of mirrors whose 'echoes' were produced through optical mimesis" (Findlen 1996, 47). Jesuits were as devoted to placing themselves at the frontiers of scientific knowledge as at those of geographic and ethnographic space (Livingstone 1992, 63ff). Their College in Rome was a key center of seventeenth-century European scientific learning, the clearing house for the Society's global information network. Indeed, one of Kircher's most ambitious projects was the reform of geography, to be achieved by means of gathering information on local magnetic variation from Jesuit houses across the globe in order to solve the problem of the longitude (Gorman 1998, 5–11). The Jesuits' task of planetary evangelism demanded extensive language skills and with these came both knowledge of and admiration for linguistic, philosophical, and intellectual traditions beyond those of the Latin West: Byzantine, Coptic, Islamic, Jewish, and Chinese. The appeal of cosmography, in scope and syncretism, to the Jesuit vision was a predictable outcome of its missionary objectives. Matteo Ricci's 1584 Ortelian world map with Chinese phonetic equivalents and his application of the Lullian mnemonic tradition to his Chinese scholarship are an early expression of this attraction (Spence 1983, 64–65, 93–107).

Neoplatonic and Pythagorean appeals to a single creative reason, a primordial *nous* which gives unity and harmony to the world's bewildering variety, had attracted cosmographers and encyclopedists since the time of Macrobius. Since its foundation in the writings of Plotinus, Neoplatonism had emphasized *light* as the primary expression of the unitary creative source, and this had always provided a stimulus and justification for Christian studies of optical phenomena, especially of solar and celestial light.[5] Such ideas held particular attraction for Jesuit thinkers. Since the late sixteenth century it had become common to inscribe the Hebrew Tetragrammaton (script and number of creation) onto cosmographic images and world maps, signifying divine illumination parting the clouds of material unknowing in order to illuminate the true nature of the cosmos. The Tetragrammaton could readily be supplemented, or even replaced, on geographic maps by the Jesuit IHS, signifying the Son's task of bringing light to a world fallen into darkness. Jesuit emblemata also commonly incorporated images of celestial and terrestrial globes, maps, representations of light and shadow, and the IHS inscribed onto the Sacred Heart of Jesus, the Christ-Apollo (*Imago Primi Saeculi Societatis Jesu* 1640). For Kircher, working within this syncretic tradition, enlightenment was at once a spiritual and material imperative, and illumination offered a coherent mode of organizing, rationally and metaphorically, the otherwise bewildering assortment of phenomena pouring into his collection from the four corners of the Jesuit missionary empire.

Kircher's science commanded enormous respect, even among critics such as Galileo Galilei and members of the Royal Society in London. As Findlen points out, while Kircher dutifully attacked heliocentrism, he was no Catholic bigot, learning from his many Protestant correspondents. Indeed, as an enthusiastic experimenter, Kircher was in many respects himself a proponent of the new philosophy. "Yet he diverged from it, as practised by Galileo, Descartes, and many members of the Royal Society, in one important respect: his insistence that wonder was a category of analysis rather than simply a tool to lead men to the contemplation of higher truths" (Findlen 1996, 93). This embrace of wonder connects directly to Kircher's attachment to graphic images, for the primary function of encyclopaedic knowledge, empirical investigation and experimentation was the identification of signs.

It is impossible to summarize here the scope and range of Kircher's publication. His logic is perhaps best summarized in *Ars Magna Sciendi* (Kircher 1669), a summary of the Lullian combinatorial art, based according to Kircher on the two great principles of *combination* (synthesizing a unity from the multitude of diverse phenomena that initially proceeded from unity) and *analogy* (seeing every individual phenomenon ultimately as a sign reflected in every other).[6] Ultimately Kircher sought a grand synthesis of Aristotelian natural philosophy, based on a combined hermeneutics of ancient textual authority and global empirical observation (Findlen 1996, 83ff). The application of these principles is apparent in his most specifically cosmographic text: the *Iter Extaticum Celeste* (Kircher 1671), in which the Jesuit imagines himself as a hermetic traveler on a physical journey through the spheres, examining visually all the planets, including the Sun itself (whose surface distribution of volcanic fires he maps)[7] before returning towards earth and viewing from space its varied zones and regions. Among the terrestrial arcana for which Kircher offers explanations are the phenomena of volcanic activity, studied at first hand in his 1637–38 descent into the craters of Etna and Vesuvius and to which he later devoted a two-volume book, *Mundus Subterraneus* (1665, 1668), tides, and water flows, including the explanation of Scylla and Charybdis, from which Coronelli explicitly drew.

While not explicitly acknowledged by Coronelli, Kircher seems also to have been a source for the former's ideas about the Nile. Kircher was seventeenth-century Europe's leading authority on ancient Egyptian hieroglyphs, long believed to contain the secrets of humanity's origins (Rowland 1995). Such fascination with "pagan" chronologies also surfaces in Isaac de la Peyrère; see Livingstone's, chapter 4 below. The Jesuit applied his theories of linguistic and symbolic correspondence to the hieroglyphic inscriptions on the obelisk raised by Gianlorenzo Bernini above the fountain of the four world rivers in the Piazza Navona, on which the sculpted face of the Nile remains shrouded to signify the mystery of its origin (Eco 1996; Schama 1995, 302–305). For hermeticists, hieroglyphics represented the ultimate sign: a symbolic unity of word and image, standing at the origins of human knowledge, articulating the universal wisdom of the creative 'One.'[8] Both Egyptian land and Egyptian civilization were creations of the Nile River, whose flood and whose source remained the most enduring of the West's geographic arcana. Pursuing a conventional link that connected cosmography and eschatology, Kircher believed that if Egypt represented the origins of human time, it must also be connected to the origins of geographic space. The source of the Nile therefore represented not only a natural mystery as the physical point of origin of the globe's greatest river, but a source of original language, knowledge, and wisdom: it was a physical and metaphysical end of the earth, a "heart of darkness" awaiting and offering illumination (see Helms 1988, and Withers, chapter 3, for attention to the sites of paradise). Kircher claimed that the Jesuit father, Pedro Pais had actually reached the river's source in 1618, in the company of the Ethiopian emperor. Kircher's *Oedipus Aegyptiacus* (1652–54) was almost certainly one of the sources consulted by Coronelli in his cartouche inscription on the Nile referred to earlier, where the cartographer claims that "geographers have so distanced the sources of the Nile from their true position that it has come to occupy a greater part of Ethiopia, which remains currently vacant. We have rather chosen to fill that space with this discourse" (Jacob 1992, 339).

The title page of Kircher's *Ars Magnae Lucis et Umbrae* (1671b) graphically summarizes the global scope of Kircherian science and the centrality of illumination as its metaphor (fig. 2.3). Four sources of knowledge and wisdom proceed from the divine origin of light signified by the Tetragrammaton to illuminate the world. These are *sacred authority* inscribed in divine texts; *reason,* whose light proceeds from the inner eye of intellect; the *senses,* whose illumination is reflected through the figure of Apollo and enhanced by human technology in the form of the telescope; and *profane authority,* furthest removed from the divine source and represented by the impoverished light of a candle. The figures of Apollo and Diana are here used to illustrate the passage of direct and reflected light from its source to a material world awaiting illumination. The text itself covers astronomy, astrology, and optics (in Kircher's words, the relations between solid bodies and all forms of simulacra, many of them produced by manipulating rays of light). It is structured around Kircher's universal instrument of observation, his *horolabiorum,* of which he describes different forms all based ultimately on the astrolabe. Kircher also gives instructions on the manufacture of celestial and terrestrial globes, and the spherical form of the globe is the recurrent motif of his text's elaborate illustrations. Among these, the theme of global illumination is best captured in the "Universal Jesuit Horoscope" (fig. 2.4), which graphically summarizes the imperial reach of Ignatian geography. An olive tree emerges from the haloed head of St. Ignatius, kneeling on the meridian line of Rome as ships depart for new worlds. Its branches hold the names of each Jesuit provincial house, and the leaves indicate each region of the Society's provinces, "diffused throughout the whole terrestrial orb." For every location the hours of sunrise and sunset are shown by a calendrical diagram, arranged to form the IHS. The image illustrates a device that Kircher actually constructed and displayed for visitors to his museum. As Michael John Gorman (1998, 11) points out, "when a stylus was placed in each Province, and the device positioned vertically so that the Roman time was correctly given, the clock allowed the time in all the Jesuit provinces to be read correctly . . . the olive tree sundial has been designed so that the shadows of the small gnomons, when aligned, spell the abbreviated name of Jesus, IHS, which appears to 'walk over the world' with the passing of time." At the four corners of the printed image (in conventional

Fig. 2.3. A. Kircher: *Artis magnae lucis et umbrae . . .* (1671, frontispiece)

Fig. 2.4. A. Kircher: Universal Jesuit horoscope (*Artis magnae lucis et umbrae . . .*, 1646, II, f.553)

world maps the spaces occupied either by the Sons of Noah inheriting the continents or by the trumpeting angels of Apocalypse), the four parts of the world are allocated to four Jesuit saints, and in thirty-four languages the same text is inscribed: "From East to West, praiseworthy is the name of the Lord." The Society of Jesus brings the light of Faith to the world without barrier of language: saying Mass, hearing confession, administering the sacraments of faith, and enunciating the mystical name across the planetary surface, each minute of every day, illuminating the turning globe doctrinally as the sun lights it physically.

This Roman geography of global illumination: literal, metaphysical, and intellectual, finds a perfect parallel iconographic expression in the painted ceiling of the Jesuit's collegiate church of St. Ignatius, decorated in 1685–86, very soon after Kircher's death and in the same years that Coronelli returned from France, having finished the great globes for Louis XIV, to establish his Venetian academy. The artwork by Andrea Pozzo is a spectacular representation of Jesuit globalism. Drawing on a tradition of cosmographic ceiling decoration dating back to medieval Byzantium (Emiliani 1977; Lehmann 1945),[9] Pozzo rendered the vault as a single illusionistic image, an 'infinite' perspective, which carries human vision along a vertiginous axis beyond the capitals of soaring columns, through billowing clouds to a heavenly source of light. St. Ignatius receives a single ray of divine illumination from the Trinity, via the body of the Christ-Apollo, in the same way that the Apollonian figure in Kircher's frontispiece receives and transmits the heavenly light. The ray fractures from the saint's breast into a fan of beams which bathe representative figures of all humanity, brought into the light of faith through Ignatius's saving mission. A secondary ray reflects from a great concave mirror raised up by an angel, to shine on the four continents, personified according to Ripan iconographic conventions: sceptered Europe resting on the globe of Faith. Here too, the textual key to the spectacular drama of universal illumination is Ignatius's: "Ignem veni mittere in terram." Pozzo was explicit in his description of the iconographic program: "Jesus illumines the heart of St. Ignatius with a ray of light which is then transmitted by the saint to the furthest quarters of the earth, which I have represented with their symbols in the four sections of the vault." His illusory space offers a dramatically global context for the theatre of the Baroque Mass, where the soul rises in light and sound from gross material ignorance towards the beatific vision via the mystical body and the light of Christ.

2.5. Conclusion: Kircher, Coronelli, and Global Enlightenment

I know of no recorded direct encounter between Coronelli and Kircher. Such an event was by no means impossible, since the precocious young Venetian was studying in Rome during 1672–74 when Kircher was still a center of academic gravity in the city and actively working on one of his most imaginative and geographical texts, the *Arca Noë*. Among other concerns, this work draws upon the Biblical story of the deluge and the materials in Kircher's museum to develop a taxonomy of animal life based on spatial allocation of species within Noah's Ark. Coronelli's own quasi-taxonomic illustrations—of ship types, for example, positioned on his globes according to their originating geographical locations—reveal a similar encyclopedic mentality. Kircher's text incorporates maps of Mesopotamia, including Eden, illustrating the location of the various materials used in the construction of the Ark, and a world map comparing the distribution of continents and oceans before and after the deluge (cf. fig. 3.1 of Moxon 1671 in chap. 3 below). Another map illustrates the division of peoples and languages descended from the Sons of Noah. Given the centrality of Kircher's museum collection and publications within the intellectual life of 1670s Rome, and the young Coronelli's focus on theology and astronomy, which he continued to study after returning to Padua in 1674, it seems impossible that he could remain entirely unaffected by Kircherian science. Coronelli certainly drew heavily on Jesuit geography in later life. His maps of China in the *Atlante Veneto,* reproduced, complete with their Jesuit insignia, IHS, were provided by Antonio Baldigiani, S.J., Kircher's successor as Professor of Mathematics at the Roman College, and himself a member of the Argonauti academy.[10]

Had these two Baroque figures interacted directly, it might have been an awkward meeting of minds. They shared the same epistemology only to a limited degree: the Franciscan Coronelli's amassing of information via the members of the *Argonauti* apparently gave limited rise to speculative desires to reduce it to the kind of monist Neoplatonic theory Kircher sought to impose on the materials reaching him from Jesuit fathers even more widely scattered across the globe. It was spatially, through the organizing principle of the graticule, on mathematically accurate globes and maps, that Coronelli sought to control his information, his textual inscriptions acting as secondary elaborations bound into the overarching cartographic logic. Located on a globe, within the rhetorical spaces and the

DENIS COSGROVE

broader iconographic context of Versailles, Coronelli's cartography has an emblematic coherence which is lost in the encyclopedic text of his *Atlante Veneto*. Kircher's work, although primarily textual, reveals the Jesuit's knowledge of and interest in cartography, as the *Arca Noë* so clearly illustrates. His use of images too, is much more than merely incidental and illustrative, it is constitutive of his entire project through Kircher's Catholic iconosophy. Kircher was more sophisticated and consistent than Coronelli in his grasp of the emblematic mode of thought and expression, partly because the Jesuit's intellectual talents were superior to those of the Minorite, partly because the former's synthetic method for uniting textual and observational knowledge was more clearly threatened by the results of scientific experimentation by Coronelli's time. Coronelli's globes illuminate the natural and ethnographic wonder of the earth with an encyclopedism equivalent to Kircher's museum at the Roman College, but they are theoretically and methodologically cautious to the degree that they are iconographically and rhetorically audacious (Jacob 1992, 224).

In their different ways, both Kircher and Coronelli worked within an emblematic tradition which emphasized the power of visual images and the logic of imagination. Both drew for information upon a global network of information sources: geographers seeking to synthesize reported information as opposed to astronomers forcing distant information to inscribe itself directly through mechanical instruments of vision. They demonstrated their intellectual results in the highly rhetorical public spaces of palace and college, located at the heart of imperial centers, Versailles and Rome. Discussing them together thus offers some insights into the relations of geography and enlightenment in the second part of the seventeenth century.

In 1685, five years after Kircher's death, two years before Newton published his *Mathematical Principles of Natural Philosophy,* in the year when Coronelli completed the Versailles globes and returned to Venice, and while Pozzo was at work on his extraordinary images in the Church of St. Ignatius, the Edict of Nantes, legal guarantor of French religious tolerance, was revoked by Louis XIV. It was a decision vociferously promoted by Coronelli's patrons, Cardinal César d'Estrées and his Admiral brother, and for which they organized a lavish celebration at Trinità di Monte and its obelisk overlooking the Spanish Steps. Henceforth the Sun King of France would assume, with the aid of the Jesuit Order, secular responsibility for global enlightenment through Catholic faith. If Louis personally perceived his role as Catholic Apollo in terms of continental struggles against

Fig. 2.5. Aerial view of the Palace and Gardens of Versailles along the east-west axis

apostacy, Colbert and others among his ministers and advisors took a more global perspective, promoting those developments in French navigational science, technology, and cartography which sought to give Paris dominance over Amsterdam and London in cosmography and the production of maps and atlases. The monuments to this project included the Cassinis' maps, mathematics and calculations of

DENIS COSGROVE

latitude and longitude based on the meridian drawn through Paris, and the publication of the *Neptune François* in 1693. Paris was challenging not only the Protestant cities of the North Sea for commercial and colonial mastery of the globe through the secular sciences of oceanic empire, but Rome itself as Christendom's calendrical and metaphysical center of global illumination. The cosmographic iconography of Versailles into which Coronelli's globes were designed to be inserted, drew explicitly upon programs originally elaborated in Rome over the course of the preceding century for the papal imperial rhetoric of the Medici, Barberini, and Pamphili families, and by the "black" papacy of the Society of Jesus. Central to all these programs was the imagery of celestial light, originating in the Trinity, its terrestrial expression the light of Faith, and its projection across global terrestrial space through rays of intellectual and spiritual illumination flowing from the Apollonian body of pope, king, and missionary priest.

Far from collapsing under the contradictions Lestringant has identified from his examination of the intellectual career of its sixteenth-century practitioner André Thevet, cosmography was developed and elaborated into the late seventeenth century, coexisting with heliocentrism and embracing both Neoplatonic speculation and empirical rigor in collecting and reporting the variety of an expanding global nature, and even incorporating the emerging scientific method of experimentation (Grant 1994, 675–79; Panese 1997). In the work of scientists such as Athanasius Kircher and of globe makers and cartographers such as Vincenzo Maria Coronelli, cosmography's project of accurately representing terrestrial space while seeking to illuminate metaphysical space, served the rhetorical purposes of Baroque rulers bent on promoting global enlightenment by means of Catholic Faith.

Notes

1. The dedicatory text of the celestial globe is published on the two-page illustration showing its intended positioning by the author in Coronelli (1690, I)

2. Coronelli's struggles to finance his vast projects led to constant recycling of the engraved images as well as of his ideas. They may also have been in part responsible for his early removal by the Pope from the senior position in his Order after only three years. After 1700 Coronelli seems to have had few genuinely new ideas, relying on republication of existing materials.

3. The *impresa* of the Argonauti was designed by Girolamo Antonio Parisotti of Castelfranco. Significantly, Francis Bacon had used the image of

a ship sailing beyond the Pillars of Hercules in the frontispiece of his *Great Instauration* of 1621, a key text for the new science.

4. F. le Large's two volumes: "Receuil des inscriptions, des remarques historiques et géographiques qui sont sur le globe terrestre de Marly exécuté par le P. Coronelli" remain unpublished in the Bibliothèque Nationale in Paris (mss. français 13.365 and 13.366). I am grateful to Christian Jacob for allowing me to see a typescript made of these.

5. Divine light and divine love were of course closely connected phenomena in Neoplatonic philosophy, most cogently expressed by Marsilio Ficino and his Florentine academy colleagues such as Francesco Berlingueri, composer of an Italian verse translation of Ptolemy's *Geography* published in 1482, copies of which were presented to Duke Montefeltro of Urbino and the Turkish Sultan Sulemann II. For the continued influence of these ideas into the seventeenth century, see Grant (1994) and Hallyn (1993).

6. Hallyn (1993, 250–51) points out that Kircher did not take mathematics or music as logical principles underlying universal harmony, as Kepler for instance had done, but rather worked through analogy and symbol, appealing to imagination rather than to intellect: "The title says it well: *ars*. All the knowledge, all the technique of the period are put in the service of a theatre of marvels directed principally to the imagination." On Kircher's paradigm of science, see Findlen (1996, 78–96).

7. Kircher's maps of the sun and moon (represented symbolically in the frontispiece to his *Ars Magna Lucis* of 1671 as Apollo and Diana) and his diagrams of eclipses were well respected scientifically. They were based on the work of his fellow Jesuit Christoph Scheiner and widely reproduced, for example in Carel Allard's *Planispheri coelestis hemisphaerum septentrionale* published in Amsterdam about 1700 (Panese 1997, 116).

8. According to Plotinus, the hieroglyphics inscribed on obelisks expressed the "essence of things," the same intention that lay behind the emblem of device of the seventeenth century. Valeriano's *Hieroglyphica* of 1556 constituted fifty-eight books of images and symbols widely known throughout Europe. The appeal of Egyptian hieroglyphics as a mystical language continued until their deciphering with the help of the Rosetta stone in the 1820s.

9. The ceiling of Kircher's own museum was decorated with cosmographic images of planets and stars.

10. The Rome branch of the *Accademia degli Argonauti* also included Cardinal Barberini, scion of the great papal family that had patronized Bernini, and indeed a certain Pietro Filippo Bernini, whose relationship (if any) to the sculptor-planner of Baroque Rome is unclear.

References

Alpers, Svetlana. 1983. *The Art of Describing: Dutch Art in the Seventeenth Century.* London: John Murray.

Bevilacqua, Eugenia. 1984. "Geografi e cosmografi," *Storia della cultura veneta*, 3/II, "*Dal primo quattrocento al concilio di Trento*," Vicenza: Da Pozzo, 355–74.

Campanella, Tomaso. 1981. *The City of the Sun,* translated by A. M. Elliott and R. Miller. London: Journeyman Press.

Cavaillé, Jean-Pierre. 1991. *Descartes: la Fable du Monde.* Paris: Vrin.

Coronelli, Vincenzo M. 1690. *Atlante Veneto.* Vol. 1, Venezia.

———. 1692. *Corso Geografico Universale.* Venezia: Accademia degli Argonauti.

———. 1693. *Epitome Cosmographia.* Venezia: Poletti.

———. 1697. *Libro dei Globi.* Venezia: Accademia degli Argonauti.

Cosgrove, Denis E. 1990. "Environmental Thought and Action: Pre-Modern and Post-Modern." *Transactions of the Institute of British Geographers,* NS 15, 3, 344–58.

Eco, Umberto. 1995. *The Search for the Perfect Language.* Oxford: Blackwell.

Emiliani, Maria D., ed. 1977. *La Prospettiva Rinacimentale: Codificazioni e Trasgressioni.* Firenze: Centro di Studi Rinascimentali.

Findlen, Paula. 1991. "The Economy of Scientific Exchange in Early Modern Italy," in Bruce T. Moran, ed., *Patronage and Institutions: Science, Technology and Medicine at the European Court 1500–1750.* London: Boydell Press.

———. 1996. *Possessing Nature: Museums, Collecting, and Scientific Culture in Early Modern Italy.* Berkeley and Los Angeles, University of California Press.

Ginzburg, Carlo. 1986. *Myths, Emblems, Clues.* London: Hutchinson Radius.

Godwin, Joscelyn. 1979. *Athanasius Kircher: A Renaissance Man and the Search for Lost Knowledge.* London: Thames & Hudson.

Gombrich, Ernst H. 1948. "Icones Symbolicae: The Visual Image in Neo-Platonic Thought." *Journal of the Warburg and Courtauld Institutes* 11, 152–180.

Gorman, Michael John. 1998. "Athanasius Kircher and the uses of erudite correspondence in the 17th century." Unpublished paper presented at the Seminar on the History of Scholarship from the Renaissance Onwards, Warburg Institute, London.

Grant, Edward. 1994. *Planets, Stars and Orbs: The Mediaeval Cosmos 1200–1687.* Cambridge: Cambridge University Press.

Hallyn, Fernand. 1993. *The Poetic Structure of the World: Copernicus and Kepler.* New York: Zone Books.

Helms, Mary. 1988. *Ulysses' Sail.* Princeton: Princeton University Press.

Hulme, Peter, and Ludmilla Jordanova, eds. 1990. *The Enlightenment and Its Shadows.* London: Routledge.

Imago Primi Saeculi Societatis Jesu a Provincia Flandro-Belgica eiusdem Societatis Repraesentata. 1640. Antwerp: Balthasaris Moretti.

Jacob, Christian. 1992. *L'Empire des Cartes: Approche Théorique de la Cartographie à travers l'Histoire.* Paris: Albin Michel.

———. n.d. "Lire les globes," unpublished typescript.

Kircher, Athanasius. 1669. *Ars Magna Sciendi.* Amsterdam: Jannzon.

———. 1671a. *Iter Extaticum Kercherianum Praelusionibus & Scholiis illustratum* Nuremburg.

———. 1671b. *Ars Magna Lucis et Umbrae sive Astrolabio-graphia Figurata,* 2d ed. Rome.

Lehman, Kenneth. 1945. "The Dome of Heaven." *The Art Bulletin* 27, 1–24.

Lestringant, Frank. 1994. *Mapping the Renaissance World: The Geographical Imagination in the Age of Discovery.* Cambridge: Polity.

Livingstone, David. 1993. *The Geographical Tradition.* Oxford: Blackwell.

Mattelart, Armand. 1996. *The Invention of Communication.* Minneapolis and London: University of Minnesota Press.

Miscellenia Francesca. 1951. *Il P. Vincenzo Coronelli dei Frati Minori Conventuali 1650–1718.* 2 vols. Roma.

Moseley, Charles. 1989. *A Century of Emblems: An Introductory Anthology.* Aldershot: Scholar Press.

Pagden, Anthony. 1995. *Lords of All the World: Ideologies of Empire in Spain, Britain and France, c. 1500–c. 1800.* New Haven and London: Yale University Press.

Panese, Francesco. 1997. "Sur les traces des taches solaires de Galilée. Disciplines scientifiques et disciplines du regard au XVIIe siècle. *Equinoxe* 103–123.

Pelletier, Monique. 1982. "Les Globes du Louis XIV. Les Sources Françaises de l'Oeuvre de Coronelli." *Imago Mundi* 34.

———. 1993. "Les Globes de Marly, Chefs-d'ouevre de Coronelli." *Revue de la Bibliothèque Nationale* 47, 46–91.

Rowland, Diana. 1995. "Mother of the World" (review of the exhibition Egyptomania: Egypt in Western Art, 1730–1930). *New York Review of Books* 42, 13, 42–47.

Schama, Simon. 1995. *Landscape and Memory.* New York: Simon and Schuster.

Spence, Jonathan. 1983. *The Memory Palace of Matteo Ricci.* London: Faber & Faber.

Stafford, Barbara. 1994. *Artful Science.* Cambridge: Cambridge University Press.

Stevenson, Edward L. 1921. *Terrestrial and Celestial Globes: Their History and Construction Including a Consideration of their Value as Aids in the Study of Geography and Astronomy.* New York: Hispanic Society of America.

Wallis, Helen, ed. 1969. *Libro dei Globi* (facsimile of Coronelli's *Atlante Veneto* Vol. I). Amsterdam: Teatrum Orbis Terrarum.

Watson, Elizabeth S. 1993. *Achille Bocchi and the Emblem Book as Symbolic Form.* Cambridge: Cambridge University Press.

Chapter Three
Geography, Enlightenment, and the Paradise Question

CHARLES W. J. WITHERS

3.1. Introduction

For the Christian world at least, geography, as all knowledge, begins in paradise (Duncan 1972, 41–2). On the one hand, the idea of terrestrial paradise is commonly associated with gardens, notably with the Garden of Eden. Renaissance paradise gardens were seen as earthly symbols of heavenly order. Early modern botanic gardens, crucial as encyclopedia to the beginnings of modern science, were created as 'mirrors' of earthly paradise (Comito 1971; Prest 1981). In the eighteenth century, landscape gardens were seen by some as "new Edens" (Collins 1717; Schulz 1985; Switzer 1718). Terrestrial paradise is also connected, for Christians and non-Christians, with the idea of an arcadian past or lost golden age positioned at the beginning of world histories (Armstrong 1969; Cohn 1996; Heinberg 1990). On the other hand, celestial paradise as the City of God has informed both biblical criticism and architectural thinking, especially for those who equated the idea of paradise either with a state of being or utopian future (McClung 1983; Sanford 1961; Skoblow 1993).

This chapter discusses, for terrestrial paradise, the question "Where was Paradise?" I want to suggest that this question was crucial to those tensions between sacred belief and rational understanding informing Enlightenment as a *process* of reasoning. Put simply, this question was a matter of enlightenment and of geography and was unavoidable. To the churchman, if not the exegetical philosopher, paradise was textual truth, a spiritual state, and a celestial place. Belief in the first at least demanded a geography for paradise. To eighteenth-century natural historians, especially those whose explanations depended upon adherence to biblical accounts even as challenges mounted to them, the location of terrestrial paradise was central to explanation of natural diversity since it was *the* point of origin of all living creatures. To philologists and social theorists, locating paradise was a linguistic matter, part of theoretical questions about the origin of languages and the "science of Man" (Fox 1995; Moravia 1980; Wood 1996). Earlier, the question was relevant to Renaissance and early modern geographers, and in the later seventeenth century, paradise prompted numerous treatises on sacred geography in which geographical knowledge was used to discuss the Genesis account in the light of new geographical knowledge. In the eighteenth century, attention to the locational geography of paradise was less common. But, as I hope to show, the idea of paradise was by then closely bound-up with European encounters with the Pacific and with theoretical debates on the presocial state of 'primitive' peoples.

My concern, however, is not simply to recognize that the question "where was paradise?" is a key one in terms of the relationship between the scriptures and empirical science that underpins the Enlightenment. I want also to suggest something of the intellectual sites in which enlightenment as a process took place and to use the paradise question as one example of the ways in which we might conceive of the enlightenment not in terms of unsituated absolute categories but rather as a geographical enterprise in which ideas themselves were situated and had a certain geography (Livingstone 1994, 1995; Ophir and Shapin 1991). Geographical knowledge, I argue, was used differently to answer the paradise question itself differently shaped by geographical location. I am not claiming that the paradise question was a question of *the* Enlightenment. For one thing, it is as a question much older and takes a number of forms in the preenlightenment period. For another, the question, far from ceasing in the eighteenth century as has been claimed (Boies 1983; Giamatti 1966), takes different forms then. Indeed, the problems continued to occupy people in the nineteenth century (Alexandre 1988; Delitizsch 1881; Engel 1885; Hennig 1950; Keerl 1861). General Gordon, later to find glory in death as Gordon of Khartoum, was convinced in 1881 he had located Paradise in the Seychelles (Plaut 1982, 1984, 1993). In 1885, one William Fairfield Warren placed Paradise in the Arctic in his *Paradise Found,* a work designed to show "the cradle of the human race at the North Pole" (Warren 1885). More recently, paradise has been placed in the home nations of individual writers: the German von Wendrin located paradise in east Germany in his 1924 *Die Entdeckung des Paradies.* Other claims to it being eastward in Eden were, for him, the result of Jewish conspiracy. As

CHARLES W. J. WITHERS

recently as 1987, claims made on the basis of remote-sensing have located paradise in the Near East (Hamblin 1987), and others support archaeological evidence relating to early hominids by siting paradise in Kenya and Ethiopia (Ford 1981).

It is possible, however, to see this as a question *of* enlightenment, one being differently understood and permitting of different answers according both to the periods in which it was posed, and by whom. Geographical enquiry, I will suggest, is central to these different contexts. What follows is in three parts. The first summarizes the paradise question before about 1650. The second examines the question in several sacred geographies and other seventeenth-century texts, works which sought reconciliation between scriptural authority and the evidence of 'modern' science. The third part examines debates about geography, language, natural history, and paradise in the eighteenth century.

3.2. Paradise as a Geographical Problem before c. 1650

For Christians, the paradise question begins with the account in *Genesis* (chap. 2, vv. 10–14) with the description of the four rivers:

> And a river went out of Eden to water the garden; and from thence it was parted, and became into four heads. The name of the first is Pison: that is it which compasseth the whole land of Havilah, where there is gold; And the gold of that land is good: there is bdellium and the onyx stone. And the name of the second river is Gihon: the name is it that encompasseth the whole land of Ethiopia. And the name of the third river is Hiddekel: that is it which goeth toward the east of Assyria. And the fourth river is Euphrates.

For those interested in terrestrial paradise, several ideas have been advanced founded upon belief in the literal truth of this account. The first argues that paradise once embraced the whole world and that no true conception of its antediluvian and prelapsarian condition could be formed from what survived on the earth only as ruins. This idea had few formal expressions in answer to the question "Where was paradise?" since, although it presumed paradise to have had a real geography, it was now forever lost. But this idea does endure as part of belief in a lost 'golden age.' The idea of the present world as a ruination, a 'paradise lost' and of the human condition as a Fall is found in cultures other than the Christian and certainly predates its particular expression in eighteenth-century Western thought (Baudet 1965; Duncan 1972).

The second belief, more widely held, was that paradise as a site had survived the Flood and that its precise location would be found through geographical discovery. In this regard, the most popular idea from the early medieval period was that paradise lay somewhere to the east (Duncan 1972, 79–91). The precise location of paradise varies in early accounts: from Ceylon, to China, Japan, the polar regions, the Canaries, an island or mountain in the equatorial regions or torrid zones, even on the moon. The earliest map of paradise, from the sixth century, simply shows it to the east of the world, a situation repeated in many early medieval *mappaemundi*. Although the four rivers—Tigris, Euphrates, Pison, and Gihon—are usually stylized on *mappaemundi* as fanning out from paradise, they are sometimes represented as real rivers: the Tigris, Euphrates, Ganges, and Nile. Placing and naming the rivers of paradise correctly was a major geographical problem in the Middle Ages. Persistent confusion over the four rivers is evident in Columbus, for example, who, on hearing a report that men on his 1498 expedition had seen four rivers at the head of the Gulf of Paria, thought that these were the rivers of paradise (Harley and Woodward 1990, 328–330). For much of this period, of course, it was technologically impossible to search for terrestrial paradise: "But in the great age of geographical discovery . . . this attitude became out of date" (Prest 1981, 31). Paradise thus shifts from being an unattainable image to become a geographical reality capable of empirical proof. It becomes, as it were, intellectually resited, from a distant past to an as yet undiscovered present. Locating paradise was only a matter of time and of geography, and it is for this reason that Prest argues that the search for paradise was a key motivation behind Portuguese voyages of discovery.

In addition to these ideas, we may distinguish two further and related themes before about 1650: the idea of paradise as a site within the 'New World'; and the idea of the contemporary New World *as* paradise, or, at least, having paradisal qualities now lost to Europe. This second idea is related to those strands of early utopian thought in which the new world of the Americas is simultaneously positioned as an earthly paradise removed from the imperfections of western civilization and yet part of emergent theories of universal history. The Elizabethan geographer Thomas Hariot certainly had utopia in mind in his *Briefe and True Report of the New Found Land of Virginia* (1588) in describing that new colony as a pastoral paradise. But his rhetoric is also that of the courtier-adventurer keen to promote himself, his patron, and the colonial enterprise (Cormack 1991; Levin 1969, 164). Hakluyt and Samuel Purchas, on the other hand, thought

CHARLES W. J. WITHERS

paradise as a site would be found "south of the equinoctial line" (Prest 1981, 32).

Something of the complexity of the idea of paradise by the early seventeenth century—as a site, a condition of natural bounteousness, and yet, paradoxically, something which increasing geographical enquiry had failed to discover—is apparent in Sir Walter Raleigh's discussion "Of the Place of Paradise" in his 1614 *History of the World*. Raleigh's expedition to Guiana lent authority to his claim that "if there be any place upon the earth of that nature, beautie, and delight, that *Paradise* had, the same must be found within the Tropicks." The tropics were, naturally, bountiful: "so many sorts of delicate fruites, . . . at all times beautified with blossoms and fruit . . . as it may of all other parts bee best compared to the *Paradise of Eden*." Yet Raleigh does not place paradise within the tropics and, indeed, offers another perspective on the new world. Alongside "these pleasures," there are "fearefull and dangerous thunders and lightnings, horrible and frequent earthquakes, the dangerous diseases, the multitude of venimous beasts and wormes." The very bounteousness of nature there was the source of human corruption. This is one early expression of that sustaining western view concerning the environmentally determined innate sloth of "the Native." It was also for Raleigh a means to affirm the truth of the biblical account regarding the site of paradise in the east. The potential earthly paradises of the Americas were, for Raleigh, *"Vitious Countries."* He rejected the arguments of those who saw the whole earth as paradise and those who "seate Paradise under the Aequinoctiall." He argued, principally "from the testimonies of the Scriptures" but also from observations in travelers' accounts, that paradise lay in Mesopotamia (Prest 1981).

Changing forms to the paradise question are also evident in its mapping. Maps of Eden or paradise as a celestial or cosmological site appear in Bibles from about 1500. Many published in the 50 years after the Reformation present symbolic affirmation of the place of Jerusalem and of paradise at the center of the Christian world (Delano Smith and Ingram 1991, 1–24). In one sense, maps of the site of a terrestrial paradise record what Plaut terms the "vagrancy" of paradise (Plaut 1984, 6). Catalan maps of c. 1450 and c. 1502 show it in East Africa while other maps of the period locate it on India's east coast at just the time the Portuguese sited it in South America. In another sense, paradise as a terrestrial site is marginalized in seventeenth-century maps, perhaps because increased geographical knowledge of the world had failed to locate it and a certain empirically derived scepticism is being expressed cartographically. Janssonius's 1632 world map positions paradise as a decorative vignette in the frame, not as a geographical site (Plaut 1982, 31).

Yet the idea of a paradisal new world endures in a particular form in the Enlightenment. The failure of voyages of discovery to locate paradise led by about 1650 to the conviction that its site was never likely to be found. This is not to argue that the forms taken in the paradise question have neat chronologies. As Alexander von Humboldt noted in 1831, the still-current idea of paradise as somewhere in the east was part of wider Orientalist interests (Humboldt 1814–1834, III, 120). But from the early 1600s, if not before, the paradise question becomes differently understood, as a result of the failure of geography, in the sense that there was common agreement that the site had not survived the Flood. "Where was Paradise?" becomes from this period a question both of historical geography and of a quite different sense of rational enquiry (Duncan 1972, 89–90).

3.3. Paradise, Enlightenment, and Sacred Geography

The term "sacred geography" does not unequivocally define a clear type of geographical enquiry. In the early modern period, the subject embraced the geography of the scriptures, that is, described lands and countries recorded there. The term was employed by Wells (1701) and used, for example, in Diderot and d'Alembert's *Encyclopédie* as a subdivision of geography as a whole and survived until at least the end of the eighteenth century (Withers 1993). In focusing upon the scriptures at a time of emergent rationality and mounting secular scepticism (and see also Livingstone, chap. 4, on this point), sacred geographies of the seventeenth and early eighteenth centuries together with sacred theorists like Burnet were, I will argue, centrally implicated in the enlightening role of scientific knowledge.

Abraham Ortelius reviewed the paradise question in his *Thesaurus Geographicus* (1596) and in his 1606 *Parergon*. His review documents earlier writers on the paradise question, among whom was John Hopkinson, an English divine, whose *Synopsis Paradisi* (1593) had been written to refute sceptics then doubting scriptural accounts. The most complete early English language account is John Salkeld's *A Treatise of Paradise* (1617), written from a concern "to show the place, and demonstrate the grace from which we fell." The fullest expression in Europe in the early seventeenth century is Agostino

CHARLES W. J. WITHERS

Invege's 1649 *Historia Sacra Paradisi Terrestris.* Inveges, a Sicilian priest, discusses the near-eastern location of paradise, its physical characteristics, and argues, too, for the exact time and place of "the first parents' action," another of the central concerns of these sacred theorists in dating the origins of human history.

In his *Geography Delineated Forth in Two Bookes* (1625), the Oxford divine and geographer Nathanael Carpenter had no doubts about the authenticity of paradise and of the importance of the question for human and natural history. Others have seen Carpenter as a "scholarly empiricist" and this work as an isolated theoretical attempt to secure for geography "some measure of academic esteem in conjunction with the sciences of the time" (Bowen 1981, 73). Review of his attention to the paradise question would suggest a more complicated basis to that faith in reason and empiricism which underlay his geography. Carpenter considered the scriptures the basis to any explanation of how "the first inhabitants of the Earth were planted in Paradise, and thence translated to the places near adjoyning." "For the confirmation of this point," he writes, "we need no further proof then [sic] the authority of God himselfe, speaking in his word, whereon all truth is grounded." But debating the *site* of paradise was open to rational enquiry. Carpenter proceeded, first, by reference to the authority vested in others' works, "but of the place of *Paradise,* where we place the first habitation, sundry disputes have been amongst Divines sufficiently examined, of late by a judicious and worthy writer [Raleigh]." Second, he used the scriptures to deny interpretation of paradise as allegory and to affirm paradise as a garden: ". . . many reasons may be drawn to prove there was a true locall *Paradise Eastward:* . . . out of the text itselfe." The scriptures were also used to challenge the idea of the whole Earth as paradise. Third, Carpenter proceeded empirically to dismiss the notion of paradise as sited on a mountain that reached "to the Moone, or higher at least than the Middle Region of the Aire." This could not logically be so, he argued: "First, because it would be subject to the eyes and knowledge of men. Secondly, it would hide the light of the *Sunne* . . . Thirdly, it would over-poize the *Earth.*" Carpenter subscribed to the view of paradise as a past site, now lost, in Mesopotamia and did so because of place name evidence concerning Eden and because the rivers of paradise "cannot be imagined to meet in any other part of the world." Carpenter's method was, then, partly empirical, partly faith, partly a search for confirmation in contemporary others, partly an appeal to ancient authorities. There was certainly no single legitimating claim to his conclusion that "howsoever it be, certain it is, that Paradise was seated in the East, from whence mankind had its first off-spring" (Carpenter 1625, 208–212). Peter Heylyn likewise favored a site in Mesopotamia in his *Cosmographie* (1652) and drew upon the scriptures and common sense to dismiss the view of paradise as the whole world: "this cannot be: for then when *Adam* was driven out of *Paradise*, it must follow that he was driven out of the world, which were most absurd" (Heylyn 1652, 767).

These tensions between ancient and scriptural beliefs and empiricist reasoning are particularly evident in Marmaduke Carver's 1666 work, *A Discourse of the Terrestrial Paradise*. Although written many years before "for private satisfaction only," this was published by Carver in order to reclaim the paradise story from "the more Heathenish Custodians of these later times." In a lengthy introductory section, Carver reviews those who had earlier written on paradise, including Raleigh and Purchas. He is chiefly concerned to dispute those who promote the fabulous in locating paradise. He is scornful of those who have been "not making out the truth of it [the Paradise question] by the help of Geography, but blanching it over with Allegoriall, or impertinent and ridiculous Interpretations." His work discusses place-name evidence and the productive capacities of modern environments in order to situate paradise. His conclusion, that paradise was in Armenia, owes less to confirmation of the biblical story than other works. In the final part of his book, he leaves open confirmation of the *exact* site of paradise, seeing that as a matter for further work:

> And now to conclude: Though all these evidences laid together (which surely are as great as well may be expected in a Subject of this nature) have not raised our confidence to such a height as some have attained to, (and, as we suppose, upon far weaker grounds;) yet we verily believe that if they whom God hath blessed with abler parts, more skill in the Tongues, History, Geography, &c. [had] a larger freedom from other imployments and distractions, with a more plentiful supply of Books, and other accommodations for such a study, (all which we want) would resume this Argument, and apply their plans to the farther search of this *not-unnecessary question,* they might here (sooner then in any other place yet discovered) find out *the true place of the Situation of the Terrestrial Paradise.* (Carver 1666, 164–165)

Carver is certainly supposing the paradise question worth asking. Among other things, better geography and better philology may allow it to be answered. He is not, however, directly pointing to deficiencies in contemporary geography. For Carver, geography as a

CHARLES W. J. WITHERS

practical pursuit has not in the past found paradise. It yet may but as a matter of historical enquiry. He is, however, making a point about how geographical knowledge should be better used as part of rational methodologies. This shift in emphasis is echoed in Bishop Huet's 1694 *A Treatise of the Situation of Paradise* and Samuel Bochart's *Geographia Sacra* (1684). Others, like Salomon van Til in 1701 in his *Malachius Illustratus: Tractiatus de situ Paradisi Terrestris: Dissertatio singularis Geographico-Theologica* advocated a more strongly theological line. Edward Wells in his three-volume *An Historical Geography of the Old Testament* (1710–1712) was largely content to support the Mosaic account. He also reviewed the etymology of the river names Pison, Euphrates, Hiddekel (Tigris), and Gihon, and only lastly proceeded to review the "works of learned men"—he cites Schickard, Steuchus, and Athanasius Kircher, who, as Cosgrove shows in chapter 2, was likewise interested in paradise sites, as well as Samuel Bochart—before confirming both that "Terrestrial Paradise" was in Eden in Mesopotamia and that it was the model for all later gardens as well as of the "Poetical Fictions concerning the Fortunate Isles, the Elysian Fields, the Gardens of the Hesperides, & c." (Butlin 1992; Wells 1712, III, 24, 37, 45). Henry Hare's claim to have found the site of paradise—in Galilee—was part of an allegorical study proving religious tolerance: "tis an honest Policy, a Stratagem to make you in Love with Religion" (Hare 1683, 2). Joseph Moxon discusses paradise in his 1671 *Sacred Geographie* and places it in the Near East on his "Mapp of all the Earth" (fig. 3.1). This siting, though not necessarily the figurative portrayal of the Fall with which Moxon frames his map, is indicative of the many commentators who took paradise to be a single site.

Chief among theorists arguing for paradise as the whole Earth was Thomas Burnet in his 1684 *Sacred Theory of the Earth*. This has been seen by one commentator on the enlightenment as "one of the most influential early Enlightenment statements of the confluence between the History of Nature and the events of Biblical narrative" (Outram 1995, 52). Burnet's *Sacred Theory* is in four books, the second of which is entitled "Concerning PARADISE." Of interest here are Burnet's statements on method. His is a deductive enquiry: "We may in the mean time observe," he writes, "how preposterously they go to work, that set themselves immediately to find out some pleasant place of the earth to fix *paradise* in, before they have considered, or laid any grounds, to explain the general conditions of it, wheresoever it was. These must be first known and determined, and we must take our aim and directions from these, how to proceed further

Fig. 3.1.   The location of Paradise as shown in Joseph Moxon's "Map of all the Earth," from his *Paradise, or the Garden of Eden with the Countries circumjacent inhabited by the Patriarchs* (London, 1690). The site of Paradise corresponds here with the traditional siting of it in most works of sacred geography. Reproduced with the permission of the Royal Geographical Society, London.

in our enquiries after it; otherwise we sail without a compass, or seek a port and know not which way it lies." Burnet's work as a whole is framed by a concordance between God's words (the sacred texts) and his works (the objects of nature). The first he presumes true; the second must, therefore, be explained to fit this truth. Burnet argued from scriptural evidence on three conditions—human longevity, the spontaneous richness of the soil and a perpetual Spring Equinox— to deny paradise both a contemporary and a local geography. His method also allowed refutation of ancient belief, at least initially: "Who can pretend to assign any place or region in this terraqueous globe, island or continent, that is capable of these conditions, or that agrees either with the descriptions given by the ancient heathens of their *paradises,* or by the Christian fathers of scripture?" Only when scripture and reason were found wanting should traditional beliefs be accommodated: "Where the two former are silent, it seems very reasonable to consult the third (Burnet 1684, 169, 176–186, 247). Burnet, no less than Carpenter, Carver, and others, clearly treated

CHARLES W. J. WITHERS

the paradise question as a legitimate object of scientific theorization. Indeed, his theory as a whole was "writ with a sincere intention to justify the Doctrines of the *Universal Deluge,* and of *Paradise*" (Burnet 1684, *Dedication*). Unlike others, Burnet proceeded deductively and saw secular authors as sources to be consulted only when the scriptures failed to support his theorizings.

Interest in paradise continued as part of sacred geographies until at least the mid eighteenth century: several works on the geography of paradise were, for example, collated by Ugolino in his 1747 *Thesaurus Antiquitatem Sacrarum.* But, by virtue of empirical failure and because, in Bishop Huet's words, "Many learned Men of ancient times, very able in other respects, have been very ignorant of Geography" and, not least, since "The old geography is not very certain" (Huet 1694, 41–2), the paradise question understood as a matter of historical geography concerning a certain past site was not answerable. What, more generally then, are we to make of these sacred geographies?

First, these and other works were commonly concerned to establish the *place* of paradise and to *date* events there in order to provide a basis to universal histories and to legitimate scriptural authority. Second, they signal a shift away from the search for a present to a past geography for paradise. But they do so, I would argue, by appealing differently to geographical knowledge as a means to examine the tensions between scriptural and empirical authority. The fact that paradise as a fact of geography was not locatable should not be understood as the discursive 'failure' of geography as a practical subject by the late seventeenth century. In one sense, quite the contrary. Geographical exploration had been successful in showing that paradise did not still exist. Rather, the paradise question in these sacred geographies was a matter of using contemporary geographical evidence and the authority of ancient writers to determine where once paradise had been. It is possible, then, to see the shift in mapping of a vagrant but undiscovered paradise, its aesthetic marginalization in Janssonius, and its place in the imaginative maps of Moxon and others as indicative of the role of a certain sort of geographical enquiry being used to support the biblical account just as others sought to question such authority.

In Carver, Huet, and Moxon, there is clear recognition of the importance of geography to that historical method being incorporated in contemporary rational enquiry. In others, geographical knowledge was differently used in reconciling scriptural truths with the evidence of nature. For Burnet, deductive theorization about the past state of the earth was the principal means to solve the paradise question. For Carpenter, empirical observation on inductive principles about the earth's present state allowed him to deny ancient authorities. For Raleigh, personal experience of New World bounteousness allowed him to equate that paradisal *state* with the scriptures as he simultaneously adhered to the scriptural description of paradise *as a place* somewhere "to the East." The paradise question understood as a matter of sacred geography was, thus, centrally placed within enlightenment tensions between scriptural doctrine and secular rationalism, was so from at least 1614 and was so in a variety of complex ways. In the eighteenth century, the paradise question is different still.

3.4. Paradise in an Age of Enlightenment: Questions of Language and Natural History

In conjunction with these sacred texts, the late seventeenth century saw a number of works seeking to answer the question, "What language was spoken in paradise?" This question not only reflected interest in Adam's naming of creatures in *Genesis,* but also underlay attempts to document that original language as part of theories of the universal language. This was, in turn, linked with emergent theories about the historical development of peoples and of nations.

Questions of language and natural history in the Enlightenment have several features in common. First, theoreticians and practitioners in both fields shared interests in naming and recognized the need for the empirical sciences to have a precise descriptive rhetoric capable of being used in different geographies. This is evident in the Royal Society's committee on scientific terminology, for example, and in texts like the Port Royal Grammar (1660) (Aarsleff 1983; Grillo 1989; Mengham 1993; Smith 1984). It has also to do with the emphasis placed upon order, and universal systems evident, for example, in Linnaean taxonomy, the language of mathematics, universal grammars, and the culture of encyclopedism (Frangsmyr, Heilbron, and Rider 1990; Shea 1983; Withers 1996). Second, these issues had a social dimension to do with the appropriate use of language: appropriate in the scientific sense of universally understood linguistic codes for ordering nature's diversity, and in the social sense of who was doing the naming. The emergence of this more precise written and spoken language within scientific discourse is directly related to questions of social authority (Eco 1995; Fox 1995, 6).

3.5. Paradise, Naming, and the Origins of Language

The language-of-paradise question was, in part, evident in the distinction between language as a divinely given capacity and a humanly created logos. If language is divine or Adamic, then both the thing itself and our knowledge of it is in ruins, and the task for those interested in language is to restore the lost order and harmony of the original. If language is a human creation, the world is itself made by the language we create for it. These issues are not mutually separate: one could allow God to have given Adam the power of speech and Adam to then name what he saw. But the language-of-paradise question becomes crucial if we see in that scientific language used in the emergent world of empiricism and reason a precise and denotative capacity to site all knowledge, in short, a language allowing no error or misunderstanding but only accurate representation of the nature of the thing itself (Aarsleff 1982; Fish 1971; Fraser 1971; McColley 1993).

The search for the language of paradise was also a search for historical national legitimacy. One such work was *The Languages of Paradise* (1688), by the Swede Andreas Kempe. Kempe argued that the seductive serpent spoke French, Adam spoke Danish, and that God was a Swedish speaker. His was one of several such works, each positioning their own nation and language at the beginnings of human history (Elert 1978; Olender 1992). In the same way, Leibniz's early eighteenth-century treatises on language argued that Europe's languages had originated on a continent called Scythia, although he did not identify one site for their precise origin. Such works represent, however, early expressions of the idea that all languages shared a common ancestor, that there was an absolute prototype or *Ursprache* recoverable or, at least, discernable through linguistic enquiry, and that that original form had its location either in paradise itself, or in the east. Herder considered language to have originated in the Indian mountains (one of the many sites in which paradise had been situated by earlier scholars), a fact assisted by his reading the map of the world in ways which allowed him to equate the Ganges with one of the rivers of paradise. Herder clearly follows Carver and Huet in linking philology to the geography of paradise (Herder 1782, II, 444–445; Olender 1992, 3–7).

Enlightenment thinkers were fascinated with language not just because it offered, as part of theoretical histories, a means to understand the evolution of the social world. Language was also the means to appropriate communication: the basis of and the means to rational knowledge, and, in the form of "polite" discourse, a means to both order the world and avoid incivility in human affairs. Condillac, Rousseau, and Smith and Herder all understood that none of the rational sciences or, indeed, society itself could develop without language. Because social progress was not possible without the transmission of knowledge through language, the enlightenment idea of progress as a matter of theoretical history is impossible without a theoretical history of language in which the origins of language are placed. Theorists interested in the natural origin of languages were confronted with no certain evidence, however, as to the exact site of paradise, as a matter either of present or historical geography. While many theories of the natural origin of language begin with the postulation of a couple of savages without language, without culture, these theories do not begin in paradise. Rather, they situate the beginnings of language in a natural, presocial, and paradisal condition, in that uninscribed naked state which demands naming in order to know (Condillac 1736; Rousseau 1755; Herder 1772; Schreyer 1989; Smith 1761).

For those enlightenment thinkers interested in the origins of language, then, the adoption of a word-thing understanding of language and the conception of this preciseness as part of the methods and outcome of empirical science based on observation of the world represented "nothing less than a return to the linguistic purity of Paradise" (Fish 1971, 113–114). As Adam had walked in the Garden naming things, so natural historians and enlightenment geographers armed with a denotative lexicon and the social and scientific authority to use it walked either the paradise that was the 'New World' or the mirror of that world that was the botanical garden as *hortus conclusus* naming, classifying, putting things in their place. In this regard, too, the concern for the origin of language was connected to the earlier search for the perfect language, one capable of representing categorically the natural world (Eco 1995). The conception of language as a naming device first employed in paradise becomes part of the ways in which enlightenment theorizations about the "science of man" treat language as a constitutive marker of scientific knowledge, and a means to the social authority of those Enlightenment natural historians and geographers discussed throughout this volume who were doing the traveling, naming, and mapping.

CHARLES W. J. WITHERS

3.6. Paradise, Natural History, and Paradisal Geographies

Simply put, to natural historians of the enlightenment, the paradise question could be treated in one of two ways. Paradise could either be judged as mythical, by virtue of the failure to find it in the present, to situate it historically or to site it linguistically. Alternatively, the biblical description of it could continue to be regarded as the definitive description of the site from which the terraqueous globe was populated. To Carl Linnaeus, the idea of paradise, "the first and noblest botanical garden," was central to his rational scientific arguments about the Earth's history. In his reasoning, it is possible to see yet another sense in which the paradise question is reworked to suit the context in which it was posed.

Linnaeus outlined his theory on paradise in his 1744 *Oratio de Telluris Habitabilis Incremento* ("On the Growth of the Habitable Earth"). He argued that the greater part of the land surface at the beginning of the world lay under water. Only a small island stood clear of the water. This was paradise. Here God laid out the Garden of Eden, made Adam and Eve and all living things. Over time, as the waters receded, the island gradually grew, land masses appeared. The island had a mountain on it, sufficiently large to allow specific variety to occur— alpine species to the top, tropical to the foot—from which the Earth was populated with flora and fauna fitted for its place in the world by virtue of its place in paradise. The same process obtained after the Flood. His concerns in his *Oratio* were essentially to advance a theory to explain the Earth's diminishing waters. Paradise in that context could not be avoided. I do not mean that Linnaeus was concerned to site his paradise island accurately. The site of paradise is not given. Linnaeus drew from his countrymen Olaus Rudbeck and Emanuel Swedenborg who, like Thomas Burnet, argued for paradise as the whole Earth before the Flood. He favored Burnet's view that it lay in the southern hemisphere, and he was aware of the work of Huet and of his fellow Swede Olof Celsius's *De Situ Paradisi Terrestris* (1714), which argued for a location in Mesopotamia. Rather, for Linnaeus, paradise becomes differently accommodated to accord with secular theorizations and mounting natural geographical evidence, not alone to fit biblical accounts. Situating it geographically is much less important than placing it conceptually as part of emergent theories of earth history and natural variation in species between continents. I am not claiming that Linnaeus represents a radical break in what was understood by the paradise question. In writing of paradise as a mountain on an island, Linnaeus was almost certainly drawing upon earlier accounts other than Burnet: from the Swede Kioping, for example, who had argued in the later 1600s that Ceylon, with the high mountain there known in English as Adam's Peak, was paradise. Linnaeus was also proceeding empirically: he had examined a collection of plants from Ceylon in 1747 and been struck by its paradisal diversity (Frangsmyr 1983, 111–125). Linnaeus shares something with Burnet but, unlike him, employed new evidence to test existing beliefs.

It is also possible to see in his interest in the condition of Nature and the new world an awareness of earlier ideas of the whole Earth as having once been in a state of paradisal grace and of islands especially being paradisal. To this complex combination of ideas we must add that eighteenth-century fascination with theorizing social progress in which Pacific islanders were, by many, envied, if not marveled at, living as they were in that state of natural presocial grace lost to Europeans. For the French geographer and natural historian, Bougainville, paradise was certainly a south seas island. Discussing what he called "New Cythera" (Tahiti), he wrote:

> Nature had placed the islands in the most perfect climate in the world, had embellished it with every pleasing prospect, had endowed it with all its riches, and filled it with large, strong, and beautiful people. Nature herself dictated the laws. The inhabitants follow them in peace and constitute perhaps the happiest society which the world knows. Lawmakers and philosophers, come here, and look upon the establishment of what your imagination could not even conceive. Moreover, one hopes for the sake of these people that Nature has refused to produce here the objects of Europe's cupidity; they need only the fruits which the land provides in abundance, and the rest, by drawing us Europeans here, would only bring to them the evils of the iron age. Farewell, happy and wise people: remain always as you are now. I will always remember you with delight, and as long as I live I will celebrate the happy island of Cythera: it is the true Utopia. (Bougainville 1766–1769 in Hammond 1970).

The language being used here is certainly that of the enlightened *philosophe* marveling at the beauty of the people and at Nature's plenitude. But it is not a uniquely Enlightenment discourse. It shares something with Raleigh. It is also that of the practical and intellectual colonialist, keen to name newly discovered tropical islands as paradises as part of the promotion of empire and of nominative empiricism. Of course, Bougainville was not alone in seeing Tahiti as a

CHARLES W. J. WITHERS

key site in enlightenment theories about the human social condition, the presumed "natural" state of pre-European humans and the origins of language, and Heffernan discusses Turgot's importance in this respect in chapter 5. Diderot commented how "The life of savages is so simple, and our societies are such complicated machines! The Tahitian is so close to the origin of the world, while the European is close to its old age" (Diderot 1751, 194). Joseph Banks and William Marsden on Cook's voyages took a close interest in native languages, both as part of comparative theoretical projects upon Europe's languages and as part of ethnological interest in the human geography of the Pacific (Gascoigne 1994, 160). Some argued for an affinity between Polynesian languages and Hebrew and sought through such theories to confirm that "there was a time when all the inhabitants of the world spoke Hebrew" (Glass 1787, 81–84). Such a view accorded well, of course, with those like James Dunbar, who argued in his *Essays on the History of Mankind in Rude and Cultivated Ages* (1780) how "The history . . . of some of the South Sea isles, which the late voyages of discovery have tended to disclose, enables us to glance at society in some of its earlier forms" (Dunbar 1780, 141). J. R. Forster knew that naming and the history of language were crucial to making sense of what was encountered in the Pacific in his 1778 *Observations made during a Voyage round the World* (Thomas, Guest, and Dettelbach 1996). Forster was much concerned with native languages and their grammatical and lexical relationship to the cultivated languages of Europe, matters which, as Bravo also shows in chapter 7, were about the commensurability of the languages used in the geographies of ethnographic encounter. He presented what he called "a monument [a diagrammatic chart of words and meanings] . . . to the geographical ingenuity of the people in the Society Isles, and of Tupaya in particular" (Forster 1778, 213–215, 511–515). Diderot's Tahiti was in some measure, of course, a figment of Europe's imaginative geography. But for others, the island embodied the real practical and theoretical problems engendered by contract within an unknown world. Ascribing an order to the paradise encountered demanded attention to native language in relation to civilized society and to new nominative categories for the peoples and nature discovered there.

Different islands were reckoned paradisal in different ways. St. Helena, for example, was described as an earthly Paradise by Portuguese mariners as early as 1598, given its fresh water and foodstuffs and optimum position in the Atlantic. Mauritius was also described as a paradise. Mauritius was judged paradisal by the French colonial botanist Pierre Poivre, who understood the connections between the European treatment of indigenous peoples and the treatment of the environment, and saw in the natural history of the south sea islands both a commercial and a utopian vision for French colonial interests (Grove 1995). Descriptions of Mauritius had additional weight in being compared with the degradation of St. Helena, which is represented as a paradise lost precisely because of careless use of its original, almost Edenic, environment (Grove 1995). For Joseph Banks in 1771, comparing the geographical situation of the Dutch in the Cape colony with the by-then overexploited St. Helena, one idea of paradise was closely associated with the political capacities of different nations to make proper use of new colonial environments: "nor do I think I go too far in asserting that was the Cape now in the hands of the English it would be a desert as St. Helena in the hands of the Dutch would as infallibly become a Paradise" (Beaglehole 1963, 265–266).

In several ways, then, the geography and the natural history of south sea islands had a crucial role in the paradise question in the eighteenth century (Gascoigne 1994, 182–183; Hulme and Jordanova 1990, 7–9; Marshall and Williams 1982; Spate 1988). For Bougainville, Banks, Diderot, Cook, and Forster, Tahiti was the new Eden. For Poivre and others, Mauritius was paradise. Such descriptions owe something, of course, to earlier notions of paradise as an island as they were later echoed in Gordon's claim about the Seychelles. Such notions were additionally overlain by a concern to situate paradise as a *condition*, one of natural presocial grace. The more general association between New World islands and paradisal geographies was to see them as the location of peoples as yet unfallen and as sites of natural richness. The paradise question, without losing earlier arcadian connotations, thus becomes differently situated as part of an emergent environmental ethic and as part of anthropological theorizations about the place of humankind in nature. What was true of the Americas in the early 1600s was applicable to the Pacific in the later 1700s. Such new worlds were, at once, a paradise found, yet always on the verge of being a paradise lost through the misuse of nature and the natives there.

For some historians, so commonplace and enduring has been this representation of the Pacific region as paradise that it has hindered proper scientific understanding of Pacific societies (Maude 1968, xxi). Certainly, this enlightenment envisioning of paradise speaks

CHARLES W. J. WITHERS

more strongly of a particular western vision of the world than it does of a detailed understanding of how exactly Pacific peoples carefully managed the abundance of their local environments (Hoëm 1996, 311; Smith 1969). And at the same moment as the New World was seen as paradisal in these ways, so the paradise that it signified was recreated, resited practically and conceptually, in the classifying spaces of botanical gardens. Poivre's vision for his Jardin du Roi on Mauritius, for example, was practically motivated but was also strongly Arcadian. Such views built upon rather than replaced the medieval and Renaissance view of the botanical garden as "mirror" of paradise and of gardening and natural history, in the Adamic sense, as practices of tending and naming.

Even so, not all enlightenment thinkers drew upon these ideas about paradise in the same way. In his *Universal Geography* (1816–1822) Malte-Brun includes a "Digression on the terrestrial Paradise" in his discussions on Persia and the Near East, part of his universalizing attempts to describe "all the parts of the World on a New Plan." The title and language of his commentary is itself suggestive. Citing several sacred geographies, Malte-Brun considered how "these various researches afford but dubious and vague conjectures, and are perhaps worse than idle." He closed his "Digression" in order to "resume the sober details [of his description]" (Malte-Brun 1816–1822, I 201). For many others, the enlightenment as a process of knowing and revealing the world was precisely about trying to answer the paradise question, not to ignore it, and to do so according to different kinds of rationality in which geographical knowledge was centrally placed.

3.7. Conclusion

Several points may be made in review. First, in the various senses it has been used, paradise, like the Enlightenment more generally, may be considered a European idea employed to make sense of the world at other times and in other places. The paradise question, I have argued, was important to the ways in which geographical knowledge was centrally but differently placed in the process of enlightenment. Simply, addressing the question demanded a key role for geography.

Second, the forms taken by the paradise question in the eighteenth century were closely related to earlier debates, notably in the work of Carpenter, Burnet, and Carver on the situation of paradise as an expression of the tensions between faith in either the Scriptures or in empirical natural philosophy. Carpenter adopted a method that saw some validity in ancient geographies and in Church tradition. Burnet's approach was more rational and for many at the time bordered on heterodoxy. In this regard, then, differences in approach to theological argument suggest that the paradise question was not simply chronological in its modes but to an extent denominational as to how one interpreted both the Scriptures and the facts of geography. I have also tried to show how understanding the paradise question was a matter of geographical location and of historical context. What distinguished later eighteenth-century interests was not the search for a site, past or present, but the way in which paradise in the Pacific fitted theories of universal history. This is not to suppose absolute divisions between paradise's sacred theorists between 1614 and 1743 and empirical rationalists thereafter. And it is not to suggest that earlier notions have significance only because they prefigure the age of enlightenment. But it is, third, to suggest for this geographical question at least, that enlightenment understood as a situated process of critical reasoning may be earlier than is often allowed. The paradise question was no less a matter of enlightenment in the late sixteenth century—"was the site still to be found?"— or in the seventeenth century—"where had Paradise been?"—than it was in the eighteenth century in Linnaeus's secular theorizations or in the speculation of Bougainville, Forster, and Banks about the Pacific.

Before about 1600, interest in terrestrial paradise was predominantly informed by belief that the site still existed. Geographical exploration needed only to be better for the originating site of all human history to be found. This idea is not replaced by other notions and is certainly not rejected by later enlightened rationalism. No less a geographer and explorer than David Livingstone wrote in 1871 to Sir Roderick Murchison, saying that he believed that at the sources of the Nile, could he discover them, he would "stand upon the site of the primeval Paradise" (Warren 1885, 21). As the known world expanded, however, the paradise question became one of historical not contemporary geography. Allied to this was the view of terrestrial paradise as once having embraced the whole world, an idea itself closely connected with visions of celestial order, and belief in an arcadian past and in a utopian future.

Fourth, sacred theorists sought to situate paradise for two main reasons: chiefly to affirm scriptural accounts, but also to situate the origins of all natural and civil history. The language-of-paradise question was important to a sense of national identity within Europe

CHARLES W. J. WITHERS

and to the categories by which new peoples could be placed. To Herder and other theorists on the origins of language, the site of paradise was unimportant. But intellectually situated it was crucial as a question to connections between ideas on the origins of civil society, social development, and the linguistic precision needed for the rational sciences. Nowhere was this notion of Adamic purity and the logical power of language more apparent than in the related practices of geography and natural history in this extended age of enlightenment.

Last, it may, then, be possible to talk of a natural history, even a geography, for the idea of paradise, especially in the eighteenth century. One form was in the languages of paradise. Another, more figuratively, was in the botanical garden, both in the Edenic and environmental discourses of Pacific colonialism and in the ordered spaces of the western academic and gardening tradition. Even this has earlier roots in, for example, John Parkinson's *Paradisi in Sole* (1629), commonly regarded as the first modern gardening book (Prest 1981). Paradise was in the Pacific also seen as a nature bountiful and uninscribed, a sense related to the idea of paradise as a physical and moral condition lost to the Old World since before the Fall and lost because of human folly. The complexity of a paradise in one sense not locatable and in another both found and lost was, thus, an important part of that situated encounter with the world in general and with the Pacific in particular that was geography's place in enlightenment reasoning from at least 1650.

Note

For comments on an earlier draft, I am grateful to Alessandro Scafi (particularly given his own work on the early iconography of Paradise), Robert Mayhew, Paul Wood, and Nicolaas Rupke, and to seminar audiences at the universities of Cambridge, Victoria, British Columbia, and Edinburgh. I also acknowledge the financial support of the British Academy.

References

Aarsleff, Hans. 1982. *From Locke to Saussure: Essays on the Study of Language and Intellectual History.* London: Athlone.

———. 1983. *The Study of Language in England 1780–1860.* Minneapolis: University of Minnesota Press.

Alexandre, Marcel. 1988. "Entre ciel et terre: les premiers débats sur le site du Paradis (Gen. 2, 8–15) et ses receptions," in *Peuple et Pays Mythiques*, eds. Francis Jouan and Bernard Deforge. Paris: Alex, pp. 187–224.

Armstrong, John. 1969. *The Paradise Myth.* Oxford: Oxford University Press.

Baudet, Henri. 1965. *Paradise on Earth: Some Thoughts on European Images of Non-European Man.* New Haven: Yale University Press.

Beaglehole, John C., ed. 1963. *The "Endeavour" Journal of Sir Joseph Banks, 1768–1771.* Sydney: Sydney Public Library in association with Angus and Robertson.

Bochart, Samuel. 1684. *Geographia Sacra.* Amsterdam: Boutesteyn and Luchtmans.

Boies, John L. 1983. *The Lost Domain: Avatars of the Earthly Paradise in Western Literature.* New York: University Press of America.

Bougainville, Louis Antoine de Comte. 1766–1769. [1985]. *Voyage de la fregate la Boudeuse et de la flute l'Etoile autour du Monde.* Paris: Editions la Decouverte.

Bowen, Margarita. 1981. *Empiricism and Geographical Thought: From Francis Bacon to Alexander von Humboldt.* Cambridge: Cambridge University Press.

Burnet, Thomas. 1684. *Sacred Theory of the Earth: Containing an Account of the origin of the Earth and of all the General Changes which it hath already undergone, or is to undergo Till the consummation of all Things. The First 2 Books. Concerning the Deluge and Concerning Paradise.* London: Kettilby.

Butlin, Robin A. 1992. "Ideological Contexts and the Reconstruction of Biblical Landscapes in the Seventeenth and Early Eighteenth Centuries: Dr Edward Wells and the Historical Geography of the Holy Land," In Alan R. H. Baker and Gideon Biger, eds., *Ideology and Landscape in Historical Perspective.* Cambridge: Cambridge University Press, pp. 31–62.

Carpenter, Nathanael. 1625. *Geography Delineated Forth in Two Bookes.* Oxford: Henry Capps.

Carver, Marmaduke. 1666. *A Discourse of the Terrestrial Paradise, aiming at a more probable discovery of the true situation of that happy place of our First Parents Habitation.* London: James Flesher.

Celsius, Olof. 1714. *De Situ Paradisi Terrestris.* Uppsala.

Cohn, Norman. 1996. *Noah's Flood: The Genesis Story in Western Thought.* Yale: Yale University Press.

Collins, Samuel. 1717. *Paradise Retriev'd: Plainly and fully demonstrating the method of managing and improving fruit-trees against walls, or in hedges . . . Together with a treatise on melons and cucumbers.* London.

Comito, T. 1971. "Renaissance Gardens and the Discovery of Paradise." *Journal of the History of Ideas* 32: 483–506.

Condillac, Etienne Bonnet de. 1746. *Essai sur L'origine des connaissances humanines.* Geneve: Villard et Nouffer.

Cormack, Lesley. 1991. "Twisting the Lion's Tale: Practice and Theory at the Court of Henry Prince of Wales," in *Patronage and Institutions: Science, Technology and Medicine at the European Court 1500–1750,* ed. B. Moran. Woodbridge and Rochester: Boydell Press, pp. 67–83.

Delano Smith, Catherine, and Ingram, Elizabeth M. 1991. *Maps in Bibles 1500–1600: An Illustrated Catalogue.* Geneva: Droz.

Delitzsch, Friedrich. 1881. *Wo lag das Paradies?* Leipzig: W. de Gruyter.

Diderot, Denis. 1751. *Discours Préliminaire* in *Encyclopédie, ou Dictionnaire raisonné des sciences, des arts et des métiers, par une société de gens de lettres.* Paris: Gide.

Dunbar, James. 1780. *Essays on the History of Mankind in Rude and Cultivated Ages*. London: Maxwell.

Duncan, James. 1972. *Milton's Earthly Paradise: A Historical Study of Eden*. Minneapolis: University of Minnesota Press.

Eco, Umberto. 1995. *The Search for the Perfect Language*. Oxford: Blackwell. Translated by James Fentress.

Elert, Charles C. 1978. "Andreas Kempe (1622–1689) and the Languages spoken in Paradise." *Historiographia Linguistica* 5: 221–226.

Engel, Maurice. 1885. *Die Lösung der Paradiesfrage*. Leipzig: W. de Gruyter.

Fish, Stanley. 1971. *Surprised by Sin: The Reader in Paradise Lost*. Berkeley: University of California Press.

Ford, Julian. 1981. *The Story of Paradise*. New York: Harper and Row.

Forster, Johann R. 1778. *Observations Made during a Voyage round the World, on Physical Geography, Natural history and Ethic Philosophy*. London: Robinson.

Fox, Christopher. 1995. "How to Prepare a Noble Savage: The Spectacle of Human Science," in *Inventing Human Science: Eighteenth-Century Domains*, ed. Christopher Fox, Roy Porter, and Robert Wokler. Berkeley: University of California Press, pp. 1–30.

Frängsmyr, Töre. 1983. *Linnaeus: The Man and His Work*. Berkeley: University of California Press.

Frängsmyr, Töre, Heilbron, J. L., and Rider, R. E., eds. 1990. *The Quantifying Spirit in the Eighteenth Century*. Berkeley: University of California Press.

Fraser, Russell. 1971. *The Language of Adam: On the Limits and Systems of Discourse*. New York: Columbia University Press.

Gascoigne, John. 1994. *Joseph Banks and the English Enlightenment: Useful Knowledge and Polite Culture*. Cambridge: Cambridge University Press.

Giamatti, A. Bartlett. 1966. *The Earthly Paradise and the Renaissance Epic*. Princeton: Princeton University Press.

Glass, Dr. 1787. "Letter to William Marsden, Esq., on the affinity of certain words in the language of the Sandwich and Friendly Isles in the Pacific Ocean, with the Hebrew." *Archaeologia* 8: 81–4.

Grillo, Robert D. 1989. *Dominant Languages: Language and Hierarchy in Britain and France*. Cambridge: Cambridge University Press.

Grove, Richard H. 1995. *Green Imperialism: Colonial Expansion, Tropical Island Edens and the Origins of Environmentalism 1600–1860*. Cambridge: Cambridge University Press.

Hamblin, Diana. 1987. "Has the Garden of Eden Been Located at Last?" *Smithsonian* 18 (2): 127–135.

Hammond, L. Davis. 1970. *New from New Cythera: A Report of Bougainville's Voyage 1766–1769*. Minneapolis: University of Minnesota Press.

Hare, Henry [Lord Coleraine]. 1683. *The Situation of PARADISE Found Out: Being a History of a Late Pilgrimage into the Holy Land*. London: Lowndes.

Hariot, Thomas. 1588. *Briefe and True Report of the New Found Land of Virginia*. London: Robinson.

Harley, J. Brian, and David Woodward, eds. 1990. *The History of Cartography Volume 1: Cartography in Prehistoric, Ancient and Medieval Europe and the Mediterranean*. Chicago: University of Chicago Press.

Heinberg, Richard. 1990. *Memories and Visions of Paradise: Exploring the Universal Myth of a Lost Golden Age*. Wellingborough: Aquarius Press.

Hennig, Richard. 1950. *Wo lag das Paradies?* Berlin: Sturtz.

Herder, Johann. 1772. *Abhandlung über den Ursprung der Sprache.* Leipzig.

Heylyn, Peter. 1652. *Cosmographie, in Four Bookes Contayning the Geographie and Historie of the Whole World.* London: Anne Seile.

Hoëm, Ingjerd. 1996. "The Scientific Endeavour and the Natives," in *Visions of Empire: Voyages, Botany and Representations,* eds. David Philip Miller and Peter Hanns Reill. Cambridge: Cambridge University Press, pp. 305–321.

Hopkinson, John. 1593. *Synopsis Paradisi.* London: Alsop and Coe.

Huet, Pierre Daniel [Bishop of Soissons]. 1694. *A Treatise of the Situation of Paradise.* London: James Knapton.

Hulme, Peter, and Ludmilla Jordanova, eds. 1990. *The Enlightenment and Its Shadows.* London: Routledge.

Humboldt, Alexander von. 1814–1834. *L'Examen Critique de l'Histoire de la Géographie du Nouveau Continent et des Progrès de l'Astronomie Nautique aux quinzième et seizème siècles.* Paris: Gide.

Inveges, Alessandro. 1649. *Historia Sacra Paradisi Terrestris et Sanctissimi Innocentia Status.* Venice: n.p.

Keerl, Friedrich. 1861. *Lehre vom Paradies.* Basle: n.p.

Kempe, Andreas. 1688. *The Languages of Paradise.* Hamburg: n.p.

Levin, Harry. 1969. *The Myth of the Golden Age in the Renaissance.* Bloomington: Indiana University Press.

Linnaeus, Carl. 1743. *Oratio de Telturis Habitabilis Incremento.* Leiden.

Livingstone, David N. 1994. "Science and Religion: Foreword to the Historical Geography of an Encounter." *Journal of Historical Geography* 20 (4): 367–383.

———. 1995. "The Spaces of Knowledge: Contributions toward a Historical Geography of Science." *Environment and Planning D: Society and Space* 13: 5–34.

Malte-Brun, Conrad. 1816–1822. *Universal Geography, or A Description of All the Parts of the World on a New Plan.* Volume II. London: Longman, Hurst, Rees, Orme and Brown.

Marshall, Peter, and Glyndwr Williams. 1982. *The Great Map of Mankind. British Perceptions of the World in the Age of Enlightenment.* London: J. M. Dent.

Maude, Harold E. 1968. *Of Islands and Men: Studies in Pacific History.* Melbourne: Oxford University Press.

Mengham, Rod. 1993. *The Descent of Language.* London: Bloomsbury.

McClung, William A. 1983. *The Architecture of Paradise: Survivals of Eden and Jerusalem.* Berkeley: University of California Press.

McColley, Diane K. 1993. *A Gust for Paradise: Milton's Eden and the Visual Arts.* Urbana: University of Illinois Press.

Moravia, Sergio. 1980. "The Enlightenment and the Sciences of Man." *History of Science* 18: 247–268.

Moxon, Joseph. 1671. *Sacred Geographie, Or Scripturall Maps.* London: J. Moxon.

Olender, Maurice. 1992. *The Languages of Paradise: Race, Religion, and Philology in the Nineteenth Century.* Harvard: Harvard University Press.

Ophir, A. Ian, and Steven Shapin. 1991. "The Place of Knowledge: A Methodological Survey." *Science in Context* 4: 3–27.

CHARLES W. J. WITHERS

Ortelius, Abraham. 1596. *Thesaurus Geographicus*. Antwerp: n.p.

———. 1606. *Parergon* [in his *Theatrum Orbis Terrarum*. London: Norton].

Outram, Dorinda. 1995. *The Enlightenment*. Cambridge: Cambridge University Press.

Parkinson, John. 1629. *Paradisi in sole paradisus terrestris*. London: R. N. Thrale.

Plaut, Alfred. 1982. "General Gordon's Map of Paradise." *Encounter* (June–July): 20–32.

———. 1984. "Where is Paradise? The Mapping of a Myth." *Map Collector* 29: 2–7.

———. 1993. *Analysis Analysed*. London: Routledge.

Prest, John. 1981. *The Garden of Eden: The Botanic Garden and the Re-creation of Paradise*. New Haven: Yale University Press.

Raleigh, Sir Walter. 1614. *History of the World*. London: W Burre.

Rousseau, Jean-Jacques. 1755. *Discours sur l'origine de l'inégalité parmi les homes*. Paris.

Salkeld, John. 1617. *A Treatise of Paradise, and the principall contents thereof: especially of the greatnesse, situation, beautie, and other properties of that place; of the trees of life, good and evill; of the serpent, cherubim, fiery sword, mans creation, immortalitie, propagation, stature, age, knowledge, temptation, fall, and exclusion out of Paradise; and consequently of his and our originall sin; with many other difficulties touching these points; collected out of the Holy Scriptures, ancient fathers, and other both ancient and moderne writers*. London: E. Griffin for N. Butler.

Sanford, Charles. 1961. *The Quest for Paradise: Europe and the American Moral Imagination*. Urbana: University of Illinois Press.

Schreyer, Rudiger. 1989. "'Pray what Language did your Wild Couple Speak, when first they met?' Language and the Science of Man in Scottish Enlightenment," in *The "Science of Man" in the Scottish Enlightenment*, ed. Peter Jones. Edinburgh: Edinburgh University Press, pp. 149–177.

Schulz, Maximillian F. 1985. *Paradise Preserved: Recreations of Eden in Eighteenth- and Nineteenth-Century England*. Cambridge: Cambridge University Press.

Shea, Richard W., ed. 1983. *Nature Mathematized: Historical and Philosophical Case Studies in Classical Modern Natural Philosophy*. Dordrecht: Van Reinem.

Skoblow, John. 1993. *Paradise Dislocated: Morris, Politics, Art*. Charlottesville: University Press of Virginia.

Smith, Adam. 1761. *Considerations concerning the first Formation of Languages and the Different Genius of Original Compounded Languages*. London.

Smith, Berhard. 1969. *European Vision and the South Pacific 1768–1850*. Oxford: Oxford University Press.

Smith, Olivia. 1984. *The Politics of Language 1791–1819*. Oxford: Clarendon Press.

Spate, Oskar H. K. 1988. *Paradise Found and Lost*. London: Routledge.

Stewart, Stanley. 1966. *The Enclosed Garden: The Tradition and the Image in Seventeenth-Century Poetry*. Madison: University of Wisconsin Press.

Switzer, Stephen. 1718. *Directions for the General Distribution of a Country Seat into Rural and Extensive Gardens, Parks, Paddock &c.* 3 vols. London.

Thomas, Nicholas, Harriet Guest, and Michael Dettelbach, eds. 1996. *Obser-*

vations Made during a Voyage Round the World. Hawaii: University of Hawaii Press.

Ugolino, Bruno. 1747. *Thesaurus Antiquitatem Sacrarum.* Venice: Joannam Gabrielem Herthzo.

van Til, Salomon. 1701. *Malachius Illustratus: Tractiatus de Situ Paradisi Terrestris: Dissertatio singularis Geographico-Theologica.* Dordraci: Willegardum Cornelium.

von Wendrin, Friedrich. 1924. *Die Entdeckung des Paradies.* Leipzig: Braunschweig.

Warren, William F. 1885. *Paradise Found: The Cradle of the Human Race at the North Pole.* London: Sampson Low, Marston, Searle and Rivington.

Wells, Edward. 1701. *A Treatise of Ancient and Present Geography.* London: James Knapton.

———. 1710–1712. *An Historical Geography of the Old Testament in Three Volumes.* London: James Knapton.

Withers, Charles W. J. 1993. "Geography in its Time: Geography and Historical Geography in Diderot and d'Alembert's *Encyclopédie.*" *Journal of Historical Geography* 19 (3): 255–264.

———. 1996. "Encyclopaedism, Modernism and the Classification of Geographical Knowledge." *Transactions, Institute of British Geographers* 21 (1): 275–298.

Wood, Paul. 1996. "The Science of Man" in *Cultures of Natural History,* edited by N. Jardine, J. A. Secord, E. Spary. Cambridge: Cambridge University Press, pp. 197–210.

Chapter Four
Geographical Inquiry, Rational Religion, and Moral Philosophy: Enlightenment Discourses on the Human Condition

DAVID N. LIVINGSTONE

As far back as 1619, an obscure writer, P. Bergeron by name, and a little later, in 1636, Tomasso Campanella, were putting forth this sort of thing: "The exploration of the globe having resulted in discoveries that have destroyed many of the *data* on which ancient philosophy reposed, a new conception of things will inevitably be called for."

—Paul Hazard

4.1. Introduction

Two debates at either end of what has been called "the long eighteenth century" serve here as vehicles for perusing some crucial connections between geographical inquiry, rational religion, and moral philosophy during the period of the Enlightenment. In both cases, the issues on which I wish to focus surface—at least in part—in questions to do with human origins and racial variation, and thus congregate around considerations bound up in one way or another with conceptions of the human constitution. By reflecting on these two episodes, one European, the other American, we may begin to glimpse something of the broader intellectual, religious, and political contexts within which geographical inquiry has been enmeshed. At the same time we may also catch sight of different renditions of the Enlightenment *mentalité* in different places and thereby come to appreciate something of the historical *geography* of Enlightenment. Besides this, focusing on these two debates—one early, one late; one French, one American; one sceptical, the other didactic—has the added advantage of addressing the question of origins in a number of different but related ways. The two major figures to whom I call attention were deeply concerned with human origins and were therefore profoundly implicated in the establishment of the categories by which the human condition has been understood since the period of the Enlightenment. At the same time, both have also been seen as playing key roles in the origins of anthropological science in Europe and America, respectively, even though the immense significance of various forms of geographical inquiry in the treatises they produced has perhaps been less visible than it might to the scholarly eye. The fact that they mobilized geographical knowledge in massively different ways—one to undergird what was regarded as a monumental heresy, the other to shore up visions of a Christian republic—will attract our attention and disclose how profoundly implicated geographical work was in different moral discourses.

Before turning to the two cases I wish to consider, however, it is worth pausing briefly to record that consideration of the causes of human variability was certainly not unknown in geographical treatises during the Enlightenment period. Take, for example, the case of John Walker, sometime engraver, teacher, medical practitioner, abolitionist, Quaker, and advocate of vaccination. He devoted the third part of his *General Geography* of 1788 to "Picturesque and General Sketches of the Different Parts of the Earth, and the Varied Appearances and Manners of its Inhabitants, both Man and Brute." Right at the outset he declared:

> The changes that the human species have *[sic]* undergone, from the difference of climate, soil, food, manner of life, and other accidental causes, are astonishing, insomuch, that some have called in question our common consanguinity. However, when we consider how regular and slow the gradations are, from the black complexion to the brown, from this to the fair, generally varying by imperceptible degrees with the climate, we shall perhaps have but little reason to suppose it necessary, that different species of men should have been created, to produce the variety observable at this day. (Walker 1797, 153)

Walker's declaration of monogenism here is itself significant. To him, the plasticity of the human form merely attested to the flexibility of a single constitution. And this had far-reaching political implications. For in reprinting the United States Declaration of Independence he did not hesitate to interpolate—after the "self-evident" propositions that "all men are created equal, [and] that they are endowed by their Creator, with certain inalienable rights"—the comment: "How specious, yet how palpably inconsistent these declarations from a people holding slaves" (Walker 1797, 550).

Nor were such inclinations the mere personal idiosyncracies of one

DAVID LIVINGSTONE

individual. In the French National Institute, which was in existence between 1795 and 1803, for example, Human Geography was placed in the Class of Moral and Political Sciences precisely because of a perceived need to systematically investigate the presumed "influences of climate, soil, terrain, occupations, form of government, and economic life on the physical and mental qualities of peoples" (Staum 1987, 333). Certainly the nascent discipline in that institution must be adjudged a failure, but that does not detract from the recognition among Enlightenment French legislators that geography was of fundamental importance in the construction of 'a science of man.' The geographical elements of French Enlightenment thinkers is the subject of Heffernan's chapter 5. It was central too in the aspirations of Marie-Jean de Condorcet and his Idéologue heirs to establish what Staum calls "gigantic research projects that would study the effect of physical geographic conditions on the body and mind" building on the heritage of hygienists like Volney and Montesquieu and furthering the revisionist medical geography of Jean-Noël Hallé and Pierre-Jean Georges Cabanis (Staum 1987, 332). Indeed it was because of the radicalism inherent in such efforts to *earth* the human constitution in this way that Napoleon closed down the moral sciences section of the Institut on the grounds of its incipient materialism (Porter 1995; Staum 1980).

My purpose in beginning with these observations is to illustrate that the issue of human diversity, its cause, and consequence were part and parcel of geography's domain during the period of what is called the Enlightenment. Their significance was of very considerable proportions. For as Paul Wood has recently put it, "a revolution in the human sciences took place during the 'long eighteenth century' in Europe which rivaled that which shook the physical sciences in the period" (Wood 1996, 197).

4.2. Geography, Human Origins, and the Path to Modern Scepticism

On 18 June 1646, Isaac de la Peyrère put the finishing touches to a letter to François La Mothe le Vayer, a sceptical French antirationalist and member of a circle of intellectuals enjoying the patronage of the Prince of Condé. This was the second such letter Peyrère had sent to the same individual. The first one, completed in 1644, was a description of Iceland but was not published until 1663. The present treatise, on Greenland, however, appeared in Paris in 1647 (Popkin 1987). These two volumes established their author's reputation as a leading authority on these regions until well into the nineteenth century.

I want to begin my reflections by focusing initially on this Greenland volume because it hints at some of the larger issues in which la Peyrère was subsequently embroiled while at the same time establishing geography's early complicity in the path to modern scepticism (Popkin 1980; also Gay 1995, 290–295). I shall work from the English edition of the text which was published, appropriately enough, by the Hakluyt Society in 1855.[1]

It is worth attending momentarily to the map of Greenland accompanying la Peyrère's volume. Compiled in consultation with the mathematician Roberval and the geographer Sanson, it revealed his uncertainties as to the eastern and western reaches of the subcontinent. He thus portrayed the western coastline merging with modern-day Baffin Island and Cap Farvel as crucially separated by a short stretch of ocean from Newfoundland (See fig. 4.1). Thereby he revealed his disagreement with those "who think that Greenland is part of the American continent" and sided with the narrative of "a Danish captain named John Munck, who tried this passage to the East by the northwest of the Gulf of Davis, and according to what he says there is great probability that this land is entirely separated from America" (la Peyrère 1663, 183). Jens Munck had led an expedition of around thirty men in 1619 in two ships into Hudson Bay. But the mission had to be aborted since, by June of the following year, all but three of the crew had died of scurvy, though the surviving remnant managed to make it back to Denmark (Debenham 1960, 111). These cartographical conceptions, as we shall see, were crucial to la Peyrère's account of the settlement of the region.

More generally, the treatise disclosed la Peyrère's method of working; he synthesized information gathered from both ancient and contemporary sources. Two major works provided him with historical information, one of ancient Icelandic origin, the other modern Danish. From the Icelandic Chronicle, la Peyrère learned of the Skreglinguer people who inhabited the western flank of Greenland, and he reported that "Doctor Vormius,[2] the most learned of all the doctors in northern researches . . . says they were the original savages of Greenland, to whom this name was probably given by the Norwegians." The native peoples of Greenland were not, then, descended from northern Europeans. As for the origin of the Skreglingres, two possibilities suggested themselves. Either they were of American derivation or were aboriginal to Greenland itself. La Peyrère himself believed "there was no need of bringing Americans here

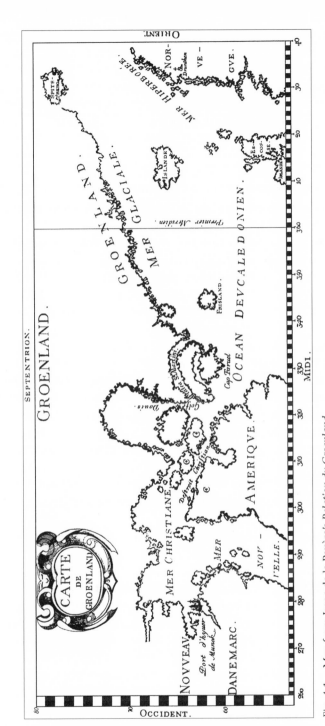

Fig. 4.1. Map from Isaac de la Peyère's *Relation du Groenland*

at all," but, either way, they predated the advent of the Norwegians and la Peyrère further suggested that "by the same reasons that Vestrebug had its original inhabitants when the Norwegians arrived there, Ostrebug had them also, and that as the eastern part was nearer the Arctic Sea, was not so fertile, and consequently less inhabited than the west, the Norwegians, who met with less resistance on that side than on the other, took possession more easily of Ostrebug than of Vestrebug." This scenario found corroboration in the Danish Chronicle which, according to la Peyrère, confirmed that Greenland "is inhabited by a variety of races, and that these races are governed by different lords, of whom the Norwegians never knew anything" (la Peyrère 1663, 193–195).

As for the characteristics of the region's indigenous peoples, la Peyrère drew on the Danish Chronicle to depict "these savages" as "deceitful and ferocious" and incapable of being "tamed, either by present or kindness": "They are fat but active, and their skins are of an olive colour; it is believed that there are black among them like Ethiopians . . . The shirts of the men and the chemises of the women are made of the intestines of fish, sewn with very fine sinews. Their clothes are large, and they bind them with straps of prepared skin. They are very dirty and filthy . . ." (la Peyrère 1663, 217).

The forces which moved la Peyrère to undertake this investigation of Greenland (and indeed his parallel study of Iceland) are crucially important to the connections I want to make between geography and Enlightenment scepticism. The Greenland work was composed during 1644–45 while la Peyrère was in Scandinavia where he met the Danish scholar Ole Worm [the Vormius cited in la Peyrère's text] with whom he subsequently sustained a lengthy correspondence. According to Popkin, this correspondence reveals la Peyrère's dominating concern to account for the origins of the American people. The problem encountered here, moreover, was similar to that of the genesis of the Inuit. Hence, as Popkin (1987, 11) puts it, "the question of whether the Bible is adequate as an account of how the world developed was challenged both geographically and anthropologically by what was then known about the Americas and about the far north . . . If Eskimos were found in Greenland by the Viking explorers, where did they come from?"

All this is fully in keeping with the final sentiments expressed in la Peyrère's Greenland excursus. Here he attacked the views of the recently deceased Dutch jurist and Swedish ambassador, Hugo Grotius[3]—as recorded in a treatise by the historian and geographer George Horn[4]—who had issued in 1643 a refutation of la Peyrère's as yet unpublished work *Praedamitae,* to which I will shortly turn, and claimed that the native peoples of America were of Norwegian descent:

> I discover at the same time the errors of the person who has written dissertations upon the origin of the people of America, whom he makes out to have come from Greenland, and makes the first inhabitants of Greenland to have come from Norway. . . . You will judge, sir, by the continuance and the reasoning of my history, that this author errs in every way. First, inasmuch as the Norwegians were not the first inhabitants of Greenland, as it appears from his narrations and the demonstrations I have given you of them; and inasmuch as that M. Vormius, who is very learned in the antiquities of the north, so far from connecting the origin of the people of America with the people of Greenland, thinks that the Skreglingres, the original habitants of Vestrebug in Greenland, came from America. Secondly, he is mistaken, inasmuch as there is little or no probability that Greenland was part of the continent of America, and that the passage from the one to the other was not so well known nor so possible as is imagined. Thirdly, he is mistaken in that which I have shown you, that there is no affinity of language or manners between Greenland and Norway; and if, as he says, the Norwegians communicated their language and manners to the Americans, they must have gone elsewhere than by Greenland to get to America. I should here have a good opportunity of showing up other errors of this dissertation, of making the author eat his own words, and of sending him to the land of visions and dreams; but as he now sleeps his last sleep, we will let him rest in quietness. (la Peyrère 1663, 249)

La Peyrère's *Relation du Groenland,* then, was a sustained effort to deploy arguments from geography, encompassing physical, demographic, linguistic, and cultural evidence, to sustain his own suspicions about the inadequacy of the traditional monogenetic story of the human race in favor of a polygenetic account.

It is for this reason that la Peyrère's tract can be regarded as a crucial moment in the modern evolution of Pyrrhonism. Taking its name from Pyrrho of Elis (c. 360–275 B.C.), this system of philosophy—technically speaking—constituted a particular formulation of scepticism emphasizing a suspension of judgment that induced a state of quietude (Popkin 1979, xv). Its recovery in the sixteenth and seventeenth centuries led Pierre Bayle to consider its reintroduction as the genesis of modern philosophy. Here I am not concerned with technicalities, however, because by the end of the seventeenth century Pyrrhonism had come to be considered as synonymous with scepticism. Scepticism of course surfaced in a variety of arenas. But in the present context I want to foreground the question of *religious* scepticism because of what I think geography's role was in the period of its modern gestation. Let us recall Paul Hazard's quip on the meaning of travel at the time. It meant "comparing manners and customs, rules of life, philosophies, religions; arriving at some notion of the relative; discussing; doubting. Among those who wandered up and down the earth in order to bring the tidings of the great unknown, there was more than one free thinker" (Hazard 1969, 27–28). Armchair geographers were no less subversive.

If the sceptical thrust of la Peyrère's Greenland studies remained thoroughly implicit, there was no mistaking its features in the grand profanity which he published in 1655 (though the work had been circulating in one form or another since 1641). The salient features are readily summarized (Livingstone 1987a, 1990, 1992; Popkin 1976a, 1987). In *Praeadamitae*, or *Men Before Adam*, la Peyrère advanced the beguilingly simple notion that human beings existed before the biblical Adam. It was, to put it another way, a polygenetic account of the human species. On the surface, the treatise was an exercise in biblical hermeneutics, constituting—as its subtitle makes plain—*a Discourse upon the Twelfth, Thirteenth, and Fourteenth Verses of the Fifth Chapter of the Epistle of the Apostle Paul to the Romans. By Which are Prov'd, That Men Were Created before Adam.* Plainly internal matters of biblical exegesis were to the forefront as la Peyrère sought to explicate St. Paul's words that "Until the law, sin was in the world; but sin was not imputed, when the law was not." Standard interpretations which took this text as referring to the Mosaic law, la Peyrère found unconvincing, and he proceeded to argue his case for it as law given to Adam. His conviction was that the history of ceremonial Judaism must be traced back beyond Moses to Adam himself. As he put it: "Long before Moses there were other Ordinances prescribed, and commended to the Jews, other Ceremonies instituted, other Laws of God decreed and confirmed for that holy and elected People. And in this place, I mean the Jews, not onely [sic] the Sons of Abraham who are called the Seed of Abraham, but also the fore-fathers of Abraham, the Posterity of Adam" (la Peyrère 1656, 4, 19). The consequences of this hermeneutic move were immediate: there must have been human beings on earth before Adam.

Yet, if internal biblical issues were the ostensible mainsprings of la Peyrère's intervention, geographical and anthropological data were even more decisive. Using material akin to that for his Greenland dissertation, he culled pagan chronicles, genealogies, and

Renaissance travel books in search of supporting testimony. *Praeada-mitae* was thus replete with details of Egyptian, Greek, Babylonian, and Chinese histories and how these challenged conventional Christian history. Taken together, the evidence of new geographical and anthropological data combined with internal biblical exegesis, helped lay the foundation of biblical criticism (see Popkin 1974; 1976b). There are, indeed, important parallels here with Withers's discussion of late seventeenth-century "sacred geographies" in the ways in which geographical knowledge was being used to question sacred authority. Extrabiblical data just had to be accorded their due role in scriptural hermeneutics. Thus Leo Strauss could comment that la Peyrère was "among the first of those who openly declared their departure from unquestioning acceptance of the Bible. This fact favours the suggestion that the recourse to Scripture, which has caused so much ridicule to be poured on la Peyrère, is no more than an adventitious trimming to a naturalistic and rationalistic theory" (Strauss 1930, 64–65). More recently, Grafton, in an exploration of "Isaac La Peyrère and the Old Testament," has made the observation that, in accounting for the intellectual transformation that was effected between the burning of Noël Journet in 1582 for his querying of scripture and Pierre Bayle's *Historical and Critical Dictionary* of 1692, "No one did more to make this revolution happen than the little-remembered French Calvinist Isaac La Peyrère" (Grafton 1991, 205). Seen in this light, the implications of la Peyrère's speculations turn out to be of epic proportions. For whether the Enlightenment's religious significance is calibrated in what Gay called "the rise of modern paganism" or along the lines of John Locke's "reasonable Christianity," the changed status of the Bible was crucially important (Gay 1995; see also Outram 1995, chap. 3). More: undermining the authority of scripture was far from merely theoretical. It "delegitimized the presumed sacredness of the foundations of civil and ecclesiastic authority" (Berti 1995, 562).

In the light of all this, it is hardly surprising that W. E. H. Lecky (1866, vol. 1, 107–108) should present la Peyrère's Preadamism as a kind of half-way house towards free thought, and that Andrew Dickson White (1894, vol. 1, 255), in his pugilistic recounting of the warfare between science and religion, should not hesitate to marshall la Peyrère in the cause of reason over against dogma. And perhaps this is not without good cause; for in the centuries following the appearance of la Peyrère's book, refutation after refutation tumbled from the pens of those alarmed at the implications of espousing

Preadamism. White (1894, vol. 2, 317), for example, noted that "his book was refuted by seven theologians within a year after its appearance, and within a generation thirty-six elaborate answers to it had appeared," while, more recently, Grafton has been able to identify a dozen repudiations in the first year (Grafton 1992, 237).

But la Peyrère's project did not stop even here. To fully appreciate its conceptual mainsprings, his Messianic vision must be taken into account.[5] For the objective of his Preadamite speculations that sets it apart from earlier inklings of the theory was his concern to separate out the Jewish experience from the rest of world history. The theological fulcrum of sacred history was, as he saw it, Judaism, and the Preadamite theory was designed to cut a deep gorge between the Jews and the Gentiles, namely, between the story of the Adamites and the Preadamites. Indeed, la Peyrère had already published a pro-Semitic work, *Du Rappel des Juifs,* in which he elaborated his vision of world history. Its thesis was that the Bible deals only with the Jews who play the leading role in Providential history. The Gentiles or, better, Preadamites merely look on as the divine drama is played out in the world, although they participate in the benefits of God's election of his own people. The political implications of this scenario for la Peyrère's France were immediate: France should admit Jews in order to hasten the nation's conversion. All in all *Du Rappel des Juifs* represented an effort to hold in balance the Bible, scientific knowledge, concern for the Jews, and French national identity (see Yardeni 1971).[6]

Taken in its full context, then, la Peyrère's project was universalistic in impulse and Messianic in character. And yet by proposing the altogether simple idea that people had existed for millennia before Adam, he was able to reconcile the shortness of biblical chronology with the latest findings of geography, anthropology, and archaeology. The polygenetic (though entirely humanitarian) thrust of his scheme earned for him an acknowledged place in the early history of the human sciences (see Allen 1963, 137; Comas 1960, 77, 80; Gossett 1963, 15; Grayson 1983, 140–142; Haddon 1910, 52; Harris 1968, 89; Hodgen 1964, 272–276; Johnson 1962; McKee 1944; Penniman 1952, 53; Rossi 1984, 132–136; Stepan 1982, 29; Trigger 1989, 112). Besides, his theory had more general biogeographical implications. Some of these were seized upon by a Dutch naturalist, Abraham van der Myl, who used it to argue that the entire fauna of the New World, animal as well as human, had been separately created (van der Myl 1667; see Browne 1983, 12–15).

Having sketched the subsequent history of the Preadamite theory elsewhere I will not dwell further upon the subject here. My aim, rather, is to establish geography's complicity in the path to modern scepticism. If indeed it is correct to say, as Popkin does, that la Peyrère "was regarded as perhaps the greatest heretic of the age, even worse than Spinoza, who took over some of his most challenging ideas," then geography was profoundly implicated in this heretical imperative. For in seeking the genesis of la Peyrère's monumental heterodoxy, Popkin himself observes that his most important evidence—above and beyond internal biblical considerations and data from classical antiquity—came from "the Voyages of Discovery" (Popkin 1987, 1, 2). Perhaps too it is significant that Anthony Grafton, in his recent examination of *New Worlds, Ancient Texts,* identifies la Peyrère's *Praeadamitae* and Blaeu's *Great Atlas* as the two key texts which widened into "frightening crevasses" the longstanding "cracks in the canon." Both "reveal in complementary ways the tremors of what would finally become an intellectual earthquake" (Grafton 1992, 237). And again, reflecting on the intellectual ferment of the early years of the Enlightenment, Paul Hazard observed: "Of all the lessons derived from the idea of space, perhaps the latest had to do with relativity. Perspectives changed. Concepts which had occupied the lofty sphere of the transcendental were brought down to the level of things governed by circumstance. Practices deemed to be based on reason were found to be mere matters of custom" (Hazard 1969, 11). Precisely this species of cross-cultural comparison, of course, was fundamental to the efforts of a succeeding century to produce stadial theories of social progress (Meek 1976) not least because, as Fox reminds us, "traveling in space also meant traveling in time" (Fox 1995, 16; also Pagden 1982).

4.3. Climate, Geography, and Visions of a Christian Republic

I want to turn now to another time, another place, and—initially at least—another exploration of racial questions. In 1787, "the first American treatise devoted to the causes of racial variation in the human species" made its appearance (Greene 1984, 323). Its author was Samuel Stanhope Smith, and the volume secured for him an acknowledged place in the annals of American anthropology (Gossett 1963, 39–40; Greene 1954) with one observer recording that his book was "the first ambitious American treatise on ethnology and long a standard work in the United States" (Stanton 1960, 4). The work was, in fact, an extended exercise in climatic determinism, so much so that Marvin Harris (1968, 86) believed that here the "ulti-mate pitch of environmentalism" had been reached. In a nutshell, Smith insisted that climatic factors predominantly, but also social conditions, could fully explain racial variation. Polygenism was thus to be rejected. According to Winthrop Jordan, "Smith's *Essay* stands as a monument to American faith in environment. It was a special faith. Environmentalism had been an underlying animus in the Revolutionary era and it towered as Americans stood poised and tri-umphant overlooking their magnificent continent" (Jordan 1965, 1).

The salient features of Smith's assessment can readily be surveyed. Climate, migration, and attendant social circumstance were more than adequate to account for human variability. The distinguishing features of human societies were thus responses to such environmen-tal conditions as "the influences of the climate, of the sterility or richness of the soil, of the elevation or depression of the face of their country, of the vicinity of seas or desarts [sic], of their insular, or con-tinental situation: or the modifications of all these, resulting from their occupation, and their habits of living" (Smith 1810, 21). By ref-erence to these geographical factors a single human constitution was preserved, flexible to be sure, but common nonetheless. One impli-cation of this commitment, of course, was that the human species had undergone some evolutionary modification and Smith plainly embraced the doctrine of the inheritance of acquired characteristics as the explanatory mechanism. Another was human cosmopolitan-ism: humankind was evidently capable of acclimatizing to the new climatic regimes which characterized zones of colonization (for a general analysis, see Livingstone 1987b).[7] Indeed, Smith went on to elaborate on the regional geography of the human complexion: "From the Baltic to the Mediterranean," he confirmed, "the different latitudes of Europe are marked by different shades of skin colour" (Smith 1810, 40). Human hair, skeletal stature, forehead shape, phys-iognomic features, even mental powers, were all correlated with "climates . . . some states of society, and modes of living" (Smith 1810, 78). Along the way, he relegated to the realm of mythology the exotic tales of travelers who were only too willing to generalize from any "little deviation from . . . ordinary standard" in order to indulge in "the most hyperbolic relations" (Smith 1810, 83, 84). Supremely, Smith's mechanistic environmental determinism—thor-oughly characteristic as it was of Newtonian science—seemed con-firmed in America's "ethnological laboratory" (Jordan 1965, xxviii).

A common origin and a shared nature, of course, did not entail

DAVID LIVINGSTONE

ethnic or cultural uniformity and Smith, evidently as accepting of cultural relativism as Montesquieu, accounted for "savage life" in terms of degeneration. Nevertheless, he remained convinced that differences between the races were literally only skin deep. He was certain, for example, that "if the Anglo-American, and the indian *[sic]* were placed from infancy in the same state of society, in this climate which is common to them both, the principal differences which now subsists between the two races, would, in a great measure, be removed when they should arrive at the period of puberty" (Smith 1810, 107). To be sure, this did not amount to modern egalitarianism; but it did confirm the unity of the human species.

By the time Smith published these sentiments in the second edition of his *Essay,* he had been President of the College of New Jersey (Princeton) for a decade and a half, having assumed the position in May 1795 (Jordan 1965, vii–liii; Noll 1989). Prior to that he had enjoyed a successful career as a Presbyterian minister, as first president of Hampden-Sydney College, and—since 1779—as professor of moral philosophy at Princeton. Soon it was clear that he was the presidential heir apparent to his father-in-law, John Witherspoon, whose daughter he had married several years before his return to Princeton. Indeed, he had been a prize student of Witherspoon at Princeton in the late 1760s and under his direction had drunk deeply at the waters of the Scottish Common Sense school of philosophy that Witherspoon championed (Sloan 1971). This brand of Enlightenment philosophy, to which I will presently return, was to dominate American college life for generations. Moreover, it undergirded Smith's personal efforts to keep physical science, moral philosophy, theology, and political ideology in conceptual tandem.

With Smith's intellectual pedigree it is scarcely surprisingly that he should be delighted to find his Baconian investigations serving "to confirm the facts vouched to us by the authority of holy writ" (Smith 1810, 149). As he had made clear on the very first page of the treatise: "The unity of the human race, notwithstanding the diversity of colour, and form under which it appears in different portions of the globe, is a doctrine, independently of the authority of divine revelation, much more consistent with the principles of sound philosophy, than any of those numerous hypotheses which have referred its varieties to a radical and original diversity of species" (Smith 1810, 7). Were there different species of humans, they would "be subject to different laws both in the physical and moral constitution of their nature" (Smith 1810, 8). The geographical facts of climate and demography, it seemed, had rescued sacred history from anthropological subversion. But even more, they had secured the intellectual integrity of moral philosophy and thereby preserved political stability:

> I must repeat here an observation which I made in the beginning of this essay, and which I trust I am now entitled to make with more confidence, that the denial of the unity of the human species tends to impair, if not entirely destroy, the foundations of duty and morals, and, in a word, of the whole science of human nature. No general principles of conduct, or religion, or even of civil policy could be derived from natures originally and essentially different from one another, and, afterwards, in the perpetual changes of the world, infinitely mixed and compounded. The principles and rules which a philosopher might derive from a study of his own nature, could not be applied with certainty to regulate the conduct of other men, and other nations, who might be of totally different species. The terms which one man would frame to express the ideas and emotions of his own mind must convey to another a meaning as different as the organization of their respective natures. But when the whole human race is known to compose only one species, this confusion and uncertainty is removed, and the science of human nature, in all its relations, becomes susceptible of system. The principles of morals rest on sure and immutable foundations. (Smith 1810, 149)

Given these assurances it is not at all surprising that Smith should find the recent polygenetic speculations of Lord Kames obnoxious. In his *Sketches of the History of Man* Kames had put forward the idea that the human race was actually composed of a variety of distinct species, each fitted to the environment within which it found itself (see Lehmann 1971; Ross 1972, 333–348). These speculations brought him to the brink of multiple creations, and though he actually drew back from that heterodoxy it was plain that he was attracted to polygenism, as indeed was David Hume (see discussions in Stocking 1975; Wokler 1988). Smith found such suggestions not only morally repugnant but also politically subversive, and he appended "Strictures on Lord Kaims' Discourse on the Original Diversity of Mankind" to his *Essay*. Thus, while the issue of biblical revelation certainly obtruded, there is little doubt that the matter of moral philosophy was the moving force behind Smith's assault. Without the unity of the human species, as he put it in the first edition of the work, any "science of morals would be absurd; the law of nature and nations would be annihilated" (Smith 1787, 48).[8] Certainly in pursuing this course, Smith was siding with Buffon who

DAVID LIVINGSTONE

followed Montesquieu's dictum that "the empire of climate is the first of all empires" despite the highly critical reception Buffon received from the Scottish moralists (Wood 1987). But whatever materialist inclinations they had detected in the Frenchman's *Histoire Naturelle,* these were less worrisome to Smith than a Kames-inspired polygenism that subverted the very possibility of a common human nature even though for Kames like "most enlightenment Scots the history of the species illustrated the workings of Divine Providence" (Wood 1996, 207).

If environmental factors thus secured the constitutional unity of the human species, they equally ensured the possibility of an inductive mental science. For it was precisely because there *was* a common human nature that it could be the object of empirical scrutiny. Indeed, the language of human nature was foundational to eighteenth-century moral philosophy and "had the status of being the shared ground on which writers of many persuasions erected moral standards of universal validity" (Smith 1996, 101). In this way, Christian moralists could co-opt scientific method and make human ethics a science even while remaining convinced that the findings of this new Baconian venture would confirm Christian virtue. If, as Witherspoon was certain, a republic was right to break both with its monarchical past *and* religious establishment; and if republics, dependent as they were on the performance of public virtue, were the fragile thing Montesquieu had shown them to be, then Christian morality could be preserved in the public square only if a universal ethical sense could be extracted from human nature by the methods of secular science. In this way, Protestant moral philosophers in America, under the leadership of Witherspoon and Stanhope Smith, domesticated Enlightenment values at the same time as they milked metaphysical meaning out of mundane science, transcendental truths out of the mere material (Guelzo 1999). Indeed, the empiricism of Smith and his successors was such that their moral project has been seen as the headwaters of the New Psychology that surfaced in late nineteenth-century America (Richards 1992, 1995). The traditional historiography which dismissed the relatively short life of American moral philosophy as 'anemic,' 'wormy,' outrageously *outré* and intellectually moribund, I suggest, is quite misconceived.

Taken overall, Samuel Stanhope Smith is best intellectually domiciled in the tradition of didactic enlightenment which was widespread in eighteenth-century Protestant America (Noll 1993, 1995). For America, as Henry May has demonstrated, was witness to several different Enlightenments. "The Enlightenment," he confirmed, already twenty years ago, ". . . has been too often homogenized" (May 1976, xv). American Protestants admired from afar the *moderate* variety as exemplified by Newton and Locke; the *sceptical* type emanating from the likes of Voltaire and Hume, and the *revolutionary* version characteristic of Rousseau and Tom Paine, were typically repudiated. But the *didactic* Enlightenment which owed its conceptual mainsprings to the Scottish philosophical moralists—Francis Hutcheson, Thomas Reid, Adam Smith, and Dugald Stewart—permitted a Protestant *rapprochement* with Enlightenment ideals even while opposing scepticism and revolutionary radicalism. The principles of Scottish Common Sense philosophy provided principles for higher education at Harvard, Brown, Yale, and Princeton. The underlying assumption was that human beings possessed a common nature whose capacities, both epistemological and ethical, could be uncovered by scientific investigation. So long as a common human nature was preserved, Protestant intellectuals were enabled to press natural science and Enlightenment rationality "into the service of traditional values" (Meyer 1976, xxvi). According to May, the official culture of the didactic Enlightenment was crucially dependent on three articles: "the essential reality and dependability of moral values, the certainty of progress, and the usefulness and importance of 'culture' . . . Every article of it was supported by reference to the Scottish Enlightenment" (May 1976, 358).

The didactic Enlightenment was thus religiously congenial, *and* it was politically comforting too. It permitted public virtue to be established in a society "that was busily repudiating the props upon which virtue had traditionally rested—tradition itself, divine revelation, history, social hierarchy, an inherited government, and the authority of religious denominations" (Noll 1995, 100). Scottish moral philosophy was thus "uniquely suited to the needs of an era still strongly committed to traditional religious values and yet searching for alternative modes of justification for those values" (Fiering 1981, 300). For after all, James Madison (himself a former student of Witherspoon at Princeton) and the other architects of the Constitution required a common human nature in order to undergird public moral order (Adair 1974; Branson 1979; Noll 1989, 123). Otherwise the "science of politics" that Madison adumbrated in the famous *Tenth Federalist,* showing his wholehearted endorsement of the Enlightenment style, would dissolve. From Witherspoon, Stanhope Smith knew how the Scottish philosophy, articulated by the Irish-born Francis Hutcheson, had developed a scientific and optimistic social theory that did not need to resort to the dictates of tradition, and cultivated a moral Newtonianism that rendered ethics a species of physical science by appealing to a moral sense intrinsic to a universal human nature.[9] It was precisely this common human nature that Smith's geographical discourse was designed to preserve in the face of seeming anthropological disruption.

There were, of course, other ways in which geography could be deployed in the service of the new republic. If Smith, the putative "father of American anthropology," secured the moral foundations of the polity by recourse to the power of environment, Jedidiah Morse, Congregationalist clergyman, staunch anti-Unitarian, and the "father of American geography" (Brown 1951; Greene 1984, 193; Wright 1959), saw to it that regional recitation could be every bit as politically serviceable. Morse's various geographies, particularly *The American Geography; or, A View of the Present Situation of the United States of America,* which first appeared in 1789, have been the subject of scrutiny by a number of the subject's historians. Unfortunately they have, generally speaking, treated his work in a lamentably internalist fashion, one dismissing his "long popularity" as "indicative of the lack of vigour in American geography at this time" (Bowen 1981, 171). Here I do not wish to review, still less evaluate, the contents of the various hugely successful school and college-level geographies he produced (see Brown 1941; Warntz 1964), so successful in fact that the name "Morse" became synonymous with geography in the period (Moss 1990, 214); most homes had a well-thumbed copy of their own "Morse." What concerns me, rather, are his mobilizing of geographical inquiry in the interests of the new republic and the revolutionary rhetoric within which his geography was domesticated. For it was precisely Morse's reproduction of the new Federal Constitution and extracts from various State Constitutions in *The American Geography* that attests to its intellectual and political mainsprings, however much Ralph Brown and other geographical commentators may have deplored "the inclusion of such a miscellany in a book entitled 'geography'" (Brown 1941, 171).

Morse, I think, can be broadly located within the same intellectual orbit as Samuel Stanhope Smith of whom he was a confidant, though perhaps with a touch less enthusiasm for even the didactic Enlightenment than other members of his circle. Nevertheless, he was closely associated with figures like Timothy Dwight, another advocate of didactic Enlightenment values in the Scottish vein, and received substantial encouragement in his geographical labors from both Witherspoon and Smith at Princeton (Brown 1941, 158). Besides he was, at least in its early phases, an advocate of the French Revolution and it seems (in keeping with other New England Calvinists like Samuel Miller, Stanhope Smith, and Ashbel Green) suspicious of revivalist excess. One of his correspondents, for example, assured him in 1799 that a local religious revival was pleasingly bereft of "any degree of enthusiasm" (cited in May 1976, 318). At the same time, convinced about the conspiratorial activities of the Bavarian Illuminati—a radical group of sceptics dedicated to rationalistic Enlightenment thought and the destruction of Christianity—he was joined by both Dwight and Smith in 1789–99 in urging that this group was engaged in fomenting political unrest in America.[10] Thus even while castigating French Jacobinism and deism, such figures did much to ensure the assimilation of enlightenment values in the immediate postrevolutionary period. Enlightenment, in both its Scottish and American articulations, flourished after all in "an environment shaped by Calvinism" (May 1976, 342).

To understand Morse's *Geography* it is vital to recognize that he had what Alan Snyder (1983) called "a dream of a republic operating upon and being sustained by Christian ideals." He found the idea of hereditary aristocracy utterly repugnant and elevated liberty to the acme of political philosophy, though he was certainly fully alert to its abuses. Indeed, it was for this reason that, in a sermon preached during the mid 1790s, he lauded the way in which the French had "burst the chains of civil and ecclesiastical tyranny" (cited in Snyder 1983, 387), even though he later came to repudiate what he saw as their intellectual and social excesses. In *The American Geography* itself he voiced strong antislavery sentiments and looked forward to a time when all distinctions between master and slave would be abolished and all such differences subsumed under the inclusivist rubric "American." Throughout his life, then, Morse articulated his own version of conservative republicanism and did everything in his power to inculcate in the new republic's citizens a sense of public virtue. It was a stern virtue, imbued with prohibitionist sentiments, supportive of abolition, and tied to an ideology of deference, not of course to monarchy, but to rank and station in life. Though Morse's personal life scarcely instantiated these values, they were the animating force behind his geographical excursions (Moss 1995; Phillips 1983).

The American Geography was authored in days of intense political tension with the passage of the Constitution and soon became, after its appearance in 1789, "the most widely read description" of the new nation (Moss 1995, 38). Within five years it had reportedly sold over 20,000 copies. In fact, the text was, as Richard Moss has put it, "a jeremiad surveying the state of the nation and calling Americans back to a vision of moral perfection and simplicity, a vision rooted in an image of New England and most specifically of Connecticut" (Moss 1990, 215). For it was here that that perfect citizen—the yeoman farmer—was most in evidence. The jeremiad indeed was already a well-established idiom and deployed by none other than Witherspoon in his fast day sermon of May 1776 to support the American cause (Sher 1985, 207). It is not surprising, therefore, that "Morse equated the New England way with the American way. If America were to be a republic, then the nation should look to Connecticut as a model" (Moss 1990, 215). And republic it certainly must be, for the "God of nature never intended that some of the best parts of his earth should be inhabited by the subjects of a monarch, 4000 miles from them" (Morse 1789, 469). Accordingly, Morse elaborated what can only be called a moral topography of the States of the Union. Maryland, for example, was characterized by "indolence and inactivity"; the South—the victim of slavocracy and excessive heat—was no less lethargic; Virginia had too many backgammon boards and was afflicted with a "gambling gentry" (Morse 1789, 352, 390, 445). By contrast, Morse envisioned New England as a near-perfect community enjoying the fruits of liberty, democratic learning, and what he termed a happy mediocrity productive of industrious labor. The moral economy and the social economy evidently mapped straight on to each other. As Richard Moss nicely summarizes it: "Morse, in effect, painted a moral vision in *The American Geography*, dominated by the enterprising New Englander who was ambitious and inventive, who reaped the profits of his labor, but who rejected the luxuries and diversions that follow" (Moss 1995, 44). Morse's text thus conveyed to the nation the contagion of his vision that it was only through inventiveness, enterprise, and moral virtue that the future of the republic could be secured.

4.4. Geographies and Enlightenments

In this essay I have been considering two significant Enlightenment moments and seeking to show how geographical motifs were pressed into the service of radically differing intellectual and political projects. The figures I have chosen to work with both occupy strategic positions in the early development of the human sciences, one in Europe, the other in America. But the fundamentally *geographical* nature of their projects has largely been ignored. By addressing these, the wider implications of the links between geographical thinking and Enlightenment discourse begin to crystallize. To a figure like la Peyrère, the opening up of the globe to European eyes meant that geographical data, cartographic representation, and spatial thinking were marshalled to advance a sceptical subversion of traditional religious and civil authority. In his hands, demographically inspired cartography articulated an heretical inevitability. By contrast, for Smith and Morse in America, geographical inquiry—either in the form of environmental necessitarianism or regional inventory—was used to demonize radicalism, to domesticate moderate Enlightenment values by preserving a common human nature, and to shore up Christian visions of liberal republicanism by compiling moral gazetteers of the newborn nation. Geographical circumstance thus provided Smith and Morse with an occasion to underwrite a Protestant vision of a public moral order in the new American Republic.

The contrast I have just been elaborating resonates with Mark Noll's (1995, 108) comment that in "France, liberty, the people, the Enlightenment, and the new sense of French national destiny stood against the church; in the United States, Protestantism identified itself with the people, the Enlightenment, democracy, republicanism, economic liberalism, and the sense of American manifest destiny." Accordingly, to tell the story of the fate of conjectural human history during the Enlightenment—as Wokler does with much conviction —from the time of la Peyrère's contemporary Samuel Pufendorf via Montesquieu to Rousseau as a tale of the relegation of God to the wings of society's self-understanding is at best an attenuated Continental European narration (Wokler 1995). As Smith's contemplations demonstrate, America's Enlightenment drew heavily on its Scottish counterpart where, as David Allan has forcefully revealed, the relations between Kirk and Enlightenment were characterized by anything but the implacable antagonism delivered to us by traditional historiography. Clearly those historians of human science, intent only on finding the roots of modern disciplines, have remained too insensitive to that enterprise's "theological parentage" (Allan 1993, 10; also Sher 1985).

Implications of even wider dimensions are, I believe, also advertised here. The fact that geography was profoundly implicated in different renditions of an Enlightenment *mentalité* means that it is misleading to seek for a universal portrait that fits everywhere from John

Witherspoon's Princeton to Voltaire's Geneva. Not only were the French and American experiences of Enlightenment hugely different, but different kinds of geographical inquiry were suited to different Enlightenment emphases and mobilized in the service of different moral projects. Plainly, future histories of geography and Enlightenment will need to be far more attentive to the historical *geography* of both geography and Enlightenment. This will certainly mean attending to the different national expressions of an Enlightenment *Zeitgeist;*[11] but it will also mean taking with far greater seriousness spatial and temporal migration of ideas and practices, the significance of geographical crafts like mapping, exploring, and regionalizing, and the ways in which Enlightenment motifs were customized to meet the needs of different ideological agendas.

To the degree that these suspicions are well founded and the "presumption" dangerous that the Enlightenment is "a single unitary process, displaying a uniform set of characteristics" (Pocock 1987, 48), then the fashionable critiques to which the so-called "Enlightenment project" has been subjected must fail for want of attending to the "ruptures," "discontinuities," and "extraordinary diversity" (Fox 1995, 3, 5) now increasingly seen as diagnostic of the Age of Enlightenment itself.

Notes

I am most grateful to Mark Noll for assistance in various ways with the preparation of this paper and to Derek Gregory, Charles Withers, and Nuala Johnson for several helpful observations on an earlier draft.

1. This translation was from the 1663 edition of *Relation du Groenland* published in Paris by Thomas Jolly.

2. Olaus Wormius—Ole Worm—with whom la Peyrère was in communication. Worm was a physician and gifted polymath at the University of Copenhagen with substantial interests in Danish antiquities (Daniel 1981, 31; Trigger 1989, 49, 54).

3. Hugo Grotius was the author of a treatise entitled *Dissertatio Altera de Origine Gentium Americanarum Adverses Obstrectatorem* (1643).

4. In *De Originibus Americanis Libri,* Horn presented his own conjectures on the subject. See "Horn," *Biographie Universelle, Ancienne et Moderne,* vol. 19 (Paris, 1817), q.v.

5. The more general "rationalistic and scathing utilization of the polemic spirit of Judaism" in the cause of anti-Christian sentiments during the period is discussed in Berti (1995).

6. Such speculations have lead several scholars to insist that la Peyrère was a Marrano, but conclusive evidence for this is apparently lacking. See Grafton (1988) and Popkin (1973).

7. In support of this position Smith called upon the medical testimony of Dr. Carl Strack, a German physician at Mainz (Smith 1787, 34). He also

directed his readers' attention to the significance of animal acclimatization (Smith 1787, 48).

8. Stocking (1975, 85) claims that Kames "did not see any contradiction between arguing on the one hand that men were not all members of the same species, and on the other, that they were all equal and shared a common human nature." Wood (1996, 204) comments, however, that Kames's contemporaries among the Scottish men of letters rejected his polygenetic inclinations "as subverting the foundations of religion and morality."

9. It is worth noting that Thomas Reid, a central figure within Scottish Common Sense philosophy, felt the need to resort to the writings of naturalists and travelers in the development of his views on human nature (Wood 1996, 210). This, moreover, was entirely in keeping with his suspicion that it might not be too long before the philosophy of the human mind would have its own Newton (Hatfield 1996). It is not surprising, therefore, that the outline of his Philosophy Course, taught in 1752, includes the study of "the Elements of Geography." Aberdeen University Library, MS 2131/8/V/I. I am most grateful to Charles Withers for drawing this item to my attention.

10. In this he was influenced by John Robison, author of *Proofs of a Conspiracy Against all the Governments and Religions of Europe,* who explained the French Revolution as the product of the Bavarian Illuminati (Wright 1983).

11. The need for locating Enlightenment in its various national contexts is apparent in Porter and Teich (1981) and Crocker (1991).

References

Adair, Douglass. 1974. "'That Politics May Be Reduced to a Science': David Hume, James Madison, and the Tenth Federalist," in Trevor Colbourn, ed., *Fame and the Founding Fathers: Essays by Douglass Adair.* New York: Norton, pp. 93–106.

Allan, David. 1993. *Virtue, Learning and the Scottish Enlightenment.* Edinburgh: Edinburgh University Press.

Allen, D. C. 1963. *The Legend of Noah: Renaissance Rationalism in Art, Science, and Letters.* Urbana, IL: University of Illinois Press.

Berti, Silvia. 1995. "At the Roots of Unbelief." *Journal of the History of Ideas* 56: 555–575.

Bowen, Margarita. 1981. *Empiricism and Geographical Thought: From Francis Bacon to Alexander von Humboldt.* Cambridge: Cambridge University Press.

Branson, Roy. 1979. "James Madison and the Scottish Enlightenment." *Journal of the History of Ideas* 40: 235–250.

Brown, Ralph H. 1941. "The American Geographies of Jedidiah Morse." *Annals of the Association of American Geographers* 31: 145–217.

———. 1951. "A Letter to the Reverend Jedidiah Morse: Author of the American Universal Geography." *Annals of the Association of American Geographers* 41: 188–198.

Browne, Janet. 1983. *The Secular Ark. Studies in the History of Biogeography.* New Haven and London: Yale University Press.

Comas, Juan. 1960. *Manual of Physical Anthropology.* Springfield, IL: Charles C. Thomas.

Crocker, Lester G. 1991. "Introduction," in John W. Yolton et al., eds., *The Blackwell Companion to the Enlightenment*. Oxford: Blackwell, pp. 1–10.

Daniel, Glyn. 1981. *A Short History of Archaeology*. London: Thames and Hudson.

Debenham, Frank. 1960. *Discovery and Exploration: An Atlas-History*. New York: Doubleday.

Fiering, Norman. 1981. *Moral Philosophy at Seventeenth-Century Harvard: A Discipline in Transition*. Chapel Hill: University of Carolina Press.

Fox, Christopher. 1995. "Introduction. How to Prepare a Noble Savage: The Spectacle of Human Science," in Christopher Fox, Roy Porter, and Robert Wokler, eds., *Inventing Human Science: Eighteenth-Century Domains*. Berkeley: University of California Press, pp. 1–30.

Gay, Peter. [1966] 1995. *The Enlightenment: An Interpretation. The Rise of Modern Paganism*. New York: Norton.

Gossett, Thomas F. 1963. *Race: The History of an Idea in America*. Dallas: Southern Methodist University Press.

Grafton, Anthony. 1988. "A Vision of the Past and Future." *Times Literary Supplement* February 12–18: 151–152.

———. 1991. *Defenders of the Text. The Traditions of Scholarship in an Age of Science, 1450–1800*. Cambridge, MA: Harvard University Press.

———. 1992. *New Worlds, Ancient Texts. The Power of Tradition and the Shock of Discovery*. Cambridge, MA: The Belknap Press of Harvard University Press.

Grayson, Donald K. 1983. *The Establishment of Human Antiquity*. New York: Academic Press.

Greene, John C. 1954. "The American Debate on the Negro's Place in Nature, 1780–1815." *Journal of the History of Ideas* 15: 384–96.

———. 1984. *American Science in the Age of Jefferson*. Ames: Iowa State University Press.

Guelzo, Allen C. 1999. "'The Science of Duty': Moral Philosophy and the Epistemology of Science in Nineteenth Century America," in David N. Livingstone, Darryl G. Hart, and Mark A. Noll, eds., *Evangelicals and Science in Historical Perspective*. New York: Oxford University Press, pp. 267–89.

Haddon, A. C. 1910. *History of Anthropology*. London: Watts & Co.

Harris, Marvin. 1968. *The Rise of Anthropological Theory: A History of Theories of Culture*. New York: Thomas Y. Crowell.

Hatfield, Gary. 1996. "Remaking the Science of Mind: Psychology as Natural Science," in Christopher Fox, Roy Porter, and Robert Wokler, eds., *Inventing Human Science: Eighteenth-Century Domains*. Berkeley: University of California Press, pp. 184–231.

Hazard, Paul. 1969. *The European Mind: 1680–1715*. Cleveland and New York: Meridian Books. Originally published as *La Crise de la Conscience Europeene*. Paris: Boivin, 1935.

Hodgen, Margaret T. 1964. *Early Anthropology in the Sixteenth and Seventeenth Centuries*. Philadelphia: University of Pennsylvania Press.

Johnson, James. 1962. "Chronological Writing: Its Concept and Development." *History and Theory* 2: 124–145.

Jordan, Winthrop. 1965. "Introduction" to Samuel Stanhope Smith, *Essay on the Causes of the Variety of Complexion and Figure in the Human Species*. Cambridge, MA: Belknap Press of Harvard University Press.

la Peyrère, Isaac de. 1656. *Men Before Adam. Or a Discourse upon the Twelfth, Thirteenth, and Fourteenth Verses of the Fifth Chapter of the Epistle of the Apostle Paul to the Romans. By Which are Prov'd, That Men Were Created before Adam*. London.

————. 1663. *Relation du Greonland*, in Adam White, ed., *A Collection of Documents on Spitzbergen and Greenland, Comprising a Translation from F. Martens' Voyage to Spitzbergen: A Translation from Isaac de la Peyrère's Histoire de Groenland* . . . London: Hakluyt Society, 1855.

Lecky, William Edward Hartpole. [1866] 1910. *History of the Rise and Influence of the Spirit of Rationalism in Europe*. 2 vols. London: Watts & Co.

Lehmann, William, C. 1971. *Henry Home, Lord Kames, and the Scottish Enlightenment: A Study in National Character and in the History of Ideas*. The Hague: Martinus Nijhoff.

Livingstone, David N. 1987a. "Preadamites: The History of an Idea from Heresy to Orthodoxy." *Scottish Journal of Theology* 40: 41–66.

————. 1987b. "Human Acclimatization: Perspectives on a Contested Field of Inquiry in Science, Medicine and Geography." *History of Science* 25: 359–394.

————. 1990. "Preadamism: The History of a Harmonizing strategy." *Fides et Historia* 22: 25–34.

————. 1992. *The Preadamite Theory and the Marriage of Science and Religion*. Philadelphia: American Philosophical Society.

May, Henry F. 1976. *The Enlightenment in America*. New York: Oxford University Press.

McKee, D. R. 1944. "Isaac de la Peyrère, A Precursor of Eighteenth-Century Critical Deists." *Publications of the Modern Language Association* 56: 456–485.

Meek, Ronald L. 1976. *Social Science and the Ignoble Savage*. New York: Cambridge University Press.

Meyer, D. H. 1976. *The Democratic Enlightenment*. New York: G. P. Putnam's Sons.

Morse, Jedidiah. 1789. *The American Geography; or, A View of the Present Situation of the United States*. Elizabethtown, NJ.

Moss, Richard J. 1990. "Republicanism, Liberalism, and Identity: The Case of Jedidiah Morse." *Essex Institute Historical Collections* 126: 209–236.

————. 1995. *The Life of Jedidiah Morse: A Station of Peculiar Exposure*. Knoxville: University of Tennessee Press.

Noll, Mark A. 1989. *Princeton and the Republic, 1768–1822. The Search for a Christian Enlightenment in the Era of Samuel Stanhope Smith*. Princeton, NJ: Princeton University Press.

————. 1993. "The American Revolution and Protestant Evangelicalism." *Journal of Interdisciplinary History* 23: 615–638.

————. 1995. "The Rise and Long Life of the Protestant Enlightenment in America," in William M. Shea and Peter A. Huff, eds., *Knowledge and Belief in America. Enlightenment Traditions and Modern Religious Thought*. New York: Cambridge University Press, pp. 88–124.

Outram, Dorinda. 1995. *The Enlightenment*. Cambridge: Cambridge University Press.

Pagden, Anthony. 1982. *The Fall of Natural Man: The American Indian and the Origins of Comparative Ethnology.* Cambridge: Cambridge University Press.

Penniman, T. K. 1952. *A Hundred Years of Anthropology.* 2d ed. London: Duckworth.

Phillips, Joseph W. 1983. *Jedidiah Morse and New England Congregationalism.* New Brunswick, NJ: Rutgers University Press.

Pocock, J. G. A. 1987. "Enlightenment and Revolution: The Case of English-speaking North America." *Seventh International Congress on the Enlightenment: Introductory Papers.* Oxford: Voltaire Foundation, pp. 45–57.

Popkin, Richard H. 1973. "The Marrano Theology of Isaac La Peyrère." *Studi Internazionali di Filosofia* 5: 97–126.

———. 1974. "Bible Criticism and Social Science," in Robert S. Cohen and Marx W. Wartofsky, eds., *Methodological and Historical Essays in the Natural and Social Sciences,* Boston Studies in the Philosophy of Science, Vol. XIV. Dordrecht: Reidel, pp. 339–360.

———. 1976a. "The Pre-Adamite Theory in the Renaissance," in Edward H. Mahoney, ed., *Philosophy and Humanism: Renaissance Essays in Honor of Paul Oskar Kristeller.* Leiden: E. J. Brill, pp. 50–69.

———. 1976b. "The Development of Religious Scepticism and the Influence of Isaac La Peyrère's Pre-Adamism and Bible Criticism," in R. R. Bolgar, ed., *Classical Influences on European Culture, A.D. 1500–1700.* Cambridge: Cambridge University Press, pp. 271–280.

———. 1979. *The History of Scepticism from Erasmus to Spinoza.* Berkeley: University of California Press.

———. 1980. *The High Road to Pyrrhonism.* San Diego: Austin Hill Press.

———. 1987. *Isaac La Peyrère (1596–1676). His Life, Work and Influence.* Leiden: E. J. Brill.

Porter, R. 1995. "Medical Science and Human Science in the Enlightenment," in Christopher Fox, Roy Porter, and Robert Wokler, eds., *Inventing Human Science: Eighteenth-Century Domains.* Berkeley: University of California Press, pp. 53–87.

Porter, R., and M. Teich, eds. 1981. *The Enlightenment in National Context.* Cambridge: Cambridge University Press.

Richards, Graham. 1992. *Mental Machinery, Part One 1600–1850.* London: Athlone Press.

———. 1995. "'To Know Our Fellow Men to Do Them Good': American Psychology's Enduring Moral Project." *History of the Human Sciences* 8: 1–24.

Ross, Ian Simpson. 1972. *Lord Kames and the Scotland of His Day.* Oxford: Clarendon Press.

Rossi, Paolo. 1984. *The Dark Abyss of Time: The History of the Earth and the History of Nations from Hooke to Vico,* translated by Lydia G. Cochrane. Chicago: University of Chicago Press.

Sher, Richard B. 1985. *Church and University in the Scottish Enlightenment. The Moderate Literati of Edinburgh.* Edinburgh: Edinburgh University Press.

Sloan, Douglas S. 1971. *The Scottish Enlightenment and the American College Ideal.* New York: Teacher's College Press.

Smith, Roger. 1996. "The Language of Human Nature," in Christopher Fox,

Roy Porter, and Robert Wokler, eds., *Inventing Human Science: Eighteenth-Century Domains*. Berkeley: University of California Press, pp. 88–111.

Smith, Samuel Stanhope. 1787. *Essay on the Causes of the Variety of Complexion and Figure in the Human Species*. Philadelphia: Robert Aitkin.

———. [1810] 1965. *Essay on the Causes of the Variety of Complexion and Figure in the Human Species*. Cambridge, MA: Belknap Press of Harvard University Press.

Snyder, K. Alan. 1983. "Foundations of Liberty: The Christian Republicanism of Timothy Dwight and Jedidiah Morse." *New England Quarterly* 56: 382–397.

Stanton, William. 1960. *The Leopard's Spots: Scientific Attitudes Toward Race in America, 1815–59*. Chicago: University of Chicago Press.

Staum, Martin S. 1980. *Cabanis: Enlightenment and Medical Philosophy in the French Revolution*. Princeton, NJ: Princeton University Press.

———. 1987. "Human Geography in the French Institute: New Discipline or Missed Opportunity?" *Journal of the History of the Behavioural Sciences* 23: 332–340.

Stepan, Nancy. 1982. *The Idea of Race in Science: Great Britain 1800–1960*. London: Macmillan.

Stocking, George W., Jr. 1975. "Scotland as the Model of Mankind: Lord Kames' Philosophical View of Civilization," in Timothy H. H. Thoresen, ed., *Toward a Science of Man: Essays in the History of Anthropology*. The Hague: Mouton Publishers, pp. 65–89.

Strauss, Leo. [1930] 1982. "Isaac de la Peyrère," in *Spinoza's Critique of Religion*. New York: Schocken Books.

Trigger, Bruce G. 1989. *A History of Archaeological Thought*. Cambridge: Cambridge University Press.

van der Myl, Abraham. 1667. *De Origine Animalium et Migratione Populorum*. Geneva.

Voget, Fred W. 1975. *A History of Ethnology*. New York: Holt, Rinehart and Winston.

Walker, John. 1797. *Elements of Geography, and of Natural and Civil History*. Dublin, 3d ed.

Warntz, William. 1961. *Geography Now and Then: Some Notes on the History of Academic Geography in the United States*. New York: American Geographical Society.

White, Andrew D. [1894] 1955. *A History of the Warfare of Science with Theology in Christendom*. New York: George Braziller.

Wokler, Robert. 1988. "Apes and Races in the Scottish Enlightenment: Monboddo and Kames on the Nature of Man," in Peter Jones, ed., *Philosophy and Science in the Scottish Enlightenment*. Edinburgh: John Donald, pp. 145–168.

———. 1995. "Anthropology and Conjectural History in the Enlightenment," in Christopher Fox, Roy Porter, and Robert Wokler, eds., *Inventing Human Science: Eighteenth-Century Domains*. Berkeley: University of California Press, pp. 31–52.

Wood, Paul B. 1987. "Buffon's Reception in Scotland: The Aberdeen Connection." *Annals of Science* 44: 169–190.

———. 1996. "The Science of Man," in N. Jardine, J. A. Secord, and E. C.

Spary, eds., *Culture of Natural History.* Cambridge: Cambridge University Press, pp. 197–210.

Wright, Conrad. 1983. "The Controversial Career of Jedidiah Morse." *Harvard Library Bulletin* 31: 64–87.

Wright, J. K. 1959. "Some British 'Grandfathers' of American Geography," in R. Miller and J. Wreford Watson, eds., *Geographical Essays in Memory of Alan G. Ogilvie.* London: Thomas Nelson and Sons, pp. 144–165.

Yardeni, Myriam. 1971. "La Religion de La Peyrère et 'Le Rappel des Juifs.'" *Revue d'Histoire et de Philosophie Religieuse* 51: 245–259.

Mappings

Both metaphorically and materially, mapping had a crucial role to play during the period of the Enlightenment. On some occasions, this manifested itself in the passion to reduce the world's geographical complexity to cartographic clarity through the exercise of surveying crafts in conjunction with expeditionary travel. On others, it surfaced in the array of schemes produced to "map" accumulated knowledge, newly encountered peoples, and the interior reaches of mind and morality alike. Either way, the mapping motif has frequently been emblematic of attempts to take the measure of enlightened *thinking*. Accordingly, the chapters that follow here dwell on questions of mapping in one form or another.

In his scrutiny of what he calls "Historical geographies of the future," Michael Heffernan considers the differing ways in which stadial theories of human progress were mapped by a variety of Enlightenment commentators. What emerges from his analysis is the way that, despite the strongly geographical thrust of the distinctive schemes advanced by Turgot, Volney, and Saint-Simon, they remained resolutely opposed to environmental determinism. Instead they chose to emphasize the capacity of the human mind to liberate society from nature's tyranny. Throughout, Heffernan seeks to locate these particular renditions of conjectural history in the respective circumstances within which their architects worked. In so doing, he highlights the irony inherent in theories of progress which were sensitive to geographical variation and yet intent on annihilating differences in space. The cumulative effect of this chapter is to retrieve the geographical significance of these major Enlightenment thinkers and to reinterpret the relationship between geographical reasoning and discourses about human progress.

Matthew Edney then turns to the map as an encyclopedic repository of global reconnaissance. Here we discover some of the ways in which the mapping of archived information was bound up with the rationalization of social hierarchy. The apparently scientific cast of cartography during the eighteenth century is thus exposed as a suite of rhetorical practices which masked the wide range of ideological interests that were embodied in the very fabric of the map. At one level, given its capacity to impose coherence on complexity, the map was emblematic of the Enlightenment ideal of encyclopedic knowledge. And yet when it came to actual mapping exercises, what Edney calls "the conceptual unity of the geographical archive" was routinely compromised by technical and other imprecisions. Using the particular example of the abbé Barthélemy's *Voyage du jeune Anacharsis* and its accompanying set of thirty-one maps by Barbié du Bocage, Edney shows how the map as a form of visualization turns out to be a "rhetorical construct" laced with "ideology and ambiguity."

Just precisely what happens during the map-making process—in those moments of cross-cultural encounter and their translation into cartographic representation—is the focus of Michael Bravo's chapter. Taking issue with Bruno Latour's actor-network account of Lapérouse, Bravo dwells upon such matters as negotiation, interpretation, and visualization as crucial to the production of cross-cultural knowledge. In Bravo's telling, Lapérouse's cartographic ventures are best understood as ethnographic strategies designed to transform indigenous spatial knowledge into acceptable enlightenment science. What is involved is a scrutiny of the ways in which practitioners of enlightened science adjudicated local knowledge claims. Bravo thus uses this study of "ethnocartography" to uncover something of the mutuality involved in the encounter between different cultures' ways of knowing. Thus the transfer of geographical information from the anonymous fishermen's scratchings on the sand to Lapérouse's map is far from the speedy thing it was long assumed to be. Rather it was effected only through recurrent negotiation. Indeed, the very same lines on the map are read differently as the encounter between the two cultures develops. One result of this investigation is to redraw attention to the immense significance of ethnography in the actual practices of enlightenment geography and thereby to affirm the intimate historical connections between the anthropological and the geographical. Place-knowing, we might say, was inextricably intertwined with people-knowing.

Anne Godlewska's consideration of Alexander von Humboldt's visual thinking completes this section. A transitional figure between the classical and the modern, Humboldt's own synthetic way of thinking serves as a means of identifying one dominant Enlightenment way of integrating cosmic holism with empirical investigation. What facilitated this aspiration was Humboldt's use of cartographic methodology as a central means of representing nature. Through a detailed examination of his cartographic manoeuvres and innovations, Godlewska reveals the dynamic character of his maps and uses these reflections to illustrate how Humbolt helped move the earth and life sciences from description towards explanation. For him, graphic expression provided a scientific language capable of revealing connections, associations, and interrelations in space; the Humboldtian map thus facilitated the shift from recital to analysis. Thus, even though Humboldt was at best on the margins of the institutional geography of his day, his imaginative cartography contributed mightily to the discourse of enlightened geography.

Cumulatively, these four chapters serve to redraw attention to the immense power of the map—both as analogy and craft—to express a variety of Enlightenment motifs. As the meeting point between theory and practice, history and geography, the explorer and the aboriginal, the ontological and the epistemological, the archive and the field, mapping emerges as one crucial element in the evolution of Enlightenment thinking.

Chapter Five

Historical Geographies of the Future: Three Perspectives from France, 1750–1825

Michael Heffernan

5.1. Introduction

[I]n the midst of their ravages manners are softened, the human mind becomes enlightened, separate nations are brought closer to one another. Finally, commercial and political ties unite all parts of the globe, and the whole human race, through alternate periods of rest and unrest, of weal and woe, goes on advancing, although at a slow pace, towards greater perfection.

> —A.-R.-J. Turgot, *A Philosophical Review of the Successive Advances*
> *of the Human Mind,* 1750 (quoted variously in Meek, 1973a, 41,
> and Manuel and Manuel 1979, 475)

Here [at Palmyra] . . . an opulent city once flourished; this was the seat of a powerful empire . . . And now a mournful skeleton is all that subsists . . . [N]othing remains of its powerful government but a vain and obscure remembrance . . . What glory is here eclipsed, and how many labours are annihilated. Thus perish the works of men and thus do nations and empires vanish away!

> —C.-F.-C. Volney, *The Ruins,* 1792a, 5

The poetic imagination places the Golden Age in the cradle of humanity amidst the ignorance and brutality of the earliest times; yet this was, in fact, the 'Iron Age.' The Golden Age of the human species is not behind us, it is before us; it lies in the perfection of the social order. Our fathers did not witness it, our children will one day arrive there. It is our task to clear the path.

> —C.-H. de Rouvroy Saint-Simon and A. Thierry,
> *De la réorganisation de la Société Européenne,* 1814, 96–7

This essay is concerned with the debate about the nature and potential of human progress and the possibility of a 'knowable' future. This was a quintessential preoccupation of the 'enlightened' eighteenth century and has been an enduring subject of philosophical discussion over the last two hundred years. The literature on the idea of human progress, and on the opposing concepts of decline and fall, is impressively large.[1] Virtually all accounts agree that the modern idea of human progress was first advanced in a systematic way in the mid eighteenth century. It was only at this juncture that a secular view of progress arose to challenge earlier, generally Christian theories of the human condition based on various forms of divine providence. F. C. Green's words, spoken at the University of Oxford in 1950, are entirely characteristic:

> The idea of progress, in less than two centuries, has become one of the ideas which form the basis of world civilization. Yet this idea has not essentially changed, although it has rapidly proliferated in the last fifty years owing to the spectacular achievements of technical science. It rests today, as in the eighteenth century, upon man's faith in the unlimited potentialities of his intelligence. It is the belief, grounded on history, that the human reason, scientifically employed, is constantly extending man's knowledge and control of the universal mechanism and, indeed, of his own nature; that this process, moreover, is accompanied by a steady improvement in the material and moral conditions of human life on this earth. (Green 1950, 1)

Despite a continuing strand of pessimistic thinking, the idea of progress became an article of faith during the nineteenth century, acquiring along the way an evolutionary and scientific gloss which shone with a special luster in the aftermath of Darwin (Bowler 1974, 1989; Buckley 1966; Gould 1988). Although severely tested by the global warfare, genocides, cyclical economic crises, and environmental degradation of this century, many still insist that an optimistic faith in the potential of human improvement has remained the dominant intellectual conviction of our age, articulated through new terms such as 'modernization' or 'development,' which now occupy comparable positions in economic, political and social debate (Binder 1986; Cowen and Shenton 1996; Iggers 1982).

This is the broad context within which this essay is set. It considers three French writers who contributed to this debate between 1750 and 1825: Anne-Robert-Jacques Turgot, the Baron de l'Aulne (1727–81); Constantin-François Chassebeouf, the Comte de Volney (1757–1820); and Claude-Henri de Rouvroy, the Comte de Saint-Simon (1760–1825). It must immediately be acknowledged that these three names, though widely discussed by historians of ideas, are rarely accorded much space in the major histories of geography. This may be because all three writers belong to a strand of eighteenth-century thinking which was strongly opposed to environmental explanations of the human condition and particularly towards any form of deterministic reasoning.

MICHAEL HEFFERNAN

This derived from their shared belief in the empiricist philosophy of Francis Bacon and John Locke, mediated in this French context through the mid eighteenth-century writings of Condillac (Bowen 1981; Condillac 1924; Locke 1975). According to both Locke and Condillac, the human mind is not a manifestation of an innate, transcendent metaphysical force; rather, it is the product of the various sensations which are brought to bear upon it. Nothing can exist in the mind that has not first passed through the body's visual, aural, olfactory, or tactile sense channels.

Expressed in these bald terms, Lockean 'sensationalism' implies little more than a physical environmental determinism. Yet this was precisely the opposite of what Locke and Condillac believed. Their project was fundamentally educational and moral. As such, they drew a clear distinction between change in the natural world and change within the human mind. While the natural order is restricted by permanent, unchanging limits—the endless circles of birth, growth, decay, death, and rebirth—the human mind has the potential for infinite variability and development. Properly tutored, the mind constantly expands; wisely directed, human societies continually improve on all fronts. The natural world of flora, fauna, and the elements, for all its constant motion and awesome power, is ultimately unchanging. By underscoring the distinction between the human mind and the physical environment, Turgot, Volney, and Saint-Simon rejected environmental causation, particularly the deterministic thinking which became so common during the late eighteenth and nineteenth centuries (Peet 1985). Rather than interpret human societies as trapped by their physical environments, these three commentators emphasized the dynamism of the human mind and its power to overcome, and ultimately to control, the forces of nature.

Despite their common philosophical impulse and their similar interpretation of the relative power of humankind and the natural world, Turgot, Volney, and Saint-Simon developed quite distinctive conclusions on the nature of human progress. To some extent, their differences reflect the historical contexts in which they wrote. Taken together, their lives span nearly a century of French history (1727–1825) from the 'high Enlightenment' of the mid eighteenth century, through the upheavals of the Revolution and the Napoleonic Empire, to the onset of the industrial age. Though Volney and Saint-Simon were near-contemporaries, the three can be connected to separate 'moments' in the genealogy of French thinking about the nature of human progress. Turgot's boundless optimism produced "the most sophisticated French analysis of general progress prior to the 1790s" (Spadafora 1990, 384–5), the apotheosis of 'enlightened' eighteenth-century faith in the potential of modernity. Volney's musings, mostly published before the end of the Napoleonic Empire, suggest a darker and more somber interpretation of the human condition as befits one who had experienced at first hand the shock of the Revolution and its bloody aftermath. Saint-Simon's utopianism, published mainly in the years immediately after the Napoleonic age, was crafted in the shadow of the Industrial Revolution and points towards a characteristically nineteenth-century faith in the power of science and technology to drive the engine of human progress to new heights.

It is important to consider these otherwise unfamiliar figures precisely because their geographical ideas are not those normally emphasized in the standard accounts. This has more than passing significance for historians of ideas because the literature on the idea of progress generally fails to consider the geographical dimensions of these debates. It is important too for historians for geography. As the editors of this volume make clear in their introduction, modern geography cannot be regarded as following a simple disciplinary trajectory; as taking flight like the owl of Minerva in the middle of a neatly defined historical period which we now call 'the Enlightenment' and swooping in a single path to the present. Both 'geography' and 'Enlightenment' are complex discursive matrices which cut across the linear narratives imposed by historians projecting present ideas into the past. Other approaches to intellectual history, particularly Michel Foucault's consideration of the 'archaeology' or 'genealogy' of ideas in which emphasis is placed not on continuities and traditions but on ruptures and unexpected lateral maneuvers, offer fruitful alternative perspectives (Foucault 1970). There are, in short, as many different 'geographies' as there are 'Enlightenments.' What follows is an account of a hitherto uncharted 'enlightened geography' across the tumultuous and revolutionary decades before, during, and after the French Revolution.

5.2. Turgot: The Geography of a "Cosmic Tory"[2]

Turgot's place in the history of the eighteenth century was secured during his brief incumbency as Louis XVI's ill-fated first finance minister (1774–6). This extraordinary period witnessed the most ambitious attempt to modernize the economic and political structures of the *Ancien Régime* before the entire system was swept aside by the whirlwind of 1789. The details of Turgot's doomed battle to transform the French economy and overturn a quasi-feudal social and political order do not bear repetition here.[3] Steven Kaplan (1986, 220) sums up the episode well: "[t]he word reform is too frail to capture the seismic energy of Turgot's vision: it was a radical assault upon the elementary structures of the Old Regime."

The scale of Turgot's ambitions in government reflected philosophical and moral beliefs which had been formulated and refined over the preceding 25 years. As a theology student at the Sorbonne (1749–51), Turgot delivered two highly original lectures before his teachers in 1750, one on Christianity's role in the development of humanity, the other on "the successive advances of the human mind" (Meek 1973a, 41–59). In the second of these lectures, Turgot underlined the Lockean division between an ultimately unchanging natural world and an endlessly expanding human mind. The former was

> Governed . . . by constant laws . . . confined within a circle of revolutions which are always the same. All things perish, and all things spring up again . . . in these acts of generation through which plants and animals reproduce themselves, time does no more than restore continually the counterpart of what it has caused to disappear.

The latter, on the other hand,

> Affords from age to age an ever-changing spectacle. Reason, the passions, and liberty ceaselessly give rise to new events . . . [S]peech and writing, by providing men with the means of securing their ideas and communicating them to others, have made all individual stores of knowledge a common treasure-house which is always being enlarged by the discoveries of each age. (Manuel and Manuel 1979, 466; Meek 1973a, 41)

A few lines later, Turgot confidently asserted the possibility of human perfectibility, quoted at the beginning of this chapter.

The transformation of this belief into a more systematic theory of human progress emerged from Turgot's engagement with Montesquieu, whose *De l'esprit des lois* had recently appeared (Montesquieu 1748; Shackleton 1961; Shklar 1987). This work provided the seeds from which a dynamic theory of human progress developed. The important passage is Montesquieu's discussion of the "general relation of the laws" in Chapter VIII, Book XVIII. This stated that

> The laws are very closely related to the way that various peoples procure their subsistence. There must be a more extensive code of

laws for a people attached to commerce and the sea than for people satisfied to cultivate their lands. There must be a greater one for the latter than for a people who live by their herds. There must be a greater one for these last than for a people who live by hunting. (quoted in Jones 1992, 746)

The idea that economic systems produce their own moral and legal structures was an important observation, although not entirely unheralded (Hunt 1988). It did not, however, imply a sequential or historical chronology. According to Montesquieu, peoples of differing modes of subsistence simply co-existed, their differences being environmentally determined. Elsewhere in Book XVIII, in his consideration of the "laws in relation to the nature of the terrain," Montesquieu famously asserted his belief in the overwhelming importance of the physical environment, particularly climate, on the nature of a given society (Mercier 1953). Montesquieu's view of the human condition was, then, a relatively static one which emphasized preordained environmental limits (Glacken 1967, 551–622).

Turgot was convinced by Montesquieu's socioeconomic materialism but was repelled by his environmental materialism, the latter running counter to a Lockean position on the potential of the human mind. In an important "digression on climate," Turgot attacked the notion that a region's climate determines the nature of its human inhabitants, insisting on the "need to have exhausted all moral causes [of a people's character] before asserting the physical influence of climate" (Turgot 1913, I, 262).

The result was a modification of Montesquieu's typology, initially developed in notes Turgot wrote as preparation for two anticipated treatises on "universal history" and "political geography."[4] These documents have been discussed in detail elsewhere (Heffernan 1994a). The essential point is that in rejecting Montesquieu's static environmental typology, Turgot was able to develop an historicized version in its place. By stacking Montesquieu's categories into a dynamic historical sequence, Turgot argued that all peoples were advancing along a 'ladder' of progress in the same upward direction. They did so, however, at different rates. This fact explained the geographical variations in the levels of development between different peoples around the world.

Turgot's account of the changing global pattern of social, economic, and political development, beginning in the earliest, postdiluvian times, was explicated through a sequence of seven "mappemondes politiques" discussed in his notes on political geography. The existing human geography of the globe was, he claimed, a "freeze-frame" in an evolving process. Assuming, as Turgot did, that there was a common origin for all humankind, the complex social, economic, and political geography of the world was itself evidence of human progress. Here was the logical extension of Locke's famous observation: "In the beginning, all the world was America" (Locke 1967, 343). As earlier writers had claimed, notably Lafitau (1724), the 'savage' peoples of the Americas were 'survivors' of a primeval order, vestiges of the original peoples who emerged after the Biblical flood (Dickason 1983; Hulme 1990). By considering the different peoples of the earth, the philosopher was, in effect, considering the past, the historical sequence of human development from the most primitive and barbaric to the most enlightened and refined.

This argument culminated in Turgot's notes on universal history with the familiar four-stage model of human progress—the age of hunters, the age of shepherds, the age of husbandmen, and the age of commerce. Each stage was associated with distinct forms of language, ideas, institutions, laws, property systems, governments, customs, manners, and morals. The engine driving advancement from one stage to the next was innovation in the arts and, especially, the sciences (Cohen 1976). These innovations were rooted in the material economic sphere.[5]

Turgot's scheme is one of the first statements of a stadial, materialist theory of human history, although a few contemporaries, notably Adam Smith, were thinking along remarkably similar lines (Meek 1971, 1976, 99–130). Its impact has been enormous, not least in geography (Bobek 1962). Turgot's idea of human progress was developed by ordering the observable human geographical variations into an historical sequence as a challenge to Montesquieu's static environmental explanation of these same geographical variations. It is for this reason that Manuel and Manuel (1979, 475) refer to Turgot's work as a "world historical geography," a fusion of history and geography to create a general theory of human progress.

Turgot's historico-geographical method was clarified in some revealing comments on the nature of history and geography. In his view, history and geography were virtually identical intellectual pursuits, concerned with exactly the same range of social, economic, and political questions. "Geography," he wrote, "the description of the present, varies unendingly as the present becomes the future; history, the record of past times, is the sequential description of past geographies" (Turgot 1913, I, 257). Today's geography is simply tomorrow's history.

It is worth underlining the novelty of this perspective. Previous theories of the human condition were usually couched in metaphors of endless circularity: the Renaissance notions of organic cycles of birth, growth, decay, and rebirth, for example, or the even older idea of 'fortune's wheel' (Manuel 1965). Giambattista Vico's discussion of the three ages in his *New Science* (1725)—the age of men, the age of heroes, the age of gods, followed by a *ricorso* or decline—is the most striking example of circular thinking in the early eighteenth century.

There were, of course, linear conceptualizations of the human condition before c. 1750 but these took a rather different form. The Christian eschatological view (the idea of humankind moving from a state of original sin at birth to a final salvation in heaven at the end of a virtuous life) certainly implies linear progression but this was an entirely nonmaterialist, spiritual perspective; an ecstatic, other-worldly resolution to life on earth. The millenarian thinkers of the Middle Ages were also largely mystical and nonmaterialist (Cohn 1970). Earlier formulations of linear progression, as developed by Bacon, Pascal, Locke, Fontenelle, and others, were also different as they focused exclusively on the development of knowledge without connecting this to social, economic, or political advance (Sorel 1969, 1–29).

Turgot's idea of human progress was novel in being linear, mate-rialist, and earth-bound. It also suggested that movement through this recurring historical sequence was accompanied by (indeed de-rived from) an increasing domination of the natural world by hu-mankind. This process represented, moreover, a steady improve-ment in the material and moral conditions of human existence. The distinctly sacrilegious possibility of achieving earthly human per-fection was implicit, as was a degree of historical inevitability. Rates of progress may vary or go temporarily into reverse but ultimately humanity was destined to improve. Human history was the upward march of human progress. Human geography provided the evidence that human progress was taking place.

The emergence of a new version of human progress in the eigh-teenth century was part of a wider reconfiguration and secularization of the language of political and philosophical debate across Europe from the seventeenth century onwards (Clark 1994; Pagden 1988). The modern idea of Europe as a political expression was itself largely an invention of the late 1600s (Burke 1980; Hay 1968, 118–9; Haz-ard 1990; Schmidt 1966; Wintle 1996).[6] Older definitions of Europe, notably 'Christendom,' derived from an earlier, pre-Reformation age when it was still possible, at least within political and intellectual rhetoric, to define 'Europe' as uniformly and uniquely Christian.

MICHAEL HEFFERNAN

The transformations of the early-modern era—the Reformation and Counter-Reformation, the rise of modern nation-states, the dramatic commercial, demographic and military expansion of Europe overseas—created an entirely new global geopolitics which demanded a new vocabulary; a vocabulary which, within a generation or so, would be reified in the great dictionaries and encyclopedias of the eighteenth century (Darnton 1979). The rise of Europe as a geopolitical expression reflected the triumph of European nation-states within the wider world.

Similar arguments have been made in respect of other 'modern' concepts. In an intriguing analysis of 'governmentality,' Michel Foucault (1991) suggests that the enhanced power of the eighteenth-century nation-state was based partly on the invention of 'population' as a scientific category in the late seventeenth century. The abstract idea of 'population,' an entity which could be enumerated and analyzed in the new language of statistics, reflected a realization that the people themselves had become both the form and instrument of government, the ultimate source of national power.[7]

The invention of the idea of progress was directly linked to these earlier formulations. It also spawned other, closely related words and concepts, notably 'culture' and 'civilization.' The changing meanings attached to these two words have been explored by several historians, the general consensus being that, despite their different usage in English, French, and German, 'culture' and 'civilization' acquired their modern, dynamic meanings in the 1750s and 1760s (Baumann 1985; Elias 1994; Febvre 1974; Kroeber and Kluckhohn 1952; Vogt 1996; Williams 1976).

As the editors show above, these abstract concepts—'progress,' 'culture,' and 'civilization'—taken together provided a new secular language with which to define the geographical limits of 'Europe' based on the supposed characteristics of the population (Wolff 1994). The same terms could also describe and explain the hierarchy of peoples around the world and the seemingly undeniable fact that all Europeans had progressed further than other peoples, had developed the most advanced economic and political system, and possessed the highest level of culture and civilization. The European characteristics of dynamism, energy, and development mixed with reason, restraint, and logic set its peoples apart from the 'savage' or 'primitive' populations of the Americas, Africa, and Asia (Jacques 1997). The deployment of a new language of the Enlightenment— 'progress,' 'culture,' and 'civilization'—was thus predicated on the invention of its own opposite—'backwardness,' 'barbarism,' and 'primitivism' (Droixhe and Gossiaux 1985; Springborg 1992). If Europe was 'enlightened,' other regions had, by definition, to be shrouded in 'darkness.'

Advocates of human progress did not regard these distinctions as permanent. The 'darker' places would ultimately arrive at the same level of enlightenment and civilization as the privileged places whose own progress would ultimately end in perfection. The relative differences between regions of the globe were important in stimulating overall progress[8] but the ultimate utopian resolution was a single civilized world.

This sublime optimism intensified in some quarters during the late eighteenth century, reaching its apogée in the writings of Condorcet in France (Baker 1975) and Joseph Priestley and William Godwin in Britain. Condorcet's Girondist revolutionary zeal, his hostility to religion, and his materialist rejection of any form of other-worldly spirituality were all part of his unlimited faith in the curative potential of science and reason. The advent of the Revolutionary Terror, when he was condemned to death, apparently did nothing to dampen his supremely optimistic outlook. As Bury (1932, 207) puts it, whereas "Turgot wrote with the calm spirit of an inquirer; Condorcet wrote with the verve of a prophet" (Condorcet 1787; Ebenstein 1982; Manuel 1962; Manuel and Manuel 1979, 461–518). In *The Sketch of an Historical Picture of the Human Mind*, composed by Condorcet while in hiding from Robespierre's henchmen, we learn that:

> [N]ature has set no term to the perfection of human faculties . . .
> the perfectibility of man is truly infinite; and . . . the progress of
> this perfectibility, from now on independent of any power that
> might wish to halt it, has no other limit than the duration of the
> globe upon which nature has cast us. (Condorcet 1794, 4; quoted
> in Glacken 1967, 634–5)

Unbounded cheerfulness was far from universal, of course, and many dismissed Turgot and Condorcet as wishful thinkers. The fact that neither could offer a convincing nonenvironmental explanation as to why some peoples had advanced further and more quickly than others from a common starting point was a serious flaw in their argument. Many were convinced that the external environment exerted a pervasive influence and that human development was ultimately limited, whether by God, nature or both. Montesquieu refused to accept the idea of progress and stuck to his belief in

MICHAEL HEFFERNAN

the older notion of rise and decline within preset limits, an argument he had expounded both in the *Lettres Persanes* (1721) and in his essay on the 'grandeur' and 'decadence' of the Romans (1734) (Chinard 1942; Montesquieu 1721, 1734; Oake 1955; Shklar 1987, 49–66). In the former publication, a series of witty observations of the different customs of Europe and Asia written by imaginary Persians visiting Paris, Montesquieu rejected the idea of European superiority, insisting that "Almost all the nations of the world travel this circle: to begin with, they are barbarous, they become conquerors and well-ordered nations; this order permits them to grow, and they become refined; refinement enfeebles them, and they return to barbarism" (Vyverberg 1958, 155).

Voltaire was also of a more pessimistic inclination, famously referring to the insane idea of being perfectly reasonable (Duchet 1971, 285). Rousseau, for his part, developed a very different view of progress by the mid-1750s following his Pauline conversion on the road to Vincennes. His was the exact opposite of the view proposed by Turgot: a plea for a return to nature rather than a conquest of it, a passionate assertion of natural human goodness perverted by social progress and civilization (Cro 1990; Green 1950; LaFrenière 1990; Rousseau 1973; Shklar 1969). Despite their differences and eventual acrimonious split, Rousseau's views on the 'nobility' of the 'savage' also influenced Diderot, notably in his *Supplément au voyage de Bougainville* (Duchet 1971, 322–375, 407–475; Krakeur 1938).

A more pessimistic view was also evident elsewhere in Europe, notably in Britain (though see Spadafora 1990). Thomas Malthus offered his first *Essay on Population* (1798) as a formal refutation of Condorcet's atheistic and sacrilegious perspective (Glacken 1967, 632–7; Kumar 1978, 21). For Malthus, there was no progress, only perpetual oscillation, homeostatically sealed in an endless cycle of misery, wretchedness, and death.

Yet despite continuing disagreement about the ultimate destiny of humankind on earth, the historico-geographical component of progressive thought—the belief that the different peoples of the earth were somehow representative of different stages of human development—won widespread support, even from those whose views about the future of humanity were anything but optimistic. The stadial idea was a central preoccupation of many eighteenth-century Scottish philosophers, beginning with Adam Smith in his lectures at Glasgow in the early 1750s and culminating with the writings of John Millar and Adam Ferguson (Meek 1976, 99–129, 150–5, 60–

77). Edmund Burke, not a man given to naive optimism, felt moved to observe in a much-quoted letter to William Robertson that

> We possess at this time very great advantages towards the knowledge of human Nature. We need no longer go to History to trace it in all its stages and periods. History from its comparative youth, is but a poor instructour . . . But now that the Great Map of Mankind is unrolld at once; and there is no state or Gradation of barbarism, and no mode of refinement which we have not at the same instant under our View. The different Civility of Europe and China; The barbarism of Tartary and arabia. The Savage State of North America and of New Zealand. (Gutteridge 1961, III, 350–1; Marshall and Williams 1982; Meek 1976, 173)

The identical point was nicely expressed in 1800 by the Baron Degérando, a prominent member of the revolutionary *idéologues,* in his extraordinary essay on the methodology of the infant science of anthropology:

> The philosophical traveller, sailing to the ends of the earth, is in fact travelling in time; he is exploring the past; every step he makes is the passage of an age. Those unknown islands that he reaches are for him the cradle of human society. (Degérando, 1969, 63; Stocking 1968)

By the turn of the eighteenth century, then, despite considerable opposition to the belief in inevitable earthly progress, the historical and geographical reasoning which underpinned this idea had won widespread support and was directly informing the conduct of science.

5.3. Volney: The Lessons of History and Geography

Of the three writers under consideration here, Volney is the most obviously geographical, the only one to be considered in these terms by historians of ideas (Broc 1969, 69–73; 1974, 353–61; Glacken 1967, 690–2; Moravia 1967, 1974, 533–671; 1980, 257; 1982, 247–70; Staum 1996, 154–71; Vallaux 1938). His reputation as a 'geographer' rests on his rigorous scientific descriptions of the Middle East (following a long sojourn in Egypt and Syria from 1783 to 1786) and of the United States of America (after a comparable period of self-imposed exile from 1795 to 1798) (Volney 1787; 1803; 1825, II, 1–432; 1825, IV, 1–478). His most successful book, however, was a hauntingly evocative work of political philosophy and comparative religion, *The Ruins, or a survey of the Revolutions of Empires* (Volney 1792a; 1826, I, 1–351).

Volney's life and travels have been recounted in detail elsewhere (Breal 1899; Gaulmier 1980; Jeanvrot 1898). The salient fact here is that Volney, like Degérando, was a member of that amorphous group of intellectuals often referred to (more by their enemies than by themselves) as the *idéologues* (Welch 1984). As Terry Eagleton shows in his masterful survey, ideology (now often represented as a kind of antiscience) was first developed in the aftermath of the French Revolution as "a rational enquiry into the laws governing the formation and development of ideas." It was an attempt to theorize a new social order for the new republican age based on the rational philosophical foundations established during the preceding decades (Eagleton 1991, 64; Head 1980). The leading *idéologues* were initially Pierre-Jean-George Cabanis and Antoine Destutt de Tracy. It was the latter who first used the word 'ideology' in 1802 (Destutt de Tracy, 1801–15; Kennedy 1978).

'Ideology' represented the apotheosis of Lockean sensationalism in France. It was, according to a pioneering account (Duzer 1935, 16): "an interpretation of knowledge based on the assumption that all ideas and all faculties of human understanding—perception, memory, and will—are compounds of sensations." Like Locke, the *idéologues* held that once the laws governing the operation of the human mind were properly understood, the collective consciousness could be transformed by mass education based on science, reason, and logic; the application of pure, transcendent truth. Associated mainly with the *Sciences Morales et Physiques* class of the *Institut National des Sciences et Arts,* the *idéologues* were social engineers who sought to remake the cultural environment in order to change collective sensations and thus improve the nature of human consciousness itself. Their objective was to rescue society from the pernicious influence of metaphysical delusion and religious superstition. An overtly political as well as an intellectual project, ideology was "a revolutionary strike at priests and kings, at the traditional custodians and technicians of the 'inner life.' Knowledge of humanity was wrested from the monopoly of a ruling class and invested instead in an elite of scientific theorists" (Eagleton 1991, 65).

Volney's involvement with this 'movement' began as a young man in Paris where he studied medicine from 1774 (the year Turgot became finance minister). He was more interested in philosophical and historical questions, however, and in 1781 (the year of Turgot's death), he published a *mémoire* on Herodotus which brought him to the attention of materialist, antireligious intellectuals such as d'Holbach and Condorcet, key influences on the *idéologues* (Kors 1976; Volney 1825, V, 283–366; 1825, VI, 1–104).

Volney decided to abandon France for an extended journey to

Egypt, the Holy Land, and Syria in 1782. This was no youthful, escapist adventure or languid Grand Tour. Volney spent fully a year in detailed preparation, and his purpose was highly serious and stunningly ambitious: to discover, by means of a meticulous and integrated historical and geographical investigation, how to reconstruct a region of former glory but present decay. For Volney, Middle Eastern civilization had demonstrably declined from its glorious zenith in the classic era and had now reached the point of almost total collapse. How had this occurred, and how might it be reversed? Drawing explicitly on the ideas of Condorcet, Volney's excursion was an empirical application of the theory of human progress to a particular region that had suffered a lengthy period of decay.

The *Voyage en Egypt et en Syrie* (1787) is a remarkable document, a defining moment in the history of European Orientalism, not least because it is said to have informed Napoleon's ambitions in the region (Gaulmier 1980, 43–77; Moravia 1967, 981–1011; Said 1978, 168–70; Volney 1787; 1825, II, 1–432). The relationship between human society and the physical environment is a central theme. Volney insisted on the enormous complexity of this interaction, a view which clearly foreshadows the 'possibilism' of French regional geography under Paul Vidal de la Blache a century later. Rejecting Montesquieu's method and conclusions, Volney dismissed all forms of environmental determinism with simple historical and geographical observations: if climate and environment were determining factors, how could a single region (the Middle East) experience such dramatically changed fortunes over recorded history with no evidence of comparable environmental transformations? And how could regions with similar environments in the present exist in so many different economic, social, political, and moral circumstances? (Volney 1787, 469). As to Montesquieu's "axiom . . . that inhabitants of hot countries must necessarily be indolent, inert of body; and from an analogy, likewise inert in mind and body . . . and hence unable to resist despotic government," Volney offered a radical political alternative. Inertia and laziness were not caused by climate but by the possession of excess wealth: "[I]t is not as inhabitants of hot, but as inhabitants of rich, countries that nations are inclined to indulgence" (Volney 1787, 471).

Yet without an environmental causation, how could one explain the 'decline' of the civilizations of the Middle East? The reasons, he claimed, were "more general and compelling than the nature of the soil or the climate: they lie in the various social institutions which

MICHAEL HEFFERNAN

we call *government* and *religion*" (Volney 1787, 474). Inept, corrupt government under the Turks and an obscurantist Islamic faith were not the outcomes of a preordained environmental context; they were root causes of the ongoing process of decline. By changing these political and social regimes, the long downward movement could be halted and reversed. If necessary, this might require the reconquest of the region by a European power (Volney 1787, 284).

After the publication of his account, Volney accepted a government position as Director General of Agriculture and Commerce in Corsica. Here he began to ponder the implications of his thinking on the Middle East for the future of Europe, particularly France. If rich countries were indeed more prone to indulgence and decadence (and the corruption of *Ancien Régime* France provided ample evidence of this), then was not French civilization doomed to suffer the same fate as the civilizations of the Middle East? More important, how might this decline be avoided and the onward march of European civilization assured?

Volney's answer to this last question was both political and revolutionary. In the months before July 1789, his views grew markedly more radical. He threw himself into the acrimonious pamphlet wars of the period, publishing *La Sentinelle du peuple* in Rennes during the widespread disorders in Brittany in the autumn and winter of 1788–9. This contained swingeing attacks on the monarchy, the church and the Breton nobility (Seltzer 1980). After the Revolution, he returned to Paris as a *député* in the *États-Généraux*, famously proposing that France should formally undertake never to embark on a future war designed to increase its territory (Gaulmier 1980, 190).

The transformed circumstances did little to lift the anxiety Volney felt about the fate of humankind, particularly in Europe. In the intellectual maelstrom of late 1789 and 1790, his scrupulously positivist scientific approach was temporarily abandoned. The arguments advanced in the *Voyage* were recast in a wholly different register for *The Ruins* (1792a), his most overtly political text. Volney's objectives in *The Ruins* are made clear in a quotation from the main text reprinted on a prefatory page:

> I will dwell in solitude amidst the ruins of cities; I will inquire of antiquity what was the wisdom of former ages. . . . I will enquire of the ashes of legislators what causes have erected and overthrown empires; what are the principles of national prosperity and misfortune; what maxims upon which the peace of society and the happiness of man ought to be founded. (Volney 1792a, 24)

In twenty-four chapters, Volney carries the reader on a swirling philosophical journey across time and space. The journey begins with "sublime meditations" provoked by his encounter with the ruins of the lost cities of Middle Eastern antiquity. At Palmyra, described above, Volney was struck by "the solitariness of the situation, the serenity of the evening, the grandeur of the scene . . . The view of an illustrious city deserted, the remembrance of times past."[9] Exhausted, he sits down at the base of an ancient column and falls into a "profound revery" (fig. 5.1). Initially comforted by the knowledge that he "had found in modern Europe the past splendour of Asia," his thoughts quickly assume a more somber tone.

> Reflecting that if the place before me had once exhibited this animated picture; who . . . can assure me that their present desolation will not one day be the lot of our own country? Who knows but that hereafter some traveller will like myself sit down upon the banks of the Seine, the Thames, or the Zuyder Sea, where now, in the tumult of enjoyment, the heart and eye are too slow to take in the multitude of sensations; who knows but he will sit down solitary amid silent ruins, and weep a people broken, their greatness changed into an empty name? (Volney 1792a, 6)

At this point, a "pale apparition" appears to stir Volney from his mournful introspection: "Interrogate these ruins!" says the ghost, "read the lessons which they present to you . . . the caprice of which man complains is not the caprice of destiny . . . it resides in himself, man bears it in his heart" (Volney 1792a, 8).

There follows an extraordinary philosophical exchange between Volney and the specter. Like the ghosts in Dickens's *A Christmas Carol*, Volney's alter ego agrees to reveal "this truth of which you are in pursuit . . . the wisdom of the tombs, and the science of the eye." With an ethereal hand placed lightly on a nervous brow, Volney is "penetrated by a celestial flame" and swept up until beneath him appears the entire earthly globe "floating in empty space" (Volney 1792a, 12). The two cosmic flyers then sweep around the earth discoursing on the causes of the rise and decline of different peoples and regions; the original state of humankind; the role of genius in human progress; the "principles of society"; the sources of good and evil; the general causes of past prosperity; the origins of revolutions and the ruin of ancient states; the obstacles to improvement; the balance between rights and duties; the nature of laws; the "conspiracies of tyrants"; the relationships between productive and nonproductive classes; the possibilities of democracy; the origins and genealogy of religious beliefs, and so on (figs. 5.2a and 5.2b).

MICHAEL HEFFERNAN

Here an opulent City once flourished: this was the seat of a powerful Empire.—Yes, these places, now so desert, a living Multitude formerly animated &c.

Chap.II.

Fig. 5.1. Frontispiece from Volney's *The Ruins* (1792 English edition). Reproduced from the original text with the permission of the Cambridge University Library.

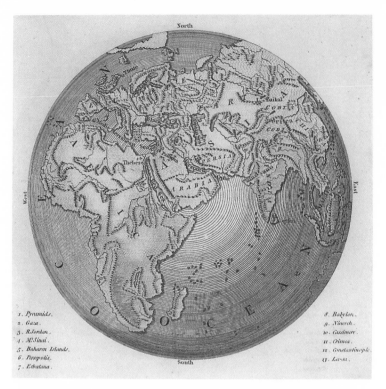

Fig. 5.2a.   The Globe from Volney's *The Ruins* (1792 English edition, opposite p. xvi). Reproduced from the original text with the permission of the Cambridge University Library.

Throughout this fantastic voyage, Volney's anxieties are constantly placated. Although everywhere constrained, restricted and downtrodden by the whims of nature and the evils of political and religious systems, humankind still possesses the capacity to break free and scale ever greater heights: "Such then is the condition of man; on the one side, subjected to the action of the elements around him; he is exposed to a variety of inevitable evils . . . [Yet] these evils [are tempered] . . . with an equal portion of benefits . . . [M]an has . . . the power of augmenting the one and diminishing the other" (Volney 1792a, 16).

Volney's message was ultimately optimistic, and his solutions to the dilemmas facing humankind were resolutely material and political. The future success of the human race lay in establishing the correct balance in all places between a generous, associative

> Self-love . . . the eternal spring of action in every individual, and a
> selfish, corrupting self-love . . . [which] . . . renders man the enemy

MICHAEL HEFFERNAN

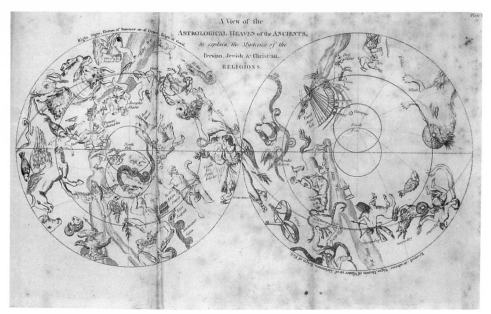

Fig. 5.2b. "A View of the Astrological Heaven of the Ancients to explain the mysteries of the Persian, Jewish & Christian Religions" from Volney's *The Ruins* (1792 English edition, opposite p. xvi). Reproduced from the original text with the permission of the Cambridge University Library.

of man, and of consequence perpetually tends to the dissolution of society. . . . It is the art of legislation . . . to temper the grasping selfishness of individuals, to keep each man's desire to possess every thing in a nice equipoise, and thus to render the subjects happy, in order that . . . all the members [of a society] should have an equal interest in the preservation and defence of the commonwealth. (Volney 1792a, 22)

The response to the anxious rhetorical question posed in the thirteenth chapter—"Will the human race be ever in a better condition than at present?" (Volney 1792a, 48)—is resoundingly positive.

The Ruins was enormously successful. Despite its rational materialism, its structure and tone had considerable impact on the subsequent rise of French Romanticism, notably via Châteaubriand (Gaulmier 1968). In Britain, cheap pocketbook translations of the whole text and of individual chapters (reprinted by the London Corresponding Society) circulated widely (Thompson 1968, 107–8; Volney 1792b, 1796). Its uncompromising hostility to all religions was denounced by the established Church. More than a decade after its appearance, the Reverend William Cockburn of St. John's College,

Cambridge was still sufficiently outraged to publish a lengthy counterblast: in his view, *The Ruins* deserved "only to be despised and forgotten" (Cockburn 1804, 5). Even those who were sympathetic to Volney's political project were dismayed by his indiscriminate dismissal of all religious faith. Joseph Priestley (an ardent supporter of the French Revolution who had been forced into exile in America following the destruction of his Birmingham home and laboratory at the hands of an antirepublican mob) penned a spirited denunciation of Volney's Godlessness, leading to a famous exchange between the two men (Fruchtman 1983; Priestley 1797; Volney 1826, I, 353–70).

Volney returned to Ajaccio in 1792 and met the young Napoleon before being forced back to Paris by Pascal Paoli's rebellion against French control of the island. Under the Terror, Volney was condemned as an opponent of the Revolution and served 10 months in prison. After his release in 1795 (Year II of the Republic), he was appointed to teach history at the newly created *École Normale* alongside other like-minded *idéologues* associated with the *Institut National*, including both Phillippe Buache de Neuville and Edme Mentelle, who together provided the lectures on geography (Broc 1974, 469–74; Moravia 1967, 949–58). Volney's *Leçons d'Histoire* were a further attempt to grapple with the problem of human progress (Gaulmier 1962; Volney 1825, VII, 1–135). The study of history in all its forms (through the written record but also through "artifacts") may well result in a depressing catalogue of crimes, vices, and errors, argued Volney, but it is necessary to know of these failings in order to avoid them in the future. Human perfectibility was the lodestar of historical investigation, and it was only by understanding the mistakes of the past that the collapse of hard-won civilization could be prevented. The past provided positive lessons too, not least by demonstrating that the revolutionary state of mind was rooted in an idealized conception of classical republicanism and Roman virtue (Parker 1937).

The closure of the *Ecole Normale* and the continuing political uncertainty in France led to Volney's second period in exile, this time in the United States. His objective was no less grandiose: to investigate the potential of the New World as a centre of a reborn civilization. If the ancient civilization of the Middle East had ceded its dominance to the modern civilization of Europe, might not Europe have to relinquish its dominance to the Americas? If this was so, what sort of civilization was destined to emerge in the Americas, and how could this be molded by actions in the present?

MICHAEL HEFFERNAN

The cyclical idea of Europe's inexorable decline and America's corresponding development was a familiar motif at this time, both in Europe and in the United States (Persons 1954). The view was elegantly, if sycophantically, expressed in Madame de Tessé's letter to Thomas Jefferson on 30 March 1787:

> I have been projected into future ages, and have seen the youth of America reading with enthusiasm and admiration all that has been collected about your travels. When the wealth of her soil and the excellence of her government have elevated North America to the summit of greatness, when the southern continent has followed her example, when you have succoured half the globe, then perhaps, people will search for the vestiges of Paris as they do today those of ancient Babylon, and the memoirs of Mr. Jefferson will guide travellers eagerly seeking the antiquities of Rome and of France, which will then be as one. (Rice 1976, 38)

Underpinning this belief was the equally familiar conviction, suggested by Thomas Jefferson and others, that the riches of the American environment, its immense resources and wide-open spaces, would generate a new race of ardent pioneers and new systems of government based on democratic principles (Chinard 1913). Volney could not accept such a view; his belief in the power of human agency to determine the course of human progress was in direct opposition to any form of historical inevitability or environmental determinism. His response to his new surroundings, articulated in his correspondence with Jefferson and others (Chinard 1923, 1925; Mathiez 1910) and in the *Tableau du climat et du sol des États Unis* (1803), was brutally realistic (Echeverria 1957, 190). Generally ambivalent about the nation's future, Volney underlined the challenges and dangers of pioneer life and made much of the poor health of Americans rather than their supposed vigor. He was especially struck, for example, by the high incidence of tooth decay (Glacken 1967, 650; Volney 1803, II, 76).

More significant was his detailed analysis of the environmental impact of American settlement, a theme which reveals the influence of Buffon's *Des époques de la nature* (1778) and Lamarck's *Hydrogéologie* (1802). In contrast to Jefferson's belief that the New World had created a more robust and morally superior people, forged in the image of its untouched, all-powerful environment, and, as Livingstone has shown, in contrast to Stanhope Smith's determinist reading of the American character and Morse's appeal to geography as a means of exploring the nature of American politics, Volney emphasized the vulnerability of American nature before the tide of humanity. Forest clearance was, he claimed, seriously affecting the climate, making it more unpredictable and prone to drought (Glacken 1967, 690–1; Williams 1989, 144–5). As Clarence Glacken has shown, this theme echoes Buffon's views on America as a degeneration of Europe as well as his belief that environmental influences could be overwhelmed—for good or ill—by the power and force of human activity (Glacken 1960, 1967, 655–705).

Back in France, Volney devoted the remainder of his life to historical and philological research (e.g., Volney 1819). His relationship with the Napoleonic regime was never easy. Napoleon grew irritated by what he saw as the impractical dreaming of the *idéologues* (he once claimed to have invented this word himself as a term of contempt), and he closed the movement's spiritual home in the *Institut National* in 1802 (Head 1980).

Although ultimately convinced that humankind would continue to better itself, Volney believed progress resulted only from constant vigilance and permanent effort. His position was thus remarkably similar to that of Edward Gibbon (Burke 1977; Furet 1977; Pocock 1977a, b, 1981; Porter 1988, 135–57). Like Gibbon, Volney recognized that great civilizations could contain the seeds of their own downfall, but he also understood the endless potential for human advancement.

5.4. Saint-Simon: The End of Geography and the Technological Sublime

Saint-Simon is by far the most difficult and complex of the three writers under consideration. His 'geographical' theories are also largely implicit, deeply embedded within his unusual philosophy. Much has been written about him and opinions differ sharply on which aspects of his sprawling, contradictory, and confused *œuvre* deserve attention. Even those who have studied his writings in detail seem uncertain that he was worth the effort. "A first-rate mind," claims one authority, "[but] a second-rate thinker, far from being original and even farther from being systematic" (Simon 1956, 311). The confusion is not merely intellectual; like Turgot and Volney, Saint-Simon's ideas were always linked to political objectives yet are difficult to categorize. His philosophy has been variously interpreted as utopian socialism,[10] radical bourgeois conservatism, and a forerunner of modern totalitarianism (Charlton 1963; Ionescu 1976; Manuel

MICHAEL HEFFERNAN

1963; Taylor 1975). The ambivalence is neatly summed up by Eric Hobsbawm (1987, 339): Saint-Simonianism "can be assigned neither to capitalism nor to socialism, because it can be claimed by both."

Only three years younger than Volney, Saint-Simon was in his mid-50s before he published his first significant text and thus belongs to later, post-Napoleonic philosophical generation. As a young man from a minor but impoverished noble family, he fought in the American War of Independence and was captured and imprisoned by the British. After his release, he traveled widely and dreamed of realizing ambitious but unlikely projects, clues to his subsequent philosophy. In Mexico, he presented a plan to the Viceroy for an interoceanic canal running through the lakes of Nicaragua; in Spain, aided by Count Barras, he proposed an equally grandiose idea for a canal linking Madrid to the sea. Neither progressed beyond his own imagination.

Back in France, Saint-Simon enthusiastically embraced the Revolution, renouncing his noble title and ostentatiously adopting the simple, peasant name of Jacques Bonhomme. Virtuous republicanism did not prevent him lining his own pockets by speculating in land and property 'liberated' from the church and the aristocracy. He even tried to buy Notre Dame cathedral to sell off the lead roofing. Arrested during the Terror (probably by mistake), he emerged a changed man, attributing his 'conversion' to a dream in which the ghost of Charlemagne, from whom he claimed direct descendence, had instructed him to devote the remainder of his life to philosophical inquiry (Heater 1992, 98).

This newfound seriousness seemed to have little impact on his lifestyle under the First Empire, which remained flamboyantly lascivious. But he did begin to publish pamphlets, usually at his own expense. Most were ignored. Though he admired the *idéologues,* he was never welcomed into their circle and was generally regarded as a crackpot. Following a mental breakdown in 1812, he settled in Paris near the *École Polytechnique,* which was to become his spiritual home. Here, in defiance of his advancing years, an attempted suicide, and persistent financial and legal difficulties, he finally began to produce work of genuine importance and originality, usually in the form of periodical publications. He was assisted by a succession of young male accomplices and students from adjacent *École,* over whom he exerted an extraordinary magnetic power, including Augustin Thierry, subsequently a major historian, and Auguste Comte, the 'father' of modern sociology and logical positivism.

Saint-Simon's project was nothing if not ambitious: he hoped to produce a general synthesis and philosophy of all human knowledge. Not surprisingly, his writings range widely in theme and method (Saint-Simon and Enfantin 1865–78). The key to his thinking was a quasi-religious positivist faith in science and technology as engines for beneficial progress; as the spiritual life force of a new utopian age of 'industrialism.' "The idea of Progress," as Simon (1956, 330) notes, "constituted the core and central inspiration of Saint-Simon's entire philosophy of history," and while there was, to be sure, more than a hint of historical predestination in his belief in the universal necessity and inevitability of human progress, Saint-Simon (like Volney) placed great store on human agency and will.

The political context in which he developed his ideas was important. To some extent, he can be seen as a "propagandist for the bourgeoisie" in its struggle to retain political power faced with a newly resurgent nobility under the restored French monarchy (Manuel and Manuel 1979, 594). Though he knew virtually no science, Saint-Simon elevated scientists and an elite class of scientifically trained entrepreneurs *(industriels)* to the status of a secular priesthood, the apex of the social, economic, and political order he foresaw. These two groups, along with the workers (sometimes also seen as *industriels*), were in his view the only productive classes and should therefore determine the nature of the new order. The *École Polytechnique,* established in 1794 as the training ground for the engineers and applied scientists of the new republican age, was precisely the sort of educational institution which would generate such an elite (Shinn 1980). The figure of Isaac Newton provided the ultimate inspiration, as he had done for many French radical thinkers (Eltin 1994, 21–3). Indeed, the Newtonian theory of gravitation was regarded by Saint-Simon as an exemplary intellectual formulation, a universal physical law which he hoped to replicate for social, economic, and political phenomena. In the meantime, a numerically small "sacerdocy of science" (Manuel and Manuel 1979, 596) would be able to adjudicate between rival classes and factions in an entirely rational and objective fashion, bringing all social and political conflict to an end. "The only useful action that man can perform," claimed Saint-Simon, "is the action of man on things. The action of man on man is always harmful to the species because of the twofold waste of energy it entails. It can only be useful if it is subsidiary and if it supplements the performance of a greater action on nature" (Manuel and Manuel 1979, 604). By turning human attention outwards towards controlling the natural world, under the benign and all-seeing

MICHAEL HEFFERNAN

direction of a scientific-technological elite, humanity's destructive passions would be transformed into constructive, peaceful conquests. Science and technology should therefore be elevated to a religious position, "the new Christianity" (Saint-Simon 1825). Although attracted to the idea of linear human progress, Saint-Simon did not see this as a smooth, gradual, or inevitable process. Rather, he believed progress was punctuated, and ultimately driven forward, by periodic crises in which the old was dramatically replaced by the new in "an ongoing pattern of universal decay and regeneration" (Haines 1978, 28).

This was one of the first philosophies of industrialism as the engine of human progress. Unlike the anti-industrial ethos of the Romantics, Saint-Simon enthusiastically welcomed the emerging industrial order, provided it was reorganized as he suggested (Crossley 1993, 105–38). Britain, the first industrial nation, was an obvious focus of his concerns, particularly its relationship with France and the rest of Europe. Foreshadowing Marx, Saint-Simon predicted that in the new industrial order, already emerging in Britain, the role of government would be massively reduced. As the state would no longer need to balance the interests of rival classes and factions, it was destined to wither away to a small, administrative rump. On an international scale, this would lead to the coalescing of existing nation-states into larger and larger confederations.

This idea was expressed most forcibly in a remarkable pamphlet on the "re-organisation of European society," co-authored with Thierry and published in October 1814, on the eve of the Congress of Vienna (Heater 1992, 97–115; Ionescu 1976, 83–98; Saint-Simon and Thierry 1814; Taylor 1975, 129–36). This laid great stress on the need for an initial Anglo-French political union (on the grounds that both nations now shared a common form of liberal government) to be followed by the creation of a general European Parliament, based on the British two-chamber model. The 'House of Commons' would be elected by a 'college' comprising the membership of new professional associations representing science and academic work, the industrial and commercial sector, the legal system, and the administration. The economic, infrastructural, and intellectual foundations for such a European union were already in place, claimed Saint-Simon, and were likely to intensify in the future. Here was a genuinely radical solution to the political problems of Europe which went far beyond the eighteenth-century blueprints for European political integration, such as those devised by the Abbé de Saint-Pierre and Rousseau (Heater 1992, 61–90). Saint-Simon's 'managerial' and

economic emphasis and his belief that communications and technology could bind together political space were entirely novel,
though widely copied, ideas (Eichthal 1840). According to Hinsley
(1967, 102), Saint-Simon's scheme was "a more far-reaching proposal for the federal organisation of Europe than anyone had ever
proposed."

After Saint-Simon's death in 1825, his ideas were developed by a
group of young disciples from the *École Polytechnique,* including
Prosper Enfantin (who emerged as the new guru of the movement),
Michel Chevalier, Charles Duveyrier, Armand Bazard, Pierre Leroux,
Gustave d'Eichthal, Émile and Isaac Periere, Léon Halévy, and Hip-
polyte Carnot. The program developed by this strange sect in numerous books, pamphlets, and newspapers (notably *Le Producteur*
and *Le Globe*) was even more wide-ranging and less coherent than
that of their eponymous mentor.[11] A central theme was their belief
in the need to overcome the physical and cultural barriers separating different classes, races, nations, regions, religions, and even the
sexes. Saint-Simonian writings on the last topic, which concerned
the need to transcend sexual divisions and forge a new androgy-
nous social order based on 'social beings' rather than traditional notions of 'men' and 'women,' proved too much for the French cen-
sors. Enfantin, Chevalier, and Duveyrier were all imprisoned in
1832 on charges of immorality, an event which precipitated the collapse of the movement (Moses 1982).

The Saint-Simonian project had a continuing influence through
the work of the group's leading members, many of whom rose to
prominence under the Second Empire. Their diverse ideas sprang
directly from a common faith in the reconstitutive power of science and technology (Carlisle 1974). Whereas Saint-Simon had focused mainly on Europe, his followers developed a global perspective, seeing Europe's colonial engagement with the non-European
world as the ultimate transcendence. The steamships, canals, rail-
ways, bridges, and telegraphs of the colonial age would, if wisely directed, bring about a new global utopia, fusing the Old World and
the New, the 'masculine' and the 'feminine,' to create a hybrid world
civilization (Émerit 1941, 1967; Enfantin 1843). The Suez Canal and
a host of other French colonial projects and exploratory missions in
Africa and Asia from the 1830s onwards were all influenced by this
utopian belief (Fakkar 1973, 1974; Heffernan 1990; Morsy 1989;
Taboulet 1968, 1971, 1973).

It is at this point that Saint-Simonian philosophy, the third incarnation of the intellectual tradition described in this chapter, begins

MICHAEL HEFFERNAN

directly to influence the emerging French geographical movement, inaugurated institutionally by the establishment of the *Société de Géographie de Paris* in July 1821, the world's first geographical society. This relationship has been widely discussed elsewhere and does not bear repetition here. Suffice it to say that the implicit geographical message of Saint-Simon's technological utopianism was to become explicit in the explorations and writings of many of France's leading nineteenth-century geographers (Émerit 1943–4, 1975; Heffernan 1989, 1994b, 1994c, 100–2; Lorcin 1995, 95–117).

5.5. Conclusion: Human Geography and Human Progress

In this chapter, I have explored three different but related perspectives on the nature and potential of earthly human progress through the writings of three French authors. The idea of human progress which these authors developed allowed the past, present, and—crucially—the future of humankind to be imagined anew. Prior to these debates, the idea of the future was still constrained by older, quasi-Medieval fears and trepidations; it was still the realm of mystics, prophets, and soothsayers whose visions were invariably apocalyptic. After these developments, a new secular language of earthly progress arose which 'tamed' the future and brought it within the orbit of 'rational' science. Prophesy gave way to prediction.

As I have tried to show, the 'progress debate' was informed by geographical reasoning. The three authors examined all emphasized the power of human ingenuity to control and dominate the natural world while welcoming this dominance as desirable evidence of human improvement. Although materialist in social, economic, and political terms, they each rejected determinist environmental explanations of the human condition.

But the three perspectives were also distinctive. Turgot's mid-eighteenth-century theory saw human geographical variation, the differences between this region and that, as the empirical proof of a stadial process of upward and onward human progress, a steady and inevitable improvement of the human condition fueled not by external forces but by intellectual advances. This view culminated with the supremely optimistic writings of Condorcet, which looked forward to a utopia of global human perfection. Volney's more anxious view, written in the wake of the French Revolution, concentrated on the regional circles of rise and decline that Turgot accepted were superimposed on global linear progression but which he choose to ignore. Volney was ultimately optimistic but he rejected historical inevitability and emphasized the importance of conscious human agency. His was also a less global perspective, more concerned with the fate of particular regions. Saint-Simonian thinking, although recognizably part of the same modernist faith in the possibility and desirability of earthly human advancement, was also distinctive. More complex than either of the other two philosophies, Saint-Simonianism shared the utopianism of Turgot but, like Volney, rejected historical inevitability. Saint-Simon and his followers, though often lapsing into a comforting belief in historical predestination, nevertheless emphasized human agency and the need to conquer space through the application of science and technology. While Turgot saw all regions advancing together under their own steam though at different rates and at different levels in a hierarchy, the Saint-Simonians spoke of lateral connections between regions and across space as the means to drive this process forward. This final perspective was, therefore, the most overtly colonial of the three.

The Saint-Simonians had a demonstrable influence on France's embryonic early nineteenth-century geographical movement. In one sense, this is perfectly logical. Their writings, though complex and often abstruse, were concerned with questions of central importance to the inchoate discipline: science, exploration, observation and the development of a new colonial policy for a new, industrial age. Yet, at a deeper level, the progressivism which the Saint-Simonians carried forward into the nineteenth century predicted a future world in which geographical divisions and barriers would be overcome. Geography, the study of spatial differentiation, would therefore seem destined to have less and less relevance. In this sense, the discipline of geography might be seen as a purely instrumental and temporary intellectual requirement which needed to exist to facilitate the eradication of its own subject matter. This is perhaps, a suitably ironic note on which to end. The rise of the modern idea of human progress has been directly associated with the contemporaneous development of the modern geographical imagination. Yet, the resolution of the progressive trajectory in some utopian future world implies a solution to the 'problem' of geography, the final conquest of space.

Notes

Though the usual disclaimers apply, I wish to thank the editors of this volume, two anonymous readers, and the other participants at the Edinburgh conference for many helpful and encouraging remarks. Tim Unwin

MICHAEL HEFFERNAN

pointed out some of the cruder assertions in an earlier draft with tact and sensitivity.

1. Too large, I think, to cite even a selection of the relevant references in the main text. Delvaille (1910) is an interesting pre-1914 survey, suggestive of the optimism and innocence of that peaceful Edwardian age. Sorel (1969, originally published in 1908) is a very different account, emphasizing the revolutionary rather then the gradual nature of progress. The eloquent inter-war considerations, written in the midst of widespread doubt that history could be equated with human advancement, have an altogether different, more modern tone. See Becker (1932); Bury (1932); Cassirer (1932); Guyénot (1941); Hubert (1923); Keohane (1982); Lovejoy (1936); Needham (1937); Wallis (1930); and Whitney (1934). The post-1945 literature, beginning with a magnificent essay by Charles Frankel (1948), reveals not only the variety of individual perspectives but also, perhaps, the general ebb and flow of optimism, pessimism, and the various shades of feeling in between. Surveys of the idea of human progress culminating in utopia include Bann and Kumar (1993); Claeys (1994); Doren (1967); Kamenka (1987); Tuveson (1949); Venturi (1971); and Wager (1967, 1969, 1972). Insights into pessimistic thinking can be gleaned from Chamberlin and Gilman (1985), Pick (1989), Starn (1975), and Vyverberg (1958). The tensions between these two positions are examined in different ways in Almond et al. (1982), Berlin (1956), Buckley (1966), Davis (1984), Gay (1966), Ginsberg (1953), Glacken (1967, 623–54), Hawthorn (1987), Horkheimer and Adorno (1973), Iggers (1965), Kumar (1978, 1987), Nef (1950), Nisbet (1980), Pollard (1968), Roger (1963), Rotenstreich (1971), Sampson (1956), Spadafora (1990), and Trevor-Roper (1963).

2. After Frankel (1948, 122).

3. Turgot's political career has been extensively analyzed. See Baker (1978); Bordes and Morange (1982); Cavanaugh (1969); Dakin (1939); Dockès (1969, 295–304); Faure (1961); Hufton (1974, 327–36); Laugier (1979); Maxwell (1933–4); White (1989).

4. Like all of Turgot's early writings, these two documents remained unpublished until early in the nineteenth century. His collected works were first compiled between 1808 and 1811 by a friend and disciple, Pierre-Samuel Dupont de Némours, who had an unfortunate habit of 'improving' on the original texts. Subsequent nineteenth-century editions perpetuated these editorial idiosyncracies. The full, unaltered material was not made available until Gustave Schelle's edition, a century after Dupont de Némours. See Turgot (1913–23).

5. Foreshadowing Auguste Comte's three-fold division of theological, metaphysical, and positive forms of understanding, developed in the 1830s, Turgot also suggested that progress through these different stages was accompanied by three different forms of philosophy, the first involving explanation through spiritual faith, the second through the ideas of "essences," and the third through mechanistic or scientific forms of explanation. See Frankel (1948, 123).

6. As several authors have noted, most recently Outram (1995, 65), the Treaty of Utrecht, signed at the end of the War of Spanish Succession in 1713, is the last accord to make reference to the pre-European term of *Respublica christiana*.

7. Turgot's brief ministry is, in this respect, highly interesting. For brilliant analyses of the impact of his political ideas on statistical theorizing, see Gillispie (1972), Hahn (1964), and, more generally, Gillispie (1981). See also Kain and Baigent (1992, 218–29) and Konvitz (1987, 34–7) in regard to his cartographic concerns.

8. This view, identical to that of Adam Smith, is articulated in Turgot's economic writings, particularly his *Réflexions sur la formation et la distribution des richesses,* published in the late 1760s during his successful period as *Intendant* for the Limousin. See Meek (1973a, 119–81), and also Meek (1971, 1973b); Groenewegen (1977); Hutchinson (1982); Kiener and Peyronnet (1979); and Lundberg (1964).

9. Volney's description of Palmyra owes more to his reading of Robert Wood's lavishly illustrated account than to his own experience. See Wood (1753).

10. The word 'socialism' appears for the first time in the early 1830s in one of the newspapers established by Saint-Simon's followers.

11. The literature on the Saint-Simonians is huge. See, for examples, Allemagne (1935), Bakunin (1976), Carlisle (1968), Charléty (1896), Charlton (1963), Derré (1986), Iggers (1958), Manuel and Manuel (1979, 615–40), Mueller (1956), Pankhurst (1957), Ratcliffe (1976), Weill (1896).

References

Allemagne, Henri-René d'. 1935. *Prosper Enfantin et les grandes entreprises du XIXe. siècle.* Paris: Librairie Gründ.

Almond, Gabriel A., Marvin Chodorow, and Roy Harvey Pearce, eds. 1982. *Progress and Its Discontents.* Berkeley and Los Angeles: University of California Press.

Baker, Keith M. 1975. *Condorcet: From Natural Philosophy to Social Mathematics.* Chicago: University of Chicago Press.

———. 1978. "French Political Thought at the Accession of Louis XVI." *Journal of Modern History* 50: 279–303.

Bakunin, Jack S. 1976. "Pierre Leroux on Democracy, Socialism, and the Enlightenment." *Journal of the History of Ideas* 37: 455–74.

Bann, Stephen, and Kristen Kumar, eds. 1993. *Utopianism and the Millennium.* London: Reaktion Books.

Baumann, Zygmunt. 1985. "On the Origins of Civilisation: A Historical Note." *Theory, Culture and Society* 2: 7–14.

Becker, C. L. 1932. *The Heavenly City of the Eighteenth-Century Philosophers.* New Haven: Yale University Press.

Berlin, Isaiah, ed. 1956. *The Age of Enlightenment.* New York: New American Library.

Binder, L. 1986. "The Natural History of Development Theory." *Comparative Studies in Society and History* 28: 3–33.

Bobek, Hans. 1962. "The Main Stages of Socio-Economic Evolution from a Geographical Point of View," in eds. P. L. Wagner and M. W. Mikesell, *Readings in Cultural Geography.* Chicago: University of Chicago Press, 218–47.

Bordes, C., and J. Morange, eds. 1982. *Turgot: Économist et administrator.*

Actes d'un séminaire organisé par la Faculté de Droit Économique de Limoges pour le bicenténaire de la mort de Turgot—8, 9 et 10 octobre, 1981. Paris: Presses Universitaires de France.

Bowen, Margarita. 1981. *Empiricism and Geographical Thought from Francis Bacon to Alexander von Humboldt.* Cambridge: Cambridge University Press.

Bowler, Peter J. 1974. "Evolutionism in the Enlightenment." *History of Science* 7: 159–83.

———. 1989. *The Invention of Progress: The Victorians and the Past.* Oxford: Blackwell.

Breal, Michel. 1899. "Volney, orientaliste et historien." *Journal des Savants* (n.v.) 98–107 and 261–71.

Broc, Numa. 1969. "Peut-on parler d'une géographie humaine au XVIIIe. siècle." *Annales de Géographie* 25: 57–75.

———. 1974. *La géographie des philosophes: géographes et voyageurs français au XVIIIe. siècle.* Paris: Éditions Ophrys.

Buckley, Jerome Hamilton. 1966. *The Triumph of Time: A Study of Victorian Concepts of Time, History, Progress and Decadence.* Cambridge, MA: Harvard University Press.

Buffon, Comte de (Georges-Louis Leclerc). 1778. *Des époques de la nature.* Vol. 5 of *Histoire naturelle, générale et particulière.* Paris: Imprimerie Royale.

Burke, Peter. 1977. "Tradition and Experience: The Idea of Decline from Bruni to Gibbon," in eds. G. W. Bowersock, John Clive, and Stephen R. Graubard, *Edward Gibbon and the Decline and Fall of the Roman Empire.* Cambridge, MA: Harvard University Press, 97–119.

———. 1980. "Did Europe Exist before 1700?", *History of European Ideas* 1: 21–9.

Bury, J. B. T. 1932. *The Idea of Progress: An Enquiry into Its Origins and Growth.* London: Macmillan.

Carlisle, Robert B. 1968. "Saint-Simonian Radicalism: A Definition and a Direction." *French Historical Studies* 5: 430–45.

———. 1974. "The Birth of Technocracy: Science, Society, and Saint-Simonians." *Journal of the History of Ideas* 35: 445–64.

Cassirer, Ernst. 1951 [1932]. *The Philosophy of the Enlightenment.* Princeton: Princeton University Press.

Cavanaugh, G. J. 1969. "Turgot: The Rejection of Enlightened Despotism." *French Historical Studies* 6: 31–58.

Chamberlin, J. Edward, and Sander L. Gilman, eds. 1985. *Degeneration: The Dark Side of Progress.* New York: Columbia University Press.

Charléty, S. 1896. *Histoire du Saint-Simonisme* (1825–1864). Paris: Hachette.

Charlton, D. G. 1963. *Secular Religions in France, 1815–1870.* Oxford: Oxford University Press.

Chinard, Gilbert. 1913. *L'Amérique et le rêve exotique dans la littérature française au XVIIe. et au XVIIIe. siècle.* Baltimore: Johns Hopkins University Press.

———. 1923. *Volney et l'Amérique d'après des documents inédits et sa correspondence avec Jefferson.* Baltimore: Johns Hopkins University Press.

———. 1925. *Jefferson et les idéologues.* Baltimore: Johns Hopkins University Press.

———. 1942. "Montesquieu's Historical Pessimism," in ed. Gilbert Chinard, *Studies in the History of Culture: The Disciplines of the Humanities.* Menasha, WI: 161–72.

Claeys, Gregory. 1994. *Utopias of the British Enlightenment.* Cambridge: Cambridge University Press.

Clark, J. C. D. 1994. *The Language of Liberty: Political Discourse and Social Dynamics in the Anglo-American World.* Cambridge: Cambridge University Press.

Cockburn, William. 1804. *Remarks on A Publication of M. Volney called "The Ruins etc."* Cambridge: Cambridge University Press.

Cohen, I. B. 1976. "The Eighteenth-Century Origins of the Concept of Scientific Revolution." *Journal of the History of Ideas* 37: 257–88.

Cohn, Norman. 1970 [1957]. *The Pursuit of the Millennium: Revolutionary Millenarians and Mystical Anarchists of the Middle Ages.* London: Pimlico.

Condillac, Etienne Bonnot de. 1924 [1746]. *Essai sur l'origine des connaissances humaines, ouvrage où l'on réduit à un seul principe tout ce qui concerne l'entendement humain* [2 vols. in 1]. Paris: Armand Colin.

Condorcet, Marquis de (Caritat, J.-A.-N. de). 1787. *Vie de Monsieur Turgot.* Berne: Kirchberger and Hatter.

———. 1794. *Esquisse d'un tableau historique des progrès de l'esprit humain.* Paris: Agasse.

Cowen, M. P., and R. W. Shenton. 1996. *Doctrines of Development.* London: Routledge.

Cro, Stelio. 1990. *The Noble Savage: Allegory of Freedom.* Waterloo, Ontario: Wilfred Laurier University Press.

Crossley, Ceri. 1993. *French Historians and Romanticism: Thierry Guizot, the Saint-Simonians, Quinet, Michelet.* London: Routledge.

Dakin, D. 1939. *Turgot and the 'Ancien Régime' in France.* London: Methuen.

Darnton, Robert. 1979. *The Business of Enlightenment: A Publishing History of the Encyclopédie 1775–1800.* Cambridge, MA: Harvard University Press.

Davis, David Brion. 1984. *Slavery and Human Progress.* Oxford: Oxford University Press.

Degérando, Joseph-Marie de. 1969 [1800]. *The Observation of Savage Peoples.* London: Routledge and Kegan Paul.

Delvaille, Jules. 1910. *Essai sur l'histoire de l'idée de progrès jusqu'à la fin du XVIIIe. siècle.* Paris: F. Alcan.

Derré, J.-R., ed. 1986. *Regards sue le Saint-Simonisme et les Saint-Simoniens.* Lyons: Presses Universitaires de Lyon.

Destutt de Tracy, Antoine. 1801–15. *Éléments d'idéologie: Projets d'éléments d'idéologie à l'usage des écoles centrales.* 4 vols. Paris: Lévi.

Dickason, O. P. 1983. *The Myth of the Savage and the Beginnings of French Colonialism in the Americas.* Edmonton: University of Alberta Press.

Dockès, P. 1969. *L'espace dans la pensée économique du XVIe. au XVIIIe. siècle.* Paris: Flammarion.

Doren, Charles van. 1967. *The Idea of Progress.* New York: F. A. Praeger.

Droixhe, D., and Pol-P. Gossiaux, eds. 1985. *L'homme des lumières et la découverte de l'autre.* Brussels: Éditions de l'Université de Bruxelles.

MICHAEL HEFFERNAN

Duchet, M. 1971. *Anthropologie et histoire au siècle des lumières: Buffon, Voltaire, Rousseau, Helvétius, Diderot*. Paris: Maspero.

Duzer, Charles Hunter van. 1935. *Contribution of the Ideologues to French Revolutionary Thought*. Baltimore: Johns Hopkins University Press.

Eagleton, Terry. 1991. *Ideology: An Introduction*. London: Verso.

Ebenstein, B. 1982. "Turgot vu par Condorcet: éléments d'une hagiographie," in Bordes and Morange, 197–204.

Echeverria, D. 1957. *Mirage in the West: A History of the French Image of American Society to 1815*. Paris: Princeton University Press.

Eichthal, Gustave d'. 1840. *De l'unité européenne*. Paris: Truchy.

Elias, Norbert. 1994 [1939]. *The Civilizing Process: The History of Manners and State Formation and Civilization*. Oxford: Blackwell.

Eltin, Richard A. 1984. *Symbolic Space: French Enlightenment Architecture and Its Legacy*. Chicago: University of Chicago Press.

Émerit, Marcel. 1941. *Les Saint-Simoniens en Algérie*. Paris: Éditions de l'Empire.

———. 1943–4. "Les explorations Saint-Simoniennes en Afrique Orientale et sur la route des Indes." *Revue Africaine* 87–8: 92–116.

———. 1967. "L'Idée de colonisation dans les socialismes français." *L'Age Nouveau* 24: 103–21.

———. 1975. "Diplomates et explorateurs Saint-Simoniens," *Revue d'Histoire Moderne et Contemporaine* 22: 397–415.

Enfantin, Prosper. 1843. *Colonisation de l'Algérie*. Paris: Bertrand.

Fakkar, Raoul. 1973. "L'Influence de Saint-Simon et ses disciples en Égypte." *Économies et Sociétés* 7: 171–200.

———. 1974. *Reflets de la socologie pre-marxiste dans le monde arabe: idées progressistes et pratique industrielles des Saint-Simoniens en Algérie et en Égypte au XIXe. siècle*. Paris: Geuthner.

Faure, Félix. 1961. *La disgrace de Turgot*. Paris: Gallimard.

Febvre, Lucien. 1974. "Civilisation: Evolution of a Word and a Group of Ideas," in ed. Peter Burke, *A New Kind of History from the Writings of Febvre*. London: Verso, 217–57.

Foucault, Michel. 1970. *The Order of Things: An Archaeology of the Human Sciences*. London: Tavistock Publications.

———. 1991. "Governmentality," in eds. G. Burchell, C. Gordon, and P. Miller. *The Foucault Effect: Studies in Governmentality*. London: Harvester Wheatsheaf, 87–104.

Frankel, Charles. 1948. *The Faith in Reason: The Idea of Progress in the French Enlightenment*. New York: King's Crown Press, Columbia University.

Fruchtman, Jack. 1983. *The Apocalyptic Politics of Richard Price and Joseph Priestley: A Study in Late Eighteenth-Century English Millenialism*. Philadelphia: Transactions of the American Philosophical Society, Vol. 73, No. 4.

Furet, François. 1977. "Civilization and Barbarism in Gibbon's History," in eds. G. W. Bowersock, John Clive, and Stephen R. Graubard, *Edward Gibbon and the Decline and Fall of the Roman Empire*. Cambridge, MA: Harvard University Press, 159–66.

Gaulmier, Jean. 1962. "Volney et ses *Leçons d'histoire*." *History and Theory* 2: 52–65.

————. 1968. "Chateaubriand et Volney." *Annales de Bretagne* 75: 570–8.

————. 1980. *L'Idéologue Volney 1757–1820: Contribution à l'histoire de l'Orientalisme en France.* Geneva and Paris: Slatkine.

Gay, Peter. 1995 [1966]. *The Enlightenment: An Interpretation.* 2 Vols. 1, *The Rise of Modern Paganism;* 2, *The Science of Freedom.* New York: W. W. Norton.

Gillispie, Charles Coulson. 1972. "Probability and Politics: Laplace, Condorcet and Turgot." *Proceedings of the American Philosophical Society* 16: 1–20.

————. 1981. *Science and Polity in France at the End of the Old Regime.* Princeton: Princeton University Press.

Ginsberg, Morris. 1953. *The Idea of Progress: A Revaluation.* London: Methuen.

Glacken, Clarence J. 1960. "Count Buffon on Cultural Changes of the Physical Environment." *Annals of the Association of American Geographers* 50, 1: 1–21.

————. 1967. *Traces on the Rhodian Shore: Nature and Culture in Western Thought from Ancient Times to the End of the Eighteenth Century.* Berkeley and Los Angeles: University of California Press.

Gould, Stephen J. 1988. *Time's Arrow, Time's Cycle: Myth and Metaphor in the History of the Geological Sciences.* Harmondsworth: Penguin.

Green, F. C. 1950. *Rousseau and the Idea of Progress.* Oxford: Clarendon Press.

Groenewegen, P. D., ed. 1977. *The Economics of Turgot.* The Hague: Martinus Nijhoff.

Gutteridge, G. H. ed. 1961. *The Correspondence of Edmund Burke.* 3 vols. Cambridge: Cambridge University Press.

Guyénot, Émile. 1941. *Les sciences de la vie aux XVIIe. et XVIIIe. siècles: l'idée d'évolution.* Paris: A. Michel.

Hahn, Roger. 1964. "The Chair of Hydrodynamics in Paris, 1775–1791: A Creation of Turgot." *Actes du dixième Congrès international d'histoire des sciences* 2: 751–54.

Haines, Barbara. 1978. "The Inter-Relations between Social, Biological, and Medical Thought, 1750–1850: Saint-Simon and Comte." *The British Journal for the History of Science* 11, 37: 19–35.

Hawthorn, Geoffrey. 1987 [1976]. *Enlightenment and Despair: A History of Social Theory.* Cambridge: Cambridge University Press.

Hay, Denys. 1968. *Europe: The Emergence of an Idea.* Edinburgh: University of Edinburgh Press.

Hazard, Paul. 1990 [1934]. *The European Mind: The Critical Years, 1680–1715.* New Haven: Yale University Press.

Head, Brian W. 1980. "The Origin of 'idéologue' and 'idéologie.'" *Studies in Voltaire and the Eighteenth Century* 183: 257–64.

Heater, Derek. 1992. *The Idea of European Unity.* Leicester: Leicester University Press.

Heffernan, Michael. 1989. "The Limits of Utopia: Henri Duveyrier and the Exploration of the Sahara in the Nineteenth Century." *Geographical Journal* 155, 3: 342–52.

————. 1990. "Bringing the Desert to Bloom: French Ambitions in the Sahara Desert during the Late Nineteenth Century—The Strange Case of 'La Mer Intérieure,'" in eds. Denis Cosgrove and Geoff Petts, *Water, Engineering and Landscape: Water Control and Landscape Formation in the Modern Period.* London: Belhaven, 94–114.

———. 1994a. "On Geography and Progress: Turgot's *Plan d'un ouvrage sur la géographie politique* (1751) and the Origins of Modern Progressive Thought." *Political Geography* 13, 4: 328–43.

———. 1994b. "An Imperial Utopia: French Surveys of North Africa in the Early Colonial Period," in ed. Jeffrey Stone, *Maps and Africa*. Aberdeen: Aberdeen University Press, 81–102.

———. 1994c. "The Science of Empire: The French Geographical Movement and the Forms of French Imperialism, 1870–1920," in eds. Anne Godlewska and Neil Smith, *Geography and Empire*. Oxford: Blackwell, 92–114.

Hinsley, F. H. 1967. *Power and the Pursuit of Peace*. Cambridge: Cambridge University Press.

Hobsbawm, Eric J. 1987. *The Age of Empire 1875–1914*. London: Weidenfeld and Nicolson.

Horkheimer, M., and Adorno, T. W. 1973. *The Dialectic of Enlightenment*. London: Allen Lane.

Hubert, R. 1923. *Les sciences sociales dans l'Encyclopédie: la philosophie de l'histoire et le problème des origines sociales*. Paris: Félix Alcan.

Hufton, Olwen H. 1974. *The Poor in Eighteenth-Century France*. Oxford: Clarendon Press.

Hulme, Peter. 1990. "The Spontaneous Hand of Nature: Savagery, Colonialism, and the Enlightenment," in eds. Peter Hulme and Ludmilla Jordonova, *The Enlightenment and Its Shadows*. London: Routledge, 16–34.

Hunt, I. 1988. "The Language of Sociability and Commerce: Samuel Pufendorf and Theoretical Foundations of the 'Four Stages Theory,'" in Pagden, 253–71.

Hutchinson, T. W. 1982. "Turgot and Smith," in Bordes and Morange, 33–45.

Iggers, Georg G. 1958. *The Cult of Authority: The Political Philosophy of the Saint-Simonians—A Chapter in the Intellectual History of Totalitarianism*. The Hague: Martinus Nijhoff.

———. 1965. "The Idea of Progress: A Critical Reassessment," *American Historical Review* 71: 1–17.

———. 1982. "The Idea of Progress in Historiographical and Social Thought since the Enlightenment," in Almond et al., 41–66.

Ionescu, Ghita, ed. 1976. *The Political Thought of Saint-Simon*. Oxford: Oxford University Press.

Jacques, T. Carlos. 1997. "From Savages and Barbarians to Primitives: Africa, Social Typologies, and History in Eighteenth-Century French Philosophy." *History and Theory* 36: 190–215.

Jeanvrot, Victor. 1898. "Volney: sa vie et ses œuvres." *La Révolution Française* 35: 278–86, 348–75.

Jones, R. 1992. "Philosophical Time Travellers." *Antiquity* 66: 744–57.

Kain, Roger J. P., and Baigent, Elizabeth. 1992. *The Cadastral Map in the Service of the State: A History of Property Mapping*. Chicago: University of Chicago Press.

Kamenka, Eugene, ed. 1987. *Utopias*. Oxford: Oxford University Press.

Kaplan, Steven L. 1986. "Social Classification and Representation in the Corporate World of Eighteenth-Century France: Turgot's 'Carnival,'" in eds. S. L. Kaplan and C. J. Koepp, *Work in France: Representations, Meaning, Organization and Practice.* Ithaca: Cornell University Press, 176–228.

Kennedy, Emmett. 1978. *A Philosopher in the Age of Revolution: Destutt de Tracy and the Origins of 'Ideology.'* Philadelphia: American Philosophical Society.

Keohane, Nannerl O. 1982. "The Enlightenment Idea of Progress Revisited," in Almond et al., 21–40.

Kiener, M. C., and Peyronnet, J.-C. 1979. *Quand Turgot regnait en Limousin: Un tremplin vers le pouvoir.* Paris: Fayard.

Konvitz, Joseph. 1987. *Cartography in France, 1660–1848: Science, Engineering and Statecraft.* Chicago: University of Chicago Press.

Kors, A. C. 1976. *D'Holbach's Côterie: An Enlightenment in Paris.* Princeton: Princeton University Press.

Krakeur, Lester Gilbert. 1938. "Diderot and the Idea of Progress." *Romantic Review* 29: 151–9.

Kroeber, A. L., and Kluckhohn, C. 1952. *Culture: A Critical Review of Concepts and Definitions.* Cambridge, MA: Harvard University Press.

Kumar, Krishan. 1978. *Prophecy and Progress: The Sociology of Industrial and Post-industrial Society.* Harmondsworth: Penguin Books.

Lafitau, J.-F. 1724. *Mœurs des sauvages amériquains comparées aux mœurs des premiers temps.* 2 vols. Paris: Saugrain l'ainé.

LaFrenière, Gilbert F. 1990. "Rousseau and the European Roots of Modern Environmentalism." *Environmental History Review* 14: 41–72.

Lamarck, Jean-Baptiste. 1802. *Hydrogéologie, ou recherches sur l'influence qu'ont les eaux sur la surface du globe terrestre.* Paris: Chez l'Auteur.

Laugier, L. 1979. *Turgot, ou le mythe des réformes.* Paris: Éditions Albatross.

Locke, John. 1967 [1690]. *Two Treatises on Government.* Cambridge: Cambridge University Press.

———. 1975 [1690]. *An Essay Concerning Human Understanding.* Oxford: Clarendon Press.

Lorcin, Patricia M. 1995. *Imperial Identities: Stereotyping, Prejudice and Race in Colonial Algeria.* London: I. B. Tauris.

Lovejoy, A. O. 1936. *The Great Chain of Being.* Cambridge, MA: Harvard University Press.

Lundberg, I. C. 1964. *Turgot's Unknown Translator: The Réflexions and Adam Smith.* The Hague: Martinus Nijhoff.

Malthus, Thomas R. 1926 [1798]. *First Essay on Population, 1798.* London: Macmillan.

Manuel, Frank E. 1962. *The Prophets of Paris: Turgot, Condorcet, Saint-Simon, Fourier and Comte.* Cambridge, MA: Harvard University Press.

———. 1963. *The New World of Henri Saint-Simon.* Notre Dame: University of Notre Dame Press.

———. 1965. *Shapes of Philosophical History.* London: Allen and Unwin.

Manuel, Frank E., and Manuel, Fritzi P. 1979. *Utopian Thought in the Western World.* Oxford: Basil Blackwell.

Marshall, Peter J., and Williams, Glyndwr. 1982. *The Great Map of Mankind: British Perceptions of the World in the Age of Enlightenment.* London: J. M. Dent.

Mathiez, Albert. 1910. "Lettres de Volney à la Revellière-Lépeaux (1795–1798)." *Annales Révolutionnaires* 3: 161–94.

Maxwell, C. 1933–4. "The Life and Work of Turgot." *History* 18: 216–29.

Meek, Ronald L. 1971. "Smith, Turgot, and the 'Four Stages' Theory," *History of Political Economy* 3: 9–17.

———. 1973a. *Turgot on Progress, Sociology and Economics*. Cambridge: Cambridge University Press.

———. 1973b. *The Precursors of Adam Smith*. London: J. M. Dent.

———. 1976. *Social Science and the Ignoble Savage*. Cambridge: Cambridge University Press.

Mercier, R.-B. 1953. "La théorie du climat des 'Réflexions critiques' à la 'L'Esprit des lois.'" *Revue d'Histoire Littéraire de la France* 53: 17–37, 159–74.

Montesquieu, Baron de (Secondat, Charles-Louis de). 1721. *Lettres Persanes*. Amsterdam. P. Brunel.

———. 1734. *Considérations sur les causes de la Grandeur des Romains et de leur Décadence*. Amsterdam: Jacques Desbordes.

———. 1748. *De l'esprit des lois*. Geneva: Barrilot et fils.

Moravia, Sergio. 1967. "Philosophie et géographie à la fin du XVIIIe. siècle." *Studies on Voltaire and the Eighteenth Century* 57: 937–1011.

———. 1974. *Il pensiero degli Idéologues: Scienza e filosofia in Francia (1780–1815)*. Florence: La Nuova Italia.

———. 1980. "The Enlightenment and the Sciences of Man." *History of Science* 18: 247–68.

———. 1982. *Filosophia e scienza umane nell'età dei lumi*. Florence: G. C. Sansoni.

Morsy, Magali, ed. 1989. *Les Saint-Simoniens et l'Orient: Vers la Modernité*. Paris: Édisud.

Moses, Claire G. 1982. "Saint-Simonian Men/Saint-Simonian Women: The Transformation of Feminist Thought in 1830s France," *Journal of Modern History* 54: 240–67.

Mueller, I. W. 1956. *John Stuart Mill and French Thought*. Urbana, IL: University of Illinois Press.

Needham, Joseph. 1937. *Integrative Levels: A Revaluation of the Idea of Progress*. Oxford: Oxford University Press.

Nef, J. U. 1950. *War and Human Progress: An Essay on the Rise of Industrial Civilization*. Cambridge, MA: Harvard University Press.

Nisbet, Robert. 1980. *History of the Idea of Progress*. London: Heinemann.

Oake, R. B. 1955. "Montesquieu's Analysis of Roman History." *Journal of the History of Ideas* 16: 44–59.

Outram, Dorinda. 1995. *The Enlightenment*. Cambridge: Cambridge University Press.

Pagden, Antony, ed. 1988. *The Languages of Political Theory in Early-Modern Europe*. Cambridge: Cambridge University Press.

Pankhurst, R. K. P. 1957. *The Saint-Simonians, Mill and Carlyle: A Preface to Modern Thought*. London: Sidgwick and Jackson.

Parker, Harold T. 1937. *The Cult of Antiquity and the French Revolution: A Study in the Development of the Revolutionary Spirit*. Chicago: University of Chicago Press.

Peet, Richard. 1985. "The Social Origins of Environmental Determinism." *Annals of the Association of American Geographers* 75: 309–33.

Persons, S. 1954. "The Cyclical Theory of History in Eighteenth-Century America." *American Quarterly* 6: 147–63.

Pick, Daniel. 1989. *Faces of Degeneration: A European Disorder, c. 1848– c. 1918*. Cambridge: Cambridge University Press.

Pocock, J. G. A. 1977a. "Between Machiavelli and Hume: Gibbon as Civic Humanist and Philosophical Historian," in eds. G. W. Bowersock, John Clive, and Stephen R. Graubard, *Edward Gibbon and the Decline and Fall of the Roman Empire*. Cambridge, MA: Harvard University Press, 103–20.

———. 1977b. "Gibbon's *Decline and Fall* and the World Views of the Late Enlightenment." *Eighteenth Century Studies* 10: 287–303.

———. 1981. "Gibbon and the Shepherds: The Stages of Society in the *Decline and Fall*." *History of European Ideas* 2: 193–200.

Pollard, Sidney. 1968. *The Idea of Progress*. Harmondsworth: Penguin.

Porter, Roy. 1988. *Edward Gibbon: Making History*. London: Weidenfeld and Nicolson.

Priestley, Joseph. 1797. *Letters to Mr. Volney occasioned by a work of his entitled Ruins, & by his letter to the author*. Philadelphia: Thomas Dobson.

Ratcliffe, Barrie M. 1976. "Saint-Simonism and Messianism: The Case of Gustave d'Eichthal." *French Historical Studies* 9, 3: 484–502.

Rice, Howard C. 1976. *Thomas Jefferson's Paris*. Princeton: Princeton University Press.

Roger, Jacques. 1971 [1963]. *Les sciences de la vie dans la pensée française du XVIIIe. siècle*. Paris: Armand Colin.

Rotenstreich, Nathan. 1971. "The Idea of Historical Progress and Its Assumptions." *History and Theory* 10: 197–221.

Rousseau, Jean-Jacques. 1973 [1755]. *Discours sur l'origine, et les fondements de l'inégalité parmi les hommes*. Paris: Presses Universitaires de France.

Said, Edward W. 1978. *Orientalism*. London: Routledge and Kegan Paul.

Saint-Simon, Claude Henri de Rouvroy. 1969 [1825]. *Nouveau Christianisme*. Paris: Seuil.

Saint-Simon, Claude Henri de Rouvroy, and Thierry, Augustin. 1814. *De la réorganisation de la Société Européenne, ou de la nécessité et des moyens de rassembler les peuples de l'Europe en un seul corps politique, en conservant à chacun son indépendance nationale*. Paris: Adrien Égron.

Saint-Simon, Claude-Henri de Rouvroy, and Enfantin, Prosper. 1865–1878. *Œuvres de Saint-Simon et d'Enfantin*, 47 vols. Paris: Dentu/Leroux.

Sampson, R. V. 1956. *Progress in the Age of Reason: The Seventeenth Century to the Present Day*. Cambridge, MA: Harvard University Press.

Schmidt, H. D. 1966. "The Establishment of 'Europe' as a Political Expression." *The Historical Journal* 9, 2: 172–8.

Seltzer, Joyce. 1980. "Volney's Science of Political Society: From Observation to Application." *Studies on Voltaire and the Eighteenth Century* 191: 774–81.

Shackleton, R. M. 1961. *Montesquieu: A Critical Biography*. Oxford: Oxford University Press.

Shinn, Thierry. 1980. *L'École Polytechnique*. Paris: Presses de la Fondation des Sciences Politiques.

MICHAEL HEFFERNAN

Shklar, Judith N. 1969. *Men and Citizens: A Study of Rousseau's Social Theory.* Cambridge: Cambridge University Press.

———. 1987. *Montesquieu.* Oxford: Oxford University Press.

Simon, Walter M. 1956. "History for Utopia: Saint-Simon and the Idea of Progress." *Journal of the History of Ideas* 17: 311–31.

Sorel, Georges. 1969 [1908]. *The Illusions of Progress.* Berkeley and Los Angeles: University of California Press.

Spadafora, D. 1990. *The Idea of Progress in Eighteenth-Century Britain.* New Haven: Yale University Press.

Springborg, Patricia. 1992. *Western Republicanism and the Oriental Prince.* Cambridge: Cambridge University Press.

Starn, Randolph. 1975. "Meaning Levels in the Theme of Historical Decline." *History and Theory* 14: 1–31.

Staum, Martin S. 1996. *Minerva's Message: Stabilizing the French Revolution.* Montreal and Kingston, Ont.: McGill-Queen's University Press.

Stocking, George W. 1955. "French Anthropology in 1800." *Isis* 4: 134–50.

Taboulet, Georges. 1968. "Aux origines du canal de Suez: le conflit entre F. de Lesseps et les Saint-Simoniens." *Revue Historique* 92, 240: 89–115.

———. 1971. "Le rôle des Saint-Simoniens dan le percement de l'isthme de Suez." *Économies et Sociétés* 5: 1295–1320.

———. 1973. "Ferdinand de Lesseps et l'Égypte avant le canal (1803–1854)." *Revue Française d'Histoire d'Outre-Mer* 60, 219–20: 61, 143–71 and 364–407.

Taylor, Keith, ed. 1975. *Henri Saint-Simon (1766–1825): Selected Writings on Science, Industry, and Social Organization.* London: Croom Helm.

Thompson, E. P. 1968 [1963]. *The Making of the English Working Class.* Harmondsworth: Penguin.

Trevor-Roper, Hugh. 1963. "The Historical Philosophy of the Enlightenment." *Studies on Voltaire and the Eighteenth Century* 27: 1667–87.

Turgot, Anne-Robert-Jacques. 1913–23. *Œuvres de Turgot et documents le concernant avec biographie et notes,* ed. Gustave Schelle. 5 vols. Paris: Félix Alcan.

Tuveson, Ernest. 1949. *Millennium and Utopia: A Study in the Background of the Idea of Progress.* Berkeley and Los Angeles: University of California Press.

Vallaux, Camille. 1938. "Deux précurseurs de la géographie humaine: Volney et Charles Darwin." *Revue de Synthèse* 15: 83–93.

Venturi, F. 1971. *Utopia and Reform in the Enlightenment.* Cambridge: Cambridge University Press.

Vico, Giambattista. 1968 [1744 ed.]. *The New Science of Giambattista Vico* (eds. T. G. Bergin and M. H. Fisch). Ithaca: Cornell University Press.

Vogt, E. A. 1996. "Civilisation and Kultur: Keywords in the History of French and German Citizenship." *Ecumene* 3, 2: 125–45.

Volney, Constantin-François-Chasseboeuf. 1787. *Travels thru' Syria & Egypt in the years 1783, 1784, and 1785 containing the present natural and political state of those countries, their productions, arts, manufactures, & commerce; with observations on the manners, customs, & government of the Turks & Arabs.* 2 vols. London: G. G. J. & J. Robinson.

———. 1792a [1791]. *The Ruins, or a Survey of the Revolutions of Empires.* London: J. Johnson.

———. 1792b. *The Torch or, a light [to] enlighten the Nations of Europe in their way towards Peace and Happiness*. London: London Corresponding Society.

———. 1796. *The Law of Nature, being an abridgement of the Law of Nature; or, catechisme of French citizens*. London: Printed for the Instruction of the People.

———. 1803. *Tableau du climat et du sol des États-Unis d'Amérique, suivi d'éclaircissements sur la Floride, sur la colonie française à Scioto, sur quelques colonies canadiennes et sur les sauvages*. 2 vols. Paris: Courcier.

———. 1819. *L'Alfabet européen appliqué aux langues asiatiques (ouvrage élémentaire, utile à tout voyageur en Asie)*. Paris: Firmin Didot.

———. 1825–6. *Œuvres*. 7 vols. Paris: Parmentier and Froment.

Vyverberg, H. 1958. *Historical Pessimism in the French Enlightenment*. Cambridge, MA: Harvard University Press.

Wager, W. Warren. 1967. "Modern Views of the Origins of the Idea of Progress." *Journal of the History of Ideas* 28: 55–70.

———. ed. 1969. *The Idea of Progress Since the Renaissance*. New York: Wiley.

———. 1972. *Good Tidings: The Belief in Progress from Darwin to Marcuse*. Bloomington, IN: Indiana University Press.

Wallis, Wilson D. 1930. *Culture and Progress*. New York: McGraw-Hill.

Weill, Georges. 1896. *L'École Saint-Simonienne: son histoire, son influence jusqu'à nos jours*. Paris: F. Alcan.

Welch, Cheryl B. 1984. *Liberty and Utility: The French Idéologues and the Transformation of Liberalism*. New York: Columbia University Press.

White, Eugene Nelson. 1989. "Was There a Solution to the Ancien Régime Financial Dilemma?" *Journal of Economic History* 44: 545–68.

Whitney, Lois. 1934. *Primitivism and the Idea of Progress*. Baltimore: Johns Hopkins University Press.

Williams, Michael. 1989. *Americans and Their Forests: A Historical Geography*. Cambridge: Cambridge University Press.

Williams, Raymond. 1976. *Keywords: A Vocabulary of Culture and Society*. Harmondsworth: Penguin.

Wintle, Michael. 1996. "Europe's Image: Visual Representations of Europe from the Earliest Times to the Twentieth Century," in ed. Michael Wintle, *Culture and Identity in Europe: Perceptions of Divergence and Unity in Past and Present*. Aldershot: Avebury, 52–97.

Wolff, Larry. 1994. *Inventing Eastern Europe: The Map of Civilization on the Mind of the Enlightenment*. Stanford: Stanford University Press.

Wood, Robert. 1753. *The Ruins of Palmyra, Otherwise Tedmor, in the Desart*. London: no publisher.

MICHAEL HEFFERNAN

Chapter Six

Reconsidering Enlightenment Geography and Map Making: Reconnaissance, Mapping, Archive

Matthew H. Edney

6.1. Introduction

The eighteenth century is widely regarded as the formative period for modern cartography. It was supposedly during the 1700s that map making was purged of its "artistic" components to leave a strictly "scientific" practice. The art/science dichotomy is, however, a false distinction promoted by Modernity's pervasive scientism. Modern culture grants maps a privileged status as objective bearers of truth, as nonindexical representations possessed of a metaphysical proximity to the world (Pickles 1992, 199; Turnbull 1993, 18–27). This status is epistemologically distinct from that accorded to other representational strategies, such as writing and landscape imagery, which are widely recognized as flexible and problematic. The current, critical reevaluation of cartography seeks to break down the false distinctions constructed between maps and texts (broadly construed) and to reconfigure maps as complex, negotiated, and ideological representations. Of particular importance to this process is the revelation of the gulf between the epistemological claims made on behalf of cartographic practices and the ability of those practices to satisfy those claims.

With this critical reevaluation in mind, I examine eighteenth-century geography as an intertextual web of discursive practices that underscored Enlightenment's ideology of encyclopedic knowledge. An idealized conception of Enlightenment map making—which might more properly be referred to as "mathematical cosmography" (Edney 1993, 1994b)—lay at the core of geographical representation and, indeed, served to epitomize Enlightenment encyclopedism. Specifically, geographical knowledge was idealized as constituting a comprehensive *archive* constructed through the geographic practices of *reconnaisance* and *mapping*. But these geographical practices proved to be epistemologically insufficient: they functioned mean-ingfully only at the small and medium scales of geographical en-quiry. They could not readily incorporate the large-scale geographies which were promoted by encyclopedism and which were increas-ingly pursued by European states after 1750. This inherent inade-quacy led to the profound reconceptualization of map making, and thus the practice of geography, in the early nineteenth century. That is, Enlightenment's idealized conceptions of space and map making before 1800 were significantly different from modern conceptions thereafter.

My periodization agrees with other studies of Enlightenment epis-temologies. An important collection of essays on Enlightenment sci-ence, for example, makes explicit the ideological intensification in the middle of the eighteenth century. Bernard de Fontanelle might have celebrated the *ésprit géometrique* in 1699 but his "quantifying spirit" did not receive widespread social support and application be-yond natural philosophy until after about 1760 (Frängsmyr, Heil-bron, and Rider 1990, 1–23). Most of the volume's contributors fur-ther suggest that significant conceptual shifts occurred in the early nineteenth century. Another recent study, of architecture and engi-neering in France, argues that the transition in architectural styles from classicism to neoclassicism and eclecticism began after 1750 and proceeded as "a slow rise culminating in a qualitative leap around the year 1800" (Picon 1992, 9).

This essay's focus is the epistemology of geographical representa-tion rather than the social conditions within which geographers functioned. I take it as understood that the Enlightenment ferment of knowledge was always a genteel intellectualism—although not adhered to by all of the gentry—to which those of lesser standing could aspire as a means of social self-advancement. The practice of geography was not only about the appropriation of knowledge and territory through the wielding of an empowered vision, it was also deeply concerned with the legitimation, reproduction, and perpet-uation of a given social order (Edney 1994b). As such, geographical enquiry was central to the era's antiquarianism (Walters 1988), Hel-lenism (Bernal 1987), and orientalism (Mackenzie 1995; Said 1978), each of which sought to configure European society in opposition both to Europe's past societies and to the contemporary societies of Asia, Africa, and the Americas (Wolff 1994; see Driver 1992).

6.2. Representing Enlightenment Map Making

The historian's usual conception of eighteenth-century map making is revealed by the frequent and almost *de rigueur* repetition of Jonathan Swift's quatrain: "So Geographers in *Afric*-maps,/ With Savage-Pictures fill their Gaps;/And o'er unhabitable Downs/Place Elephants for want of Towns" (Swift 1733, 12). Historians have interpreted Swift's derision as epitomizing the development of a new and quintessentially Enlightened sensibility that rejected the creative, imaginative, and *artistic* trappings of early maps and that promoted the pristine, empirical, and *scientific* essence of geographical fact (most recently, Hall 1992, 372; Nobles 1993, 13–15; Thrower 1996, 110). This perceived transformation is the central tenet of traditional cartographic history. It has determined the cartographic canon and accounts for the discipline's recalcitrant opposition to critical theory (Edney 1993, 1996). Ironically, the same transformation also underlies the current critique of cartography. Humanistic geographers chide the coldly scientific modern map for its apparent denial of personal experience and for its promotion of neutral and sterile "space" over emotive and vital "place." Cultural studies follow suit and promote a negative critique of modern maps as "totalizing" and "universal" representations which reject and obscure the local knowledges and nuances that are so central to postcolonial and poststructural theories (examples of many instances are Avery 1995; Duncan and Ley 1993; Ryan 1994). We might point in particular to Harvey's (1989, 249) conviction that, during the 1700s, maps were "stripped of all elements of fantasy and religious belief, as well as any sign of the experiences involved in their production," and became "abstract and strictly functional systems for the factual ordering of phenomena in space." These opinions are clearly bound up with general assumptions made of the "Enlightenment Project," whose progressive disenchantment of the world first requires the disenchantment of the mechanisms of knowledge acquisition and representation (Horkheimer and Adorno 1972, 3–42; and see the Introduction above).

It is not surprising that historians, regardless of their theoretical alignment, have fixated upon Enlightenment's disenchantment of geographical representation. The variety and scope of technological and institutional innovations in map making during the eighteenth century is overwhelming. This was, after all, the period when the problem of accurately determining longitude at sea was finally solved by John Harrison's fourth marine chronometer of 1759, ushering in a whole new era of "scientific" geographic exploration. More generally, it was the period when natural philosophy was reconfigured with an instrumentalist, and highly empiricist, quantifying spirit (see Frängsmyr, Heilbron, and Rider 1990). It was the period in which European states and their landed elites began to undertake large-scale topographic and very large-scale cadastral surveys in order to intensify their control of their territories and properties. It was the "Second Great Age of Exploration," distinguished from the preceding, post-Renaissance "Age of Reconnaisance" by its ideology of exploration in the name of science rather than for overtly mercantile or military purposes (Goetzmann 1986, 8). It was the period when European philosophers began to use the world itself as the source of ideas and data for European science, as in the French expeditions to Lapland and Ecuador to measure the size of the earth (1735–44) or the voyages to observe the transits of Venus (1761, 1769). It was the period when new map projections were designed to regulate the unavoidable distortions according to mathematical rather than aesthetic principles (Snyder 1993, 62–94). And, of course, it was the period when map makers abandoned decoration. This litany makes very clear how map making seemingly became a "science," how the responsibility for map making was apparently transferred from map makers in urban offices to surveyors in distant fields. The essential locus of map making seemingly shifted from the site of imagination and creativity constrained by commercial and political needs to the site of direct experience and pristine measurement, free from any corrupting influence.

Recent analyses of Enlightenment map making have begun to expose the cultural ideology of geographical disenchantment (Harley 1989, 1996). The adoption by explorers, travelers, and surveyors of the new geographical technologies was by no means as fast or as widespread as has been assumed. In addition to restrictions imposed by the mechanical delicacy of the new instruments and the skill needed to use them, their high cost put them well beyond the reach of all but the wealthiest individuals and state-sponsored geographers. Thus, a sufficiently simple mechanism for the marine chronometer—one which would allow economical production and widespread adoption—was not developed until 1806 (Betts 1996). The ability of the European states to engage in comprehensive cadastral surveys was limited (see Kain and Baigent 1992), while detailed topographical surveys remained little more than "magnificent military sketches" (Roy 1785, 386–87) in which "no geometrical exactness is to be expected, the sole object in view being, to shew remarkable

MATTHEW H. EDNEY

things, or such as constitute the great outlines of the Country" (Roy 1765, 2–3). Nor is it easy to identify unambiguously the presumed decline in cartographic "art." Makers of "formal" maps might have expunged mimetic icons and figures from their spatial imagery, but they persisted in providing large, allegorical title cartouches that were intended to be read as "historical narratives" (Green 1717, 155; see Clarke 1988). A thriving mode of "popular" map making that possesses all the emotive, symbolic, and overtly indexical components of premodern cartography has continued to the present, and with no end in sight, despite its exclusion from the cartographic canon and serious historical discussion. Finally, and perhaps most importantly, there was in the eighteenth century no distinct and coherent practice of "cartography" which could become less "artistic." Eighteenth-century map making existed in an interdisciplinary ferment that rejects simplification; the term "cartography" is itself a neologism coined in 1837 (Harley 1987, 12–13), and I use it in this essay to refer to practices pursued, and to conceptions held, after the middle of the nineteenth century.

The increasing plainness of formal maps during the eighteenth century did not mean that there was necessarily any significant change in the character and quality of either geographical data or map construction (Stone 1988). The plain, factual map was a rhetorical construct which has been interpreted within modern culture as signifying factuality and empirical truth but which, like any representation, is nonetheless replete with ideology and ambiguity. Swift's decorative elephants might have been slowly edged off maps of Africa, but geographical "fancy" remained. Bassett and Porter (1991), for example, explore James Rennell's cartographic creation in 1798 of a mountain chain in western Africa: the mountains of "Kong." Rennell took a single observation of distant inselbergs by the explorer Mungo Park as confirmation of his own hypothetical configuration of northern Africa. The mountains survived on maps until the 1890s, when they metamorphosed into a spurious region of northeastern Côte d'Ivoire that survives today in the nineteenth edition of the highly regarded *Goode's World Atlas* (Espenshade 1995, 210). Again, Belyea (1992) briefly examines Guillaume Delisle's 1696 "cartographic discovery" of a large sea—the "Mer de l'Ouest"—in the interior of North America; the sea subsequently remained on maps of the continent until after 1750. Or we might point to the frequent misunderstanding by European map makers of Native American information, such as the Cree description of a short river flowing into Lake Superior which was transformed in European maps into a river flowing from Lake Superior all the way to the Pacific Ocean (Lewis 1991; see Belyea 1996). And so on. Once the surface rhetoric of the map is pierced, numerous idiosyncratic and ideological statements can be identified that belie the apparent objectivity and factual character of even the most "scientific" cartography.

6.3. Enlightenment's Geographical Archive

The epistemological ideology of the geographical archive was Enlightenment's "encyclopedic mentality." This ideology was rooted in two convictions, first that "every point of view, whatever its source, could be brought into rational debate with every other," and second that "such rational debate could always, if adequately conducted, have a conclusive outcome" (MacIntyre 1990, 172). The implication of these convictions for geography was that accounts of the same phenomena from different sources could, in principle, always be reconciled, no matter the degree of difference between them. The ambiguities inherent in any set of observations could, by a process of careful and systematic comparison, be eliminated to reveal singular and unambiguous facts. The idealized product of such a process would be a corpus of data, continually growing and correcting itself, its ultimate purpose to encompass and to replicate the real world. Such an archive could be conceptualized because of the eighteenth century's conception of a mechanistic cosmos that was held to be observable, measurable, and so ultimately understandable. Unambiguous geographical data could thus be accumulated, regardless of their scales and origins, eventually producing one definitive archive of knowledge.

It proved quite impossible for western European states to construct, let alone control, actual geographical archives (Turnbull 1996). The last early modern attempt to create and maintain a global geographical database collapsed shortly before 1700. Jean Dominique Cassini's *planisphère terrestre* was engraved into the floor of the Paris Observatory in 1667 and was updated regularly, yet by the turn of the century this rigid, stone map was abandoned and allowed to wear away under the pressure of its viewers' feet; it was eventually replaced by a printed map in 1696 (Jacob 1992, 131–32). Enlightenment's universal geographical archive was, in contrast, a conceptual knowledge space within which a host of geographical descriptions could be assembled, analyzed, compared, and housed. Yet the prospect of achieving a universal archive was held out by the construction of numerous 'microarchives,' especially towards Enlightenment's

MATTHEW H. EDNEY

apogee at the end of the eighteenth century. I think here not only of the many universal geographical grammars and grand, global geographies such as Anton Friedrich Büsching's *Neue Erdbeschreibung* in eleven volumes (1754–1792), but also of more focused geographies, such as Christoph D. Ebeling's *Erdbeschreibung und Geschichte von Amerika* (1793–1816) in twenty-five volumes, although only seven were completed, or the multiauthor *Description de l'Egypte* in twenty-two volumes (1809–22).

The *Description de l'Egypte* might have been much larger than most other Enlightenment geographical works, but it was little different in the intertextual complex of representational strategies which it employed. Its twenty-two volumes brought together a vast collection of field notes, drawings, and sketch maps produced during the French occupation of Egypt in 1798–1801. The result was a coherent, comprehensive, and systematically organized body of knowledge about Egypt, both ancient and modern. Overall, this massive work rested on the power of its graphic imagery, which brought the spectacle of ancient and modern Egypt to European eyes and comprehension (Prochaska 1994). As a coherent body of knowledge—as a microarchive—the *Description* depended on the fifty-sheet *Carte topographique de l'Egypte*, published in 1825. These maps provided not only the spatial frame of reference for the reader of the *Description*, they also constructed the *Description's* own spatial structure. Even as each item of data had been observed in the field, its uniqueness was defined by its specific location in space (Godlewska 1995). It was the entire process—the field examination and the office compilation—and it was the whole representation—texts, numbers, graphics, and maps—which together encompassed Egypt, which together defined and appropriated its essence, and which together reproduced it for the understanding and edification of Europeans.

More generally, the representation of space inherent to Enlightenment's maps was the unifying force that gave substance to the conceptual potential of a universal geographical archive. For example, the construction of public works throughout *ancien régime* France produced a huge, but disparate, collection of geographical information that was nonetheless held by the civil engineers to form a coherent whole:

> The engineers conceived of the map as an image of the whole. Even if an infinity of plates were required to provide an exhaustive account of the world, cartographic depiction served to guarantee that an indissoluble link remained between words, drawings and

things. . . . [T]he word 'map' . . . signified the encounter between reality and its [archival] double. . . . (Picon 1992, 214)

The map's ideological equivalency with the world was defined by its visible graticule[1] of parallels and meridians, which replicates the invisible parallels and meridians that score the earth's surface. Tectonically determined by the graticule, the map's spatial structure was taken to be a direct reproduction of the spatial structures not only of geodesic space but also of the political and cultural spaces of human existence (Edney 1994a). The map's graticule thus established the structural correctness of the conceptual geographical archive. Graticules were uniformly precise and internally coherent; they could reconstruct the world at any scale, from the entire globe to, in principle, its tiniest portion. The knowledge space of the graticule is flexible and allows all geographical data produced by the European *reconnaissance* of the world—whatever their scale and extent—to be related and reconciled to each other, so as to recapitulate the world's geography. The scale independency of Enlightenment's archive was manifested in the reformulation of geography's subdivisions by subject (mathematical, human, physical) rather than the traditional subdivision by scale (topography, chorography, and geography; Edney 1997a, 41–46).

Indeed, Enlightenment could not conceive of maps without graticules. The graticule was absolutely necessary for each map; it alone established the epistemological truth of geographic representation. Definitions of "map" were all similar to Samuel Johnson's in 1755, of "a geographical picture on which lands and seas are delineated according to the longitude and the latitude" (quoted in Edney 1994a, 387; see Andrews 1996). The obligatory section on maps found at the start of general geographical works—so obligatory, in fact, as to suffice for the entirety of many geographies—followed a common plan: the globe was introduced first, with its lesser and greater circles, instances of which form the focus of Cosgrove's analysis of Coronelli. Only then would the map be explained as the projection of those meridians and parallels into two dimensions. We might also point to the numerous pedagogic works containing a series of maps to be copied onto blank graticules by a child, who would thereby learn his (perhaps her) geography.

The Enlightenment map was distinct from the more particular and focused large-scale "plan," "survey," or *"essai topographique"* so favored of civil and military engineers in the eighteenth century. Such works were the product not of regional abstraction but of direct

MATTHEW H. EDNEY

and highly detailed observation. Christopher Packe, for example, argued that the topographic survey and the common map presented "as manifest a Difference" between them "as there is between the Frame of any Building, and the same Finish'd into a compleat House, adorn'd with all it's [sic] Furniture." His own "philosophico-chorographical chart" comprised a topographical image of all the hills and valleys within a sixteen-mile radius of Canterbury, Kent, as seen and drawn from the vantage of the cathedral tower. Packe's purpose was thus to illuminate "the beautiful Distinction of all the Parts, and the exact Harmony of the whole Country . . . *not as a Map, but as in a Landskip* . . ." (Packe 1743, 3, original emphasis). French engineers commonly underscored the topographic survey's indebtedness to overarching and empowered vision *(surveiller)* by combining large-scale survey plans with topographic landscape views in compound images (fig. 6.1). Nonetheless, these highly localized surveys of discrete places could be, and were, incorporated into the space of the geographical archive by the provision of the extra attribute of location, specifically the place's latitude and longitude. In a practice that today strikes us as dishonest, district surveys that were constructed without reference to latitude and longitude were transformed into maps by the superimposition of spurious graticules (Laxton 1976).

Enlightenment's maps were obviously and intuitively unambiguous. Geographers might debate the actual location of a town or a river, but each location only occurs once in the world and on the map. Indeed, geographical description was Enlightenment's most successful form of knowledge creation, even recognizing, as Bravo cautions below (chap. 7) that such description was often reached only after accommodating native knowledges en route to the finished maps of Europeans. As such, the map was widely regarded during the eighteenth century as the epitome of encyclopedic knowledge. Both natural historians, whom Foucault (1970, 125–65) took to be paradigmatic of eighteenth-century classifiers of knowledge, and the encyclopedists themselves explicitly described what they did as metaphorical map making (Edney 1993, 1994b). They made maps of nature and of knowledge—plant taxonomies; trees of knowledge—and in so doing they constructed their ordered and categorical understanding of the world. By analogy, those abstract knowledge spaces would possess the same structures as the phenomena they represented. Geographical practices served to establish and legitimate Enlightenment's ideological self-image as an inquisitive, rational, knowing, and hence empowered state.

VIEW OF DELPHI AND THE TWO ROCKS OF PARNASSUS.

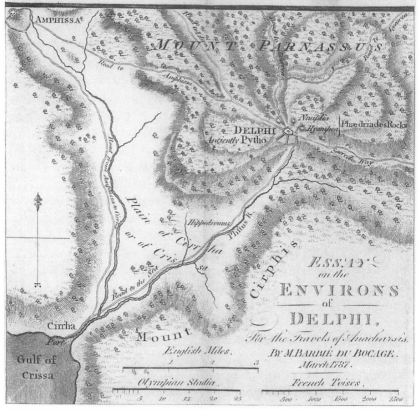

Fig. 6.1. "Essay on the Environs of Delphi. For the Travels of Anacharsis. By M. Barbié du Bocage March 1787" (Barbié du Bocage 1792, pl. 16). The combination of large-scale topographical surveys with landscape views was a common practice for French engineers and geographers; there are several instances in Choiseul-Goufier (1782). The framing of both images was carefully chosen to ensure that the landscape features were correctly aligned. The contemporary means of surveying relief was to sketch it, in a manner very much akin to the period's highly conventionalized landscape art. Such compound representations accordingly celebrated Enlightenment's visual epistemology and the power of observation. Reproduced with the permission of the Osher Map Library, University of Maine.

The map was thus the conceptual unifier of geographical knowledge: as Burke, and Diderot and d'Alembert all signaled, a means to and metaphor for global ordering. It provided the medium for rationalizing and correcting field observations; it was at once the fabric and indexing system of the geographical archive. Conversely, the otherwise mythic archive, with its promise of coherent and comprehensive knowledge, was manifested in Enlightenment's maps. It is for this reason that "geographers" were understood in the eighteenth century to be makers of maps.

There yet remains the issue of how the archive's spatial framework—such a neat and precise replication of the world's spatial structure—was filled in with geographical details. How was the archive itself constituted as a body of knowledge? The remainder of this essay discusses Enlightenment geography's two principal discursive practices that together embodied the archive: *reconnaissance* and *mapping*. It would be a mistake to configure these practices according to the ideological preconceptions prevalent after about 1800. Those preconceptions make a clear distinction between experience and reason, between empirical field work on the one hand and theoretical office work on the other. But just as the office-to-field progression of Enlightenment cartography is a fiction of Modernity—and see also Bravo's discussion below of map making in the field as a process of "ethnographic navigation"—so too is the experience/reason distinction. More properly, Enlightenment encyclopedism entailed a dialectic of experience and reason: it is experience, usually visual in nature, which produces information about the world, but it is reason which refines and structures that experience and which gives it meaning. The synthesis of this dialectic comprises the discursive practices which construct geographical knowledge. That is, I distinguish between reconnaissance and mapping not by their locus—field versus office—but by their spatial ordering of geographical description. Reconnaissance emphasizes the linear route of the geographical traveler, whereas mapping entails the systematic conceptualization and description of discrete areas across the earth's surface.

6.4. Reconnaissance

Narrative geographical descriptions were, of course, intimately related to Europe's cultural encounter with the world. The character of that encounter in the eighteenth century was markedly different from that in earlier periods. Continued economic expansion, the consolidation of the military-fiscal states, and burgeoning overseas empires together intensified Europe's examination of the peoples and lands not only of the rest of the world but also of Europe itself. The Pacific voyages of James Cook, Jean François de Lapérouse, Vitus Bering, and others, voyages which finally pushed mythical *terrae incognitae* off Europe's maps, might be taken to epitomize Enlightenment's geographical disenchantment (e.g., Goetzmann 1986), yet they comprised only a small portion of Enlightenment's geographical inquiry. More generally, reconnaissance—the purposeful movement through and examination of the world[2]—was practiced by merchants, missionaries, civil and military officials, scientists, surveyors, and even tourists who all observed, examined, and documented the constituent features of the landscapes through which they passed. Examples of such reconnaissances include William Bartram's botanizing travels in Georgia and Florida; Arthur Young's agricultural tours of England, Ireland, and France; the French and Spanish geodetic expeditions to northern Sweden and Ecuador; and, the numerous military expeditions throughout Europe and its colonies, in which officers kept records of their movements.

Implicit within the act of reconnaissance was a specific manner of structuring space and classifying geographical information. The accounts of reconnaissance are structured in a linear narrative that replicates the observer's route through the world and each of his acts of observation (reconnaissance was an activity invariably pursued by men). The conceit was that each phenomenon recorded in the narrative was indeed seen, and so known, by the traveler at a specific moment, and therefore in a specific place, in the journey. This was supported through a variety of representational strategies. Their rhetorical style turned the geographer's analytical gaze away from his subjective being and toward the objectified world around him, in contrast to less geographical accounts in which travelers intruded themselves into narratives that focus on their subjective being (see Carter 1987; Hassam 1990; and also Outram's essay in this volume [chap. 9]). Reconnaissance accounts also featured an attention to the prosaic and quotidian—even at the cost of repetition and redundancy—and the inclusion of graphics, cartographics, measurements, and statistics. All told, reconnaissance celebrated the geographer's presence in and visual examination of the landscape (Edney 1997a, 64–69). The products of reconnaissance accordingly possessed a wide variety of formats. Some reconnaissance accounts were

MATTHEW H. EDNEY

comprised of little more than a list of directions and distances; others had great swathes of descriptive text, with only brief annotations of geometric relations. While many had maps, in many more the geographer assumed that the reader would have access to the necessary imagery and did not provide them.

Yet the rhetorical emphasis on visual experience hid the manner in which the geographical traveler engaged in *reasoned* observation. Reason dictated the selection of phenomena to be examined, and reason guided the classification in situ of each phenomenon, as exemplified by the practices of natural historians (Regis 1992, 1–39). All reconnaissance accounts, even the traveler's notebook, were constructed documents (Edney 1997a, 79–85). Their goal was to enumerate the details of each place within the wider scope of geographical space represented by the general map. Reconnaissance accounts presented textual and graphic views of myriad places even as they gave the reader the tools to amalgamate those views into general, regional landscapes. Reconnaissance was accordingly the primary means of fleshing out Enlightenment's geographical archive.

That is, reconnaissance narratives were constructed about a specific epistemological ideal that was maintained in the subsequent editing and publication of geographical accounts in narrative form. Many collections of travel accounts were published throughout the eighteenth century, some of which were substantial (Crone and Skelton 1946, 78–133). These brought the world in all its exotic, fascinating, sublime, and romantic variations into the drawing room, and in doing so tamed and made agreeable the world's dangers and strangeness. When published, the most focused and highly detailed reconnaissances, such as Bartram's *Travels* (1791), were not recast into a more systematic format but were left in narrative form: the geographical facts they presented were important, but so were the spatial structure and the display of Enlightenment's epistemological ideal underlying those facts.

Indeed, reconnaissance's combination of facts and epistemology was so powerful that it was used as a means to construct geographical descriptions even in the absence of actual travel. The satirical use of the travel account, most notably in Swift's *Gulliver's Travels* (1726), is well known, as are the overtly fantastical accounts, such as Rudolph Eric Raspe's purposefully ridiculous *Adventures of Baron Munchausen* (1786). Reconnaissance was used as a model to construct more serious geographical accounts. For example, in *A New Voyage Round the World by a Course Never Sailed Before* (1725), Daniel Defoe drew on his large library of travel literature to describe the geography and economic potential of South America and the South Seas through an account of a fictitious voyage. That work was quickly exposed as a "fake," but Defoe's *Life, Adventures, and Piracies of the Famous Captain Singleton* (1720) was still thought to be a true account a hundred years later; Singleton was credited in the middle of the nineteenth century as the first European explorer of the White Nile (Adams 1962, 110–111).

Despite the popularity of reconnaissance accounts and of travel literature generally in the eighteenth century, the character of reconnaissance as geographical description is foreign to modern sensibilities. Today, we expect geographical accounts to be written in a more analytical fashion. I will therefore provide a detailed example of a narrative geography. In order to stress the created and artificial nature of such accounts—they were not rooted simply in the experience of the world—I have selected what was perhaps the most successful of Enlightenment's fictitious reconnaissances, Jean Jacques Barthélemy's *Voyage de jeune Anacharsis*, first published in 1788.

6.5. *Voyage de jeune Anacharsis:* Reconnaissance as Geographical Description

The abbé Barthélemy (1716–1795) was a renowned philologist and classical scholar. He spent some thirty years planning and writing his masterful description of Greece's last days of glory and freedom before its conquest by Philip of Macedon in 338 B.C.E. Eighteenth-century philhellenes understood the Macedonian conquest to have been the beginning of Greece's servitude to foreign powers, a servitude from which it had yet to emerge. The philhellenes were preoccupied with ending Ottoman occupation and with restoring Greece to its properly independent and glorious state. To this end, Barthélemy did not provide "any new interpretations of Greek civilization," but focused instead on making that "distant world more palpable and vivid" (Augustinos 1994, 37–38). It goes almost without saying that Barthélemy's Greece was that of grand monuments and high art, of ruins and crumbling manuscripts. A substantial work, comprising no less than eighty-two chapters and a volume of thirty-one maps and other images (Barbié du Bocage 1788; Barthélemy 1788), the book was an instant success and went into numerous editions in French and eight other languages (Badolle 1926, 227–31).[3]

As a geographical description, *Voyage du jeune Anacharsis* was constructed as the narrative of the travels through Greece by "Anacharsis the younger." Herotodus recorded the tradition of a visit to Greece in the sixth century B.C.E. by a Scythian prince, Anacharsis, whose great wisdom had so amazed the Greeks that they adopted him as one of the Seven Sages. Barthélemy now imagined that a descendent of Anacharsis had similarly lived in Athens, for three decades in the fourth century B.C.E. Barthélemy supposed that this younger Anacharsis had made periodic tours of Greece's different districts, had met the great men of the time, had seen the great sights, and had written a narrative account of Greece upon his return to Scythia. This conceit also allowed Barthélemy to explain away his incomplete knowledge of ancient Greece by having Anacharsis lose most of his notes and books in the Black Sea, when he was shipwrecked during the voyage home.

Barthélemy nonetheless made it clear that his work was constructed entirely from authentic sources and was not a work of fiction. A principal source was Pausanias's second century C.E. guidebook to Greece, the one such guide to survive into the eighteenth century (Casson 1974, 292–99). Integral to the work's factuality was Barthélemy's use of a wide array of graphics: small-scale maps; large-scale plans of towns; architectural plans and elevations; landscape views; and engravings of a few coins from the royal cabinet of medals (whose superintendent was Barthélemy) to provide actual images from the era. The graphic images were organized, in the atlas volume, in the sequence in which they were mentioned in the text; notes in the text directed the reader to the plates, while marginal comments on the plates directed the reader back to the relevant chapter, or even to the relevant page in some editions.

Most of the graphics came from materials collected by Barthélemy's former pupil, the Comte de Choiseul-Goufier, who had toured Greece and the Aegean with a large retinue of architects and engineers in 1776–79 (Augustinos 1994, 157–73; Choiseul-Goufier 1782; Constantine 1984, 173–82). Of particular importance among the products of this expedition were the architectural plans, views, and *essais topographiques* by J. Foucherot, an engineer trained at the École des ponts et chaussées (Barbié du Bocage 1792, 6). It is nonetheless clear that such graphics—at least maps—were a central component of Barthélemy's plans for his book as early as 1757. To show Anacharsis's various tours, Barthélemy had commissioned several small-scale regional maps from *premier géographe du roi* Jean Baptiste Bourguinon d'Anville, whose particular interest was the compilation of historical maps of the ancient world (Barthélemy 1792, iv). The first map, of the Cyclades, was printed in 1758 (Barbié du Bocage 1792, pl. 30; Manne and Barbié du Bocage 1802, no. 51). The detail on this map was quite sparse; overall, it appears as if there were insufficient data even for d'Anville to work with, and the project languished. When Choiseul-Goufier returned in 1779, he gave the task of arranging his collection of topographical drawings, landscape views, and architectural plans to d'Anville's star pupil, Jean Denis Barbié du Bocage (Dacier 1831). At about the same time, d'Anville also passed responsibility for Barthélemy's maps to his pupil.[4] Working under Barthélemy's close supervision, the young geographer used Choiseul-Goufier's bounty to complete d'Anville's maps of the Black Sea and the Hellespont and then to create a series of brand new maps and images (Barbié du Bocage 1792, 7).

The core rhetoric of *Voyage du jeune Anacharsis* was that it did indeed present the truth of ancient Greece. To do this, Barthélemy relied on the rhetoric of empowered vision which permeated the Enlightenment's geographical narratives. Except for one segment when Anacharsis himself acts as our guide through Athens, he was always in the company of knowledgeable locals whose information could be guaranteed. But Anacharsis is also accorded all the power and dispassion of Enlightenment vision. In chapter 1, Barthélemy leads Anacharsis across the "vast solitudes" of the steppes (Barthélemy 1792, 1:235), where there is nothing worth seeing, to the Black Sea. There, Anacharsis encounters his first civilized town, which demands to be examined. Anacharsis reveled in his newfound powers of observation and in his access to knowledge that knew no social or physical boundaries:

> I was incessantly in motion, and could never satisfy myself with viewing the citadel, the arsenal, the harbour, the vessels, their rigging, and manner of working; I entered at random into private houses, manufactories, and the most inconsiderable shops; I went out of the town, and my eyes were fixed on orchards, covered with fruit, and fields laden with rich harvests. My sensations were lively, and my narratives full of animation. (Barthélemy 1792, 1:236–37)

Similar sentiments mark Anacharsis's arrival at Athens (Barthélemy 1792, 1:293). Observation is unproblematic: preexistent objects in the world present themselves to view and are thus known. The narrative, in listing the factual results of continual observation, is a celebration of the era's visual empiricism and its certainty that truth can indeed be *seen*.

MATTHEW H. EDNEY

The empiricist textual representation of the geographer's empowered vision was intensified through the use of graphic views of, in this case, ancient Greece. That Barbié du Bocage could never have gone to ancient Greece to see the views that he engraved was not a problem; indeed, Barthélemy himself had only ever traveled as far as Rome (in 1753–57). The plans and views collected by Choiseul-Goufier were thus central to the final product. Barbié du Bocage engraved six of Foucherot's architectural plans of ruined buildings that he took to be the types of Athenian architecture in the fourth century B.C.E. Barbié du Bocage engraved several of the topographic plans made by Choiseul-Goufier's engineers, celebrating as he did so the power of Enlightenment's highly focused and localized vision. He also combined large-scale plans with landscape views to make explicit the manner in which the landscapes had been directly examined in the field (fig. 6.1).

Barthélemy and Barbié du Bocage did not merely copy Choiseul-Goufier's fortuitous bounty. For the topographic plan of the strategic pass at Thermopylæ, Barbié du Bocage used not only Foucherot's observations (Barbié du Bocage 1792, 35) but probably several other images collected by Barthélemy.[5] Most important, Barbié du Bocage created his own images. In one view, Plato is seen "discoursing to his disciples," including Anacharsis, on a promontory on the coast of Attica. Using the cartographic conventions established for large-scale survey maps (Picon 1992, 211–55), Barbié du Bocage created *essais topographiques* of the festival site at Olympia and the battleground of Platæa. The image of such maps proclaimed their basis in measurement and factual observations, yet both were "designed entirely from Ancient Authors" (Barbié du Bocage 1792, pls. 22 and 4); antiquarians did not survey Olympia, with its stadium and the temple of Jupiter, until 1787, which was too late for the results to be incorporated into the maps (Barbié du Bocage 1792, 17; Constantine 1984, 188–209).

Barthélemy's final strategy for grounding Anacharsis's narrative in truthful vision was to ensure that Barbié du Bocage's maps were explicitly tied to the chronology of the travels. In the sense of the narrative, in which each moment of observation is equated to a specific moment in the linear space-time of travel, Barbié du Bocage included Anacharsis's routes on his small-scale maps. In the historic sense, Barbié du Bocage took great pains to construct each map "for the period when Greece was free" and more especially for "the very year in which Anacharsis is supposed to have traveled through the province it represents" (Barbié du Bocage 1792, 49 and 51). For example, Barbié du Bocage used a special symbol on the map of Arcadia for those cities that had been destroyed, and their inhabitants forcibly removed to Megapolis, just fifteen years before Anacharsis would tour the region (Barthélemy 1792, 3:55).

Barbié du Bocage's maps served to connect the textual and visual explication of places to the larger context of extensive geographical space. The general maps provide the reader with that larger spatial framework within which to select and assemble the details of particular places. Barthélemy assumed that anyone who was sufficiently educated to read and appreciate his work would also be able to read a map. Within the narrative itself, Anacharsis is taught the geography of the Black Sea from a map drawn by a ship's captain. He subsequently learned the basic layout of Greece, its islands, and its colonies in the same way. Barthélemy expected his readers to follow suit, to learn the overall framework by relating the discussions in the text with the graphic images he provided (Barthélemy 1792, 1:237–38, 1:259–62).

The interdependency of the different representational strategies within Barthélemy's work is most clear in the description of Athens, a description enabled by the extensive surveys of the city's antiquities made by Europeans over the preceding century. The description is in two parts, separated temporally by almost the entire period of Anacharsis's stay in Greece, but both are explicitly cartographic in their explication. Immediately upon his arrival in the city, Anacharsis undertook a "rapid view of the curiosities it contains" (Barthélemy 1792, 1:302) and of its general layout, pushing out from the citadel (the Acropolis), to the city, to the harbors, and to the surrounding countryside (Barthélemy 1792, 1:292–302). Ultimately, this description becomes sufficiently broad that the scale of description changes and becomes an account of Attica. Once he had satisfied his immediate "delirium" (Barthélemy 1792, 1:293), Anacharsis engaged in a more leisurely exploration of the famous academy in the northwestern suburbs of the city and of the great thinkers then teaching there, notably Plato and Aristotle (Barthélemy 1792, 1:303–17). This too warranted its own map to aid the reader's comprehension of the description of the academy's topography.

In the second and far more detailed description of the city's layout, Anacharsis assumes the role of tour guide through the city as it was during the later years of his sojourn there. He has been resident long enough to have become one of the local cognoscenti, and the reader is thrust into the role of diligent pupil. Anacharsis otherwise occupies this role throughout the rest of the text but is here a passionless docent remarking on items of interest to the museum visitor. The tour begins with landfall at the harbor of Piræus and quickly establishes its instructional, and highly visual, tone: "Before we set foot on shore, let us cast our eyes on the neighboring promontory.... Behold those vessels that are arriving, those that are preparing to depart, or are already under sail" (Barthélemy 1792, 1:364–65). Once within the city, the reader is guided by the numbers on the accompanying map, which Barthélemy designed himself (fig. 6.2). Anacharsis guides us first to the Royal Portico and thence, via the street of the Hermæ, to the forum. After a circuit of the forum, we are led southward, past the Temple of Theseus, and then into a circuit of the base of the citadel. While the first description sent us spiraling outward from the city, this brings us ever inward toward its heart: the Acropolis, with its many shrines to the goddess Minerva. We mount the steps to the citadel, past the grotto of Pan, to the gateway to the Acropolis. Our guide points out several shrines and monuments before allowing the account's literal high point: "Your eyes have long been turning towards that famous temple of Minerva...," the Parthenon. From this seat of Athenian power, Anacharsis finally redirects our view outward from a series of vantage points to show the broad sweep of the city, its suburbs, and its adjacent country. And he concludes with a quotation from a comedy by Lysippus:

> Whoever does not desire to see Athens is stupid; whoever sees it without being delighted is still more stupid; but the height of stupidity is to see it, to admire it, and to leave it. (Barthélemy 1792, 1:386)

The readers have been shown the glory of Athens; presumably they are delighted, unless of course they are stupid; and now, equipped with this proxy vision, readers can revisit Athens—or at least an Athens constructed by the orientalists and philhellenes—to their hearts' content.

Sitting at their desks in Paris, Barthélemy and Barbié du Bocage thus created a geographical vision of the last days of the glory that had been Greece. They did so through the figurative use of the empowered vision wielded so effectively by Europeans in their attempts to understand and control both their home territories and their colonies. The rhetorical, visualist claims made by the text functioned through an appeal to the perspective view and its supposedly truthful presentation of places. At the same time, both were concerned with a larger conception of Greece as a territory. The uniqueness of

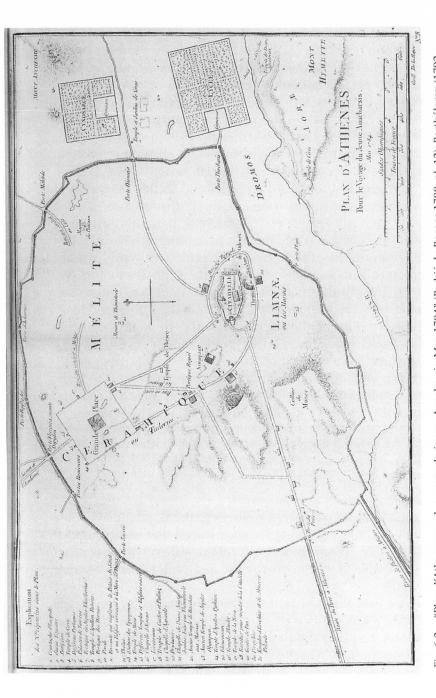

Fig. 6.2. "Plan d'Athens pour le voyage du jeune Anacharsis. May 1784" (Barbié du Bocage 1788, pl. 12). Barthélemy (1792, 1:489–500) described the construction of this plan by combining information from ancient authors, especially Pausanias, who had described his own tour of Athens, with topographical data collected by recent travelers. Barthélemy took responsibility for any errors of interpretation. Barthélemy also constructed the *essai* of Sparta (Barbié du Bocage, 1792, pl. 25; Barthélemy 1792, 2:509–12). Reproduced with the permission of Harvard University.

place—of Delphi dedicated to Apollo, of Olympia to Jupiter, and of Athens to Minerva—which was remarked in the view and summarized in the large-scale map, was eventually subsumed beneath a monolithic conception of Greece as a space of intellectual fascination and ideological legitimation manifested in the small-scale regional map. The range of images found in works such as Barthélemy's *Voyage du jeune Anacharsis* are not mere illustrations, graphics to improve the beauty of a work or to ease the mind of the dull reader, but are all integral to the construction of place and space. They simultaneously advocated the empiricist understanding of the world even as they provided the intellectual tool whereby those facts could be amalgamated with each other to form *knowledge*.

Like any reconnaissance, Barthélemy's *Voyage du jeune Anacharsis* should be understood as an orderly examination and arrangement of historical sources and contemporary travel reports. Barthélemy sifted through, judged, and rationalized the sources available to him. "I have examined and discussed them all before I have made use of them," he wrote in his *advertisement*. "I have even, on a revisal, suppressed a great part of them, and ought perhaps to have suppressed still more" (Barthélemy 1792, 1:xii–xiii). The contemporary geographical traveler had the same task of arranging the repeated observations of landscapes and peoples, of accepting or rejecting the spoken testimony of interviewees, and of relating his own tour to those of previous travelers. In this regard, the practice of reconnaissance entailed the same rational processes as those that are usually taken to characterize the practice of *mapping* and the creation of more analytical geographical descriptions. As such, it was common for the reconnaissance account to slip into a more systematic description, as when Barthélemy provided a detailed description of Attica not as a single narrative, as with other districts, but as a digest of Anacharsis's numerous visits to the country houses of Athenians (Barthélemy 1792, 3:181–213). As already noted, the difference between reconnaissance and mapping lay in their respective structuring of objects in space, between the overtly empiricist narrative of reconnaissance and the overtly rational analysis of mapping. Reconnaissance might thus be considered as the inverse of mapping, the second practice that fleshed out the geographical archive.

6.6. Mapping

The *Oxford English Dictionary* records several metaphorical usages of "map" during the seventeenth century, but one had come to dominate by 1700 and it has been prevalent ever since. This metaphorical understanding of "map" is the drawing together and presentation of the relations and hierarchies between phenomena. The space in which these phenomena are related was initially geographical space. At the start of the century, for example, Cotton Mather presented a nongraphic, "ecclesiastical map" of New England that comprised a hierarchical division of the region by colony, county, and then ministry (Mather 1702, 27–28). By the middle of the century, the tie to geographical space was increasingly irrelevant, and we can find innumerable references to abstract knowledge spaces, or "maps." Linnæus, for example, referred to his taxonomic system as a "map of nature" and Diderot and d'Alembert called their *Encyclopédie* "a kind of world map." That is, "mapping" was understood to be the essence of knowledge creation through the reconciliation of differences between observations and the laying out of the relations between individual pieces of data in a manner directly analogous to the making of a map (Edney 1994b). Mapping was the core of Enlightenment's encyclopedism; the map artifact was the key in validating Enlightenment's epistemological ideology.

Contemporary map making constituted an extended exercise in reasoned argument. The geographer progressively added data into the framework of the graticule. Only a few locations were known by observed latitudes and longitudes at the middle of the eighteenth century. Tobias Mayer published a *mappa critica Germaniae* in 1753 to advertise the lack of firm, reliable geographical locations in central Europe and to demonstrate the resultant variability of geographical outlines. Of the two hundred towns in his map, there were observed latitudes for only 33 and observed longitudes for none (Forbes 1980, 63). Like all other geographers, Mayer had to define the positions of the towns from reconnaissance accounts, either by geodetic calculations or by plotting routes between known points in the graticule. In either case, the geographer had to make many informed decisions about the quality and relative importance of each source in order to reconcile conflicts. Significantly, any conflicts could always be reconciled. Map making was thus a process of establishing the relations —the routes—between known entities and then using those relations to interpolate new entities. D'Anville referred to this process, when carried out carefully and rigorously, as "positive geography"

(Pedley 1995, 94). It accordingly served as a powerful analogy for the general processes of knowledge acquisition and as a convenient validation of Enlightenment epistemology.

The concept of "mapping" was enshrined in the memoirs written by map makers of how they teased geographic knowledge out of the mass of conflicting data. It is tempting to pursue the modern position and argue that a map's validity in the eighteenth century rested on its graphic plainness and its obviously scientific character. But the graphic map displayed only the unambiguity of geographical knowledge. Certainty and faith in a map's representation, in part or in whole, could only come from knowing the character of its data and the circumstances of its construction. As one president of the Royal Society of London had stated late in the seventeenth century, a map of eastern Asia that had been sent to the society could be interpreted only as being the product of "magic" unless some explanation was given of its sources and of the manner of its construction (Southwell 1687). At the other end of the eighteenth century, an anonymous reviewer accused James Rennell, prominent geographer and Fellow of the Royal Society, of being an "adept" of an "occult science" for turning "conjecture into certainty" in making his maps of western Africa; Rennell's failing was in not providing an explanation of how his maps were made (Anonymous 1802, 135). The particular concern of geographers was that the process of mapping entailed reasoned thought and that in turn depended on each individual's intellectual abilities. Couching his argument in general terms, John Green wrote that

> If the same materials be put into the Hands of twenty different Persons to [be worked] upon, each of them will contribute more or less, according to his Abilities, to improve the Art or Science. Some will strike Light out of bare Hints; while others will make but bad Work, though furnished with the best Help. (Green 1755, 3–4)

The purpose of the memoirs was therefore to explain how geographers arranged and combined their data in order to distinguish between geographical knowledge that was simply unambiguous and that which was *certain*. Without the memoirs, geographical knowledge was invalid.

Geographical memoirs were produced in particular by geographers as a means of legitimizing their intellectual pretensions; by the later eighteenth century they had become the hallmark of the superior geographers who were patronized by governments and who had claims to social gentility (Edney 1994b). Only a very few maps were ever accompanied by a separate memoir; the vast majority, however, featured abbreviated statements of their sources. Geographers noted their sources directly on the map, listing names, dates, and perhaps abbreviated titles of publications; usually these appear as marginalia but were sometimes engraved as small text blocks in the respective areas of the map. The larger and more expensive maps, maps which might be expected to have been constructed through a critical process but which lacked a memoir, often bore "comparative tables" of the various latitudes and longitudes which had provided the basis for the geographer's compilation. John Mitchell, for example, used several strategies for his 1755 map of eastern North America. An official statement certified that the map was based on data and special surveys collected by the Board of Trade and Plantations. Plentiful notations described the geographic character of particular locations and regions and assessed the competing territorial claims between English colonies and between the English and the French. Furthermore, Mitchell added large blocks of engraved text—a minimemoir—to the second edition (ca. 1757) in response to Green's (1755) criticism of Mitchell's sources (Edney 1997b).

Smaller and cheaper maps often announced the names of the prestigious geographers from whose works they had been copied and simplified. At the very least their makers would claim to have made the maps from the "latest" or "best" authorities. Such statements on geographical hackwork usually elicit snide remarks from historians because the map's crudeness of execution and sparseness of detail stand in stark contrast to the geographer's apparent claim to have undertaken the same careful process of evaluating all of the available data sources as was followed by the most respected and scholastic geographers. The statements are accordingly dismissed as mere commercial puffery. The geographers' claims should perhaps more properly be understood as brief announcements not that they had meticulously undertaken their own compilations but that they had derived their data from other geographers who had. That is, even the humblest hack geographers of the second half of the eighteenth century produced their maps with reference, no matter how far removed, to the larger discourse of epistemological legitimacy.

Geographical memoirs that outlined how the locations of specific places were determined, region by region, were functionally indistinguishable from the works which historians have usually considered to be geographical accounts, specifically those which enumerated and described places, region by region. Even those memoirs

MATTHEW H. EDNEY

which were highly focused on the details of map production, such as Barbié du Bocage's (1788) "critical analysis" of his maps of ancient Greece, necessarily engaged in geographical description. The larger memoirs were produced with the overt intention of serving as geographical descriptions. I think here particularly of Rennell's various *Memoirs of a Map of Hindoostan* (1783–93; see Edney 1997a, 96–104). We might also consider Lewis Evans's *Analysis* of his 1755 map of the "middle British Colonies." Evans had been helped in preparing both map and analysis by Thomas Pownall, who subsequently expanded the entire work. Pownall extended the geographic coverage to include New England, and he inserted a large amount of geographical description into the analysis, which he published in 1776 as *A Topographical Description of Such Parts of North America as are Contained in the Map of the Middle British Colonies* (Pownall 1949).

Pownall's title made explicit the mutual dependency of text and map in systematic geographical description. The map's dependency for its validation on textual explication is clear. Conversely, maps were essential to textual description. The map epitomized the rational process—mapping—by which the geographer assembled a large and disparate mass of information to create a coherent and comprehensive text. Most "special" geographies, which enumerated the world, region by region, accordingly celebrated maps and mapping in their opening chapters. The map gave spatial structure to the physical and social spaces of the world (see Edney 1994a), simultaneously arranging them in a spatial web and hierarchy of well-defined regions and subregions; it thereby permitted the relation of each section of the geographical text with all others. The map, literal and figurative, thus brought meaning to the text. This was the case both for the large "bibliographic dinosaurs" (Downes 1971) and for the numerous geographical *crambe recocta* that were little more than prose gazetteers (Bowen 1981, 144–54). It was the case both for the special geographies and for the more particular descriptions of specific regions. In this regard, Sitwell's otherwise excellent bibliography of English-language special geographies is marred by his inattention to their constituent maps (Sitwell 1993).

I do not mean to suggest that geographical works necessarily included both text and maps, as well as perhaps other graphic and numeric representations. The different mechanics and physical formats of intaglio and letterpress printing lent themselves to separation and so to commercial specialization. Notably, geographical memoirs were published as books distinct from their respective maps, no matter how the two coincided in their chronology and marketing.

Guild restrictions in *ancien régime* Paris enforced the separation of the different print technologies, with a significant impact on the character of geographical publishing (Pedley 1995). Furthermore, the complexity of incorporating maps into books, together with the high cost of engraving compared with typesetting, meant that liberally illustrated geographies were too expensive to be bought by a large readership. All in all, the actual production of geographies in the eighteenth century entailed a wide range of formats, from separately published maps and textless atlases, through complex interweavings of maps, graphics, and texts, to books and pamphlets unadorned by any prints.

Yet all of these different combinations of representational strategies functioned as a whole. Even if a geographical book lacked maps, those maps were nonetheless available and provided the geographical reader's intellectual framework. Maps were read in the context of the geographical texts available to a literate society. All the different geographical accounts constituted a vast, intertextual discursive field that was the geographical archive. The practices of which the archive was comprised—reconnaissance and mapping—were both empiricist in their enumeration of places; both were highly idealized in their representations; both entailed the reasoned manipulation of data. They are distinguished by their structuring of space and their representation of observation. The accounts of reconnaissance, with their large-scale plans and landscape views, emphasized the geographical traveler's movement through and observation of the world; the systematically constructed smaller-scale maps and texts emphasized the structured and coherent space of the world and the interrelations of phenomena within that space. Reconnaissance and mapping together replicated the world as a conceptual museum that could be indefinitely expanded to accommodate ever more extensive and intensive geographical information. In doing so, they epitomized, and legitimated, Enlightenment's encyclopedic epistemology.

6.7. The Geographical Archive Reconfigured: Enlightenment to Modernity

Despite the claims to truth with which Enlightenment ideology invested the geographical archive, the conceptual unity of that archive was undermined by the unavoidable messiness of the archive's actual implementation. The archive contained a wide range of materials housed in many different physical locations. Access to those materials was highly variable, ranging from widely available published texts to tightly controlled manuscript collections. These constituent materials could wear out, or they were discarded, mislaid, and forgotten. Ultimately, the quality of the archive was only ever as good as its indexes, and they ranged in quality from an individual's fallible memory to thorough catalogs. Moreover, eighteenth-century geographical technologies were unavoidably limited. Although Enlightenment's ideology claimed that the archive could successfully incorporate large-scale topographical surveys, the observation of latitude and longitude was still too imprecise for this actually to be the case. This would not have been a problem as long as large-scale surveys were highly focused on particular places scattered across a country. Yet a process adjunct to the increasing sophistication of Europe's military-fiscal states, and of their attempts to secure more complete and rational control over their own territories, was the mapping of those territories at ever larger scales and across ever more extensive areas. Such surveys could not be properly integrated within the archive's existing control mechanism of the graticule.

The more extensive of the new surveys, such as France's famous Cassini surveys, employed a highly precise technology for controlling and reconciling geographical observations: triangulation. The mesh of triangulation presented a far better fit to a landscape than the graticule ever could. The location of the triangle vertices—the control points—was determined by the lay of the land. Triangulation provided a far greater density of control points than observations for latitude and longitude ever could, and so could manage even the largest-scale survey. And triangulation could be carried over an entire country. It functioned simultaneously at the smallest and largest of scales. Triangulation thus represented a dramatic extension of state power, reaching deep down into different landscapes all across a state's territory. It was a far more efficacious technology than the older mapping could ever be. Triangulation's results were not only unambigious, they were *certain*.

Established as a technological fix for Enlightenment's epistemological ideals, triangulation engendered a significant shift in cultural conceptions of space and geography. Enlightenment encyclopedism, the rational reconciliation of conflicting viewpoints, was reconfigured to a new ideology of systematic and disciplined observation. The stress of the new, modern ideology was on correct and proper *observation* rather than on Enlightenment's correct and proper *reason*. Yet it was by no means easy for European states to implement triangulation-based surveys in the proper and correct manner needed to fulfill modernity's scientistic faith. Triangulations are extremely laborious and expensive, and they require substantial institutional and fiscal resources that have rarely been completely forthcoming. Although European states initiated triangulation-based topographic surveys of their territories in 1790–1820, they did not completely commit themselves to the technology until at least the 1860s, if not later. Nonetheless, the practical problems did not prevent the promulgation of the new spatial ideology after 1790.

The development of this modern conception of geography can be traced in a variety of post-1800 developments. Reconnaissance was falling out of favor as a strategy for geographic description as early as 1807, when critics objected to Francis Buchanan's extensive narrative account of Mysore, in southern India, precisely because of its narrative structure (Edney 1997a, 78). Those objections indicated an initial acceptance of Modernity's distinction between data acquisition (field/laboratory) and data analysis and presentation (office). With certainty ensured by the observational methodologies, the need for memoirs to demonstrate the certainty of geographical knowledge declined. Geographical memoirs were replaced by a new genre of text that laid out the mathematical calculations of each triangulation network; maps and texts were increasingly validated by their rhetoric of innate plainness and factuality. The popular texts on using globes and maps also disappeared in the early 1800s, as the graticule ceased to be the defining feature of maps. In a rapid reformulation of geographic pedagogy in about 1800, maps were suddenly recast as "pictures" of the landscape. No longer understood as abstractions of reality, maps were instead taken to be realist replications of each small portion of the world. The sequence of geographical primers also changed. Whereas in the 1700s they had begun with the globe and its circles of longitude and latitude, and had then proceeded to world and regional maps, very soon after 1800 they began by idealizing maps as large-scale topographic depictions of small extents of land, from which were eventually derived the small-scale maps of regions and the world (Andrews 1996). The Enlightenment's global fetish was torn down and replaced by Modernity's reductionist conviction that its knowledge of the world is detailed, particular, and exact. Significantly, it is only after 1800 that we find commentators actively engaging with the idea of maps constructed at the scale of 1:1.

Modernity's ideology has turned the two main representational strategies for geographical description away from each other. The geographical text and geographical map were increasingly distinguished in geographical discourse as maps and texts were each invested with their own rhetoric of exactness and totality. The success of the new ideology is evident in the mid-nineteenth-century recognition of a separate and autonomous discipline of "cartography." Enlightenment is central to modern cartography's self-definition as a "science" because it is then that "art" was apparently purged from maps, thereby freeing cartography's pristine, scientific core to develop and to progress. Much work remains to be done—in studying geographical memoirs, in elucidating geography's representational strategies, and in examining the survival of Enlightenment's geographical practices even into the twentieth century—but it should be clear that the historical construction of an art-to-science progression is quite misleading. Enlightenment geography was not the necessary precursor to modern geography; nor were the widespread adoption of triangulation and the intensification of Enlightenment's spatial ideology the "natural" outgrowths of Enlightenment. Instead, Enlightenment geography was a complex and dynamic intellectual activity which employed multiple representational strategies and spatial constructions to enumerate and describe the world. Complete understanding of that geography requires that we pay attention to all of those strategies, to both narrative and analytical descriptions of space, and to the ideal of the geographical archive.

Notes

A preliminary presentation of some material was delivered to the College Art Association, San Antonio, 28 January 1995. I would like to thank Joe Conforti, Denis Cosgrove, David Friedman, Anne Godlewska, Mike Heffernan, Owen Kelly, John Krygier, David Livingstone, Harold Osher, Kathy Tremper, Charlie Withers, and the anonymous reviewers for comments on and for discussions of this paper and related ideas.

1. "Graticule" is preferable to "grid" because it avoids the incorrect presumption that "projective" and finite cartographic space was, in the eighteenth century at least, infinite and somehow both Cartesian and Euclidean in nature.

2. *Oxford English Dictionary,* "Reconnaissance," definition 2b: "A survey, inspection, etc., made in order to gain information of any kind." I use the term because it lacks the problematic connotations of "voyage of discovery," "exploration," or "survey," and because its semantic openness encompasses the many different narrative forms in which geographical information has been recorded.

3. Citations are made to the 1792 English translation. With respect to the volume of graphics, page numbers refer to the preface, plate numbers to the

graphics; note also that the translation is actually quite erroneous concerning the technical details of map construction (Barbié du Bocage 1792, 9–10). In some editions (such as the 1792 German edition), the plates were tipped directly into the text volumes at the proper place.

4. Tolias (1990) is an incomplete overview of Barbié du Bocage's activities in historical geography.

5. Several plans and views of the pass are found in a collection of Barbié du Bocage's materials, now British Library Additional MS 15,326.

References

Adams, Percy G. 1962. *Travelers and Travel Liars, 1660–1800.* Berkeley: University of California Press. Reprinted, New York: Dover, 1980.

Andrews, J. H. 1996. "What Was a Map? The Lexicographers Reply." *Cartographica* 33, no. 4: 1–11.

Anonymous. 1802. "*The Journal of Frederick Hornemann's Travels,* from Cairo to Mourzouk, the Capital of the Kingdom of Fezzan, in Africa, in the years 1797–8. London: Nicols. 1802." *Edinburgh Review* 1, no. 1: 130–41.

Augustinos, Olga. 1994. *French Odysseys: Greece in French Travel Literature from the Renaissance to the Romantic Era.* Baltimore: Johns Hopkins University Press.

Avery, Bruce. 1995. "The Subject of Imperial Geography," in *Prosthetic Territories: Politics and Hypertechnologies,* edited by Gabriel Brahm, Jr., and Mark Driscoll. Boulder, CO: Westview Press, pp. 55–70.

Badolle, M. 1926. *L'abbé Jean-Jacques Barthélemy (1716–1795) et l'hellénisme en France dans la seconde moitié du XVIIIe siècle.* Paris: Thèse Paris.

Barbié du Bocage, Jean Denis. 1788. *Receuil des cartes géographiques, plans, vues et medailles de l'ancienne Grèce, relatifs au Voyage du jeune Anacharsis; précède d'une analyse critique des cartes.* With a separately paginated preface, "Analyse critique des cartes de l'ancienne Grèce, dressées pour le voyage du jeune Anacharsis." Paris.

———. 1792. *Maps, Plans, Views and Coins, Illustrative of the Travels of Anacharsis the Younger in Greece, during the Middle of the Fourth Century before the Christian Aera.* With a separately paginated preface, "Critical Observations on the Maps of Ancient Greece, compiled for the Travels of Anacharsis the Younger." London; 2d ed., 1793.

Barthélemy, Jean Jacques. 1788. *Voyage du jeune Anacharsis en Grèce, vers le milieu du quatrième siècle avant l'ère vulgaire.* 4 or 7 volumes. Paris.

———. 1792. *Travels of Anacharsis the Younger in Greece, during the Middle of the Fourth Century before the Christian Aera.* 4 volumes. London; 2d ed., 1793; 3d ed., 1796.

Bartram, William. 1791. *Travels through North & South Carolina, Georgia, East & West Florida, the Cherokee Country, the Extensive Territories of the Muscogulges, or Creek Confederacy, and the Country of the Chactaws; containing an account of the soil and natural productions of those regions, together with observations on the manners of the Indians.* Philadelphia: James & Johnson.

Bassett, Thomas J., and Philip W. Porter. 1991. "'From the Best Authorities': The Mountains of Kong in the Cartography of West Africa." *Journal of African History* 32: 367–414.

MATTHEW H. EDNEY

Belyea, Barbara. 1992. "Images of Power: Derrida, Foucault, Harley." *Cartographica* 29, no. 2: 1–9.

———. 1996. "Inland Journeys, Native Maps." *Cartographica* 33, no. 2: 1–16.

Bernal, Martin. 1987. *The Fabrication of Ancient Greece, 1785–1985.* Vol. 1 of *Black Athena: The Afroasiatic Roots of Classical Civilization.* London: Free Association Books.

Betts, Jonathan. 1996. "Arnold and Earnshaw: The Practicable Solution," in *The Quest for Longitude: The Proceedings of the Longitude Symposium, Harvard University, Cambridge, Massachusetts,* November 4–6, 1993, edited by William J. H. Andrewes. Cambridge, MA: Harvard University Press for the Collection of Historical Scientific Instruments, Harvard University, pp. 311–28.

Bowen, Margarita. 1981. *Empiricism and Geographical Thought from Francis Bacon to Alexander von Humbolt.* Cambridge: Cambridge University Press.

Büsching, Anton Friedrich. 1754–92. *Neue Erdbeschreibung.* 11 vols. Hamburg: C. E. Bohn.

Carter, Paul. 1987. *The Road to Botany Bay: An Essay in Spatial History.* London: Faber and Faber.

Casson, Lionel. 1974. *Travel in the Ancient World.* London: George Allen & Unwin. Reprinted, Baltimore: Johns Hopkins University Press, 1994.

Choiseul-Goufier, comte de. 1782. *Voyage pittoresque de la Grèce.* Volume 1. Paris.

Clarke, G. N. G. 1988. "Taking Possession: The Cartouche as Cultural Text in Eighteenth-Century American Maps." *Word & Image* 4: 455–74.

Constantine, David. 1984. *Early Greek Travellers and the Hellenic Ideal.* Cambridge: Cambridge University Press.

Crone, G. R., and R. A. Skelton. 1946. "English Collections of Voyages and Travels, 1625–1846" in *Richard Hakluyt & His Successors: A Volume Issued to Commemorate the Centenary of the Halkuyt Society,* edited by Edward Lynam. Hakluyt Society, 2d ser., vol. 93. London: Hakluyt Society, pp. 63–140.

Dacier, B. J. 1831. "Notice historique sur la vie et les ouvrages de M Barbié du Bocage." *Histoire et Mémoires de l'Institut royal de France, Académie des inscriptions et belles-lettres* 9: 132–45.

Defoe, Daniel. 1720. *The Life, Adventures, and Pyracies, of the Famous Captain Singleton.* London: J. Brotherton.

———. 1725. *A New Voyage Round the World, by a Course Never Sailed Before. Being a Voyage Undertaken by Some Merchants, Who Afterwards Proposed the Setting Up an East-India Company in Flanders.* London: A. Bettesworth and W. Mears.

Downes, Alan. 1971. "The Bibliographic Dinosaurs of Georgian Geography (1714–1830)." *The Geographical Journal* 137: 379–87.

Driver, Felix. 1992. "Geography's Empire: Histories of Geographical Knowledge." *Environment and Planning D: Society and Space* 10: 23–40.

Duncan, James, and David Ley. 1993. "Introduction: Representing the Place of Culture," in *Place/Culture/Representation,* edited by James Duncan and David Ley. London: Routledge, pp. 1–21.

Ebeling, Christoph Daniel. 1793–1816. *Erdbeschreibung und Geschichte von Amerika. Die Vereinten Staaten von Nordamerika.* Hamburg: C. E. Bohn.

Edney, Matthew H. 1993. "Cartography without 'Progress': Reinterpreting the Nature and Historical Development of Mapmaking." *Cartographica* 30, nos. 2 and 3: 54–68.

———. 1994a. "Cartographic Culture and Nationalism in the Early United States: Benjamin Vaughan and the Choice for a Prime Meridian, 1811." *Journal of Historical Geography* 20: 384–95.

———. 1994b. "Mathematical Cosmography and the Social Ideology of British Cartography, 1780–1820." *Imago Mundi* 46: 101–16.

———. 1996. "Theory and the History of Cartography." *Imago Mundi* 48: 185–91.

———. 1997a. *Mapping an Empire: The Geographical Construction of British India, 1765–1843.* Chicago: University of Chicago Press.

———. 1997b. "The Mitchell Map: An Irony of Empire." Web document, 21 April 1997. http://www.usm.maine.edu/~maps/mitchell.

Espenshade, Edward B., Jr., ed. 1995. *Goode's World Atlas,* 19th ed. Chicago: Rand MacNally.

Forbes, Eric G. 1980. *Tobias Mayer (1723–62): Pioneer of Enlightened Science in Germany.* Niedersächsische Staats- und Universitätsbibliothek Göttingen, Arbeit 17. Göttingen: Vandenhoeck and Ruprecht.

Foucault, Michel. 1970. *The Order of Things: An Archaeology of the Human Sciences.* New York: Random House.

Frängsmyr, Tore, J. L. Heilbron, and Robin E. Rider, eds. 1990. *The Quantifying Spirit in the 18th Century.* Berkeley: University of California Press.

Godlewska, Anne. 1995. "Map, Text and Image: The Mentality of Enlightened Conquerors: A New Look at the *Description de l'Egypte.*" *Transactions of the Institute of British Geographers,* ns 20: 5–28.

Goetzmann, William H. 1986. *New Lands, New Men: America and the Second Great Age of Discovery.* New York: Viking.

Green, John [*né* Bradock Mead; attrib.]. 1717. *The Construction of Maps and Globes.* London: T. Horne et alia.

———. [attrib.]. 1755. *Explanation for the New Map of Nova Scotia and Cape Britain, With Adjacent Parts of New England and Canada.* London: Thomas Jefferys.

Hall, Stephen S. 1992. *Mapping the Next Millennium: The Discovery of New Geographies.* New York: Random House.

Harley, J. B. 1987. "The Map and the Development of the History of Cartography," in *Cartography in Prehistoric, Ancient, and Medieval Europe and the Mediterranean,* edited by J. B. Harley and David Woodward, vol. 1 of *The History of Cartography.* Chicago: University of Chicago Press, pp. 1–42.

———. 1989. "Deconstructing the Map." *Cartographica* 26, no. 2: 1–20.

———. 1996. "Power and Legitimation in the English Geographical Atlases of the Eighteenth Century," in *Images of the World: The Atlas Through History,* edited by John A. Wolter and Ron E. Grim. New York: McGraw-Hill for the Library of Congress, pp. 161–204.

Harvey, David. 1989. *The Condition of Postmodernity: An Enquiry into the Origins of Cultural Change.* Oxford: Basil Blackwell.

Hassam, Andrew. 1990. "'As I Write': Narrative Occasions and the Quest for Self-Presence in the Travel Diary." *Ariel* 21, no. 4: 33–47.

Horkheimer, Max, and Theodor W. Adorno. 1972. *The Dialectic of Enlightenment*. Translated by John Cumming. New York: Herder and Herder.

Jacob, Christian. 1992. *L'empire des cartes: Approche théorique de la cartographie à travers l'histoire*. Paris: Albin Michel.

Kain, Roger J. P., and Elizabeth Baigent. 1992. *The Cadastral Map in the Service of the State: A History of Property Mapping*. Chicago: University of Chicago Press.

Laxton, Paul. 1976. "The Geodetic and Topographical Evaluation of English County Maps, 1740–1840." *The Cartographic Journal* 13: 37–54.

Lewis, G. Malcolm. 1991. "La grande rivière et fleuve de l'ouest: The Realities and Reasons Behind a Major Mistake in the 18th-Century Geography of North America." *Cartographica* 28, no. 1: 54–87.

MacIntyre, Alasdair. 1990. *Three Rival Versions of Moral Enquiry: Encyclopaedia, Genealogy, and Tradition*. Notre Dame, IN: Notre Dame University Press.

MacKenzie, John M. 1995. *Orientalism: History, Theory and the Arts*. Manchester: Manchester University Press.

Manne, L. Ch. J., and Jean Denis Barbié du Bocage. 1802. *Notice des ouvrages de M d'Anville, premier géographe du Roi . . . précède de son éloge*. Paris.

Mather, Cotton. 1702. *Antiquities. The First Book of the New-English History. . . .* Volume 1 of *Magnalia Christi Americana: or; the Ecclesiastical History of New England, from its First Planting in the Year 1620, unto the Year of our Lord, 1698*. 10 volumes. London: Thomas Parkhurst.

Nobles, Gregory H. 1993. "Straight Lines and Stability: Mapping the Political Order of the Anglo-American Frontier." *The Journal of American History* 80: 9–35.

Packe, Christopher. 1743. *ΑΝΚΟΓΡΑΦΙΑ, sive Convallium Descriptio. In which are Briefly but Fully Expounded the Origine, Course and Insertion; Extent, Elevation and Congruity of all the Valleys and Hills, Brooks and Rivers (as an Explanation of a New Philosophico-Chorographical Chart) of East Kent. Occasionally are Interspers'd some Transient Remarks that relate to the Natural History of the Country; and to the Military Marks and Signs of Cæsar's Rout thro it, to his Decisive Battle in Kent*. Canterbury, Kent: J. Abree.

Pedley, Mary Sponberg. 1995. "'Commode, complet, uniforme, et suivi': Problems in Atlas Editing in Enlightenment France," in *Editing Early and Historical Atlases: Papers Given at the Twenty-Ninth Annual Conference on Editorial Problems, University of Toronto, 5–6 November 1993*, edited by Joan Winearls. Toronto: University of Toronto Press, pp. 83–108.

Pickles, John. 1992. "Texts, Hermeneutics and Propaganda Maps," in *Writing Worlds: Discourse, Text and Metaphor in the Representation of Landscape*, edited by Trevor J. Barnes and James S. Duncan. London: Routledge, pp. 193–230.

Picon, Antoine. 1992. *French Architects and Engineers in the Age of Enlightenment*. Cambridge Studies in the History of Architecture. Cambridge: Cambridge University Press.

Pownall, Thomas. 1949. *A Topographical Description of the Dominions of the United States of America [being a Revised and Enlarged Edition of] A Topographical Description of Such Parts of North America as are Contained in the*

(Annexed) Map of the Middle British Colonies, &c. in North America. Edited by Lois Mulkearn. Pittsburgh: University of Pittsburgh Press.

Prochaska, David. 1994. "Art of Colonialism, Colonialism of Art: The *Description de l'Egypte* (1809–1828)." *L'Esprit Créateur* 34, no. 2: 69–91.

Raspe, Rudolph Eric [attrib.]. 1786. *Baron Munchausen's Narrative of his Marvellous Travels and Campaigns in Russia*. Oxford.

Regis, Pamela. 1992. *Describing Early America: Bartram, Jefferson, Crèvecoeur, and the Rhetoric of Natural History*. DeKalb: Northern Illinois University Press.

Roy, William. 1765. "Military Description of the South-East Part of England." London, July 1765. Unpublished MS. British Library, Map Library, C.7.d.12.

———. 1785. "An Account of the Measurement of a Base on Hounslow-Heath." *Philosophical Transactions of the Royal Society* 75: 385–479.

Ryan, Simon. 1994. "Inscribing the Emptiness: Cartography, Exploration and the Construction of Australia," in *De-Scribing Empire: Post-Colonialism and Textuality*, edited by Chris Tiffin and Alan Lawson. London: Routledge, pp. 115–30.

Said, Edward. 1978. *Orientalism*. New York: Random House.

Sitwell, O. F. G. 1993. *Four Centuries of Special Geography: An Annotated Guide to Books that Purport to Describe all the Countries in the World Published in English before 1888, With a Critical Introduction*. Vancouver: University of British Columbia Press.

Snyder, John P. 1993. *Flattening the Earth: Two Thousand Years of Map Projections*. Chicago: University of Chicago Press.

Southwell, Robert. 1687. "An Account of a Large and Curious Map of the Great Tartary, Lately Publish'd in Holland, by Mr. Nicholas Witsen, being an Extract of a Letter from the Author thereof, to the Honourable Sir Robert Southwell, Knt. and President of the Royal Society." *Philosophical Transactions of the Royal Society* 16, no. 193: 492–94.

Stone, Jeffrey C. 1988. "Imperialism, Colonialism and Cartography." *Transactions of the Institute of British Geographers*, ns 13: 57–64.

Swift, Jonathan. 1726. *Travels into Several Remote Nations of the World. In Four Parts. By Lemuel Gulliver, First a Surgeon, and then a Captain of Several Ships*. London: Benjamin Motte.

———. 1733. *On Poetry, A Rapsody*. Dublin.

Thrower, Norman J. W. 1996. *Maps and Civilization: Cartography in Culture and Society*. Chicago: University of Chicago Press.

Tolias, Yorgos. 1990. "The Cartographer Barbié du Bocage and the Approach to the Greek World in the Late 18th and Early 19th Centuries." *IMCoS Journal* 40: 5–9.

Turnbull, David. 1993. *Maps are Territories: Science is an Atlas: A Portfolio of Exhibits*. Chicago: University of Chicago Press.

———. 1996. "Cartography and Science in Early Modern Europe: Mapping the Construction of Knowledge Spaces." *Imago Mundi* 48: 5–24.

Walters, Gwyn. 1988. "The Antiquary and the Map." *World & Image* 4: 529–44.

Wolff, Larry. 1994. *Inventing Eastern Europe: The Map of Civilization on the Mind of the Enlightenment*. Stanford: Stanford University Press.

MATTHEW H. EDNEY

Chapter Seven
Ethnographic Navigation and the Geographical Gift[1]

MICHAEL T. BRAVO

The frontispiece seems to represent Navigation resting on a rudder, describing to History the route of Lapérouse attended by the Americans of the western coast . . . Painting is at the feet of History; and an attendant, resting on an anchor, appears to mourn the fate of the commander. The design is elegant and even beautiful.— The maps are some of the best that we have seen; and we do not perceive that any of them are carelessly or incorrectly executed.

Critical Review, New Series, vol. 26, 1799, 405

7.1. Introduction

The frontispiece to the English translation of Lapérouse's *Voyage around the World* (1785–1788) (fig. 7.1) describes the route of the *Astrolabe* and the *Boussole* which ended in the tragic loss of the captain, his ships, and most of the crew; the ships were last sighted off the Northwest coast of America and then vanished without a trace (Allen 1959; Chapin 1990; Gaziello 1984). The power of narration in this engraving resides in Navigation. It is she who speaks authoritatively about Geography, who has the privilege of knowing the truth of Lapérouse's fate, and who gives lessons to History. It is in the power of Navigation to make pronouncements about geographical knowledge.

The idea that geographical knowledge can be bequeathed as a gift is also meaningful in the context of enlightened encounters between explorers and their human Other, the peoples native to the places they visited. Whether one identifies Navigation or local people as the source of the geographical gift depends on who speaks for nature. The apparent autonomy and authority of Navigation and his geographical preeminence were derived from the institutional power of the King and Navy, the knowledge of the savants of the Academy of Sciences, and especially, the fine-precision surveying and astronomical instruments the navigators possessed as they voyaged around the world. In this sense, the navigators operated their ships and instruments like the rugged, reliable machinery that they in fact were (Bourguet 1997; Law 1987; Sorrenson 1996). However, during various landfall episodes on these voyages, the officers turned away from their instruments to consult the local people and to solicit geographical information from them in the form of hand-drawn sketches or maps (Belyea 1992; Lewis 1997; Mundy 1996). At such moments, cross-cultural encounters could break the repetitive oscillation between one's course by dead reckoning and the absolute location measured by one's instruments, to interject a distinctly different discourse. The episodes in which explorers solicited native maps, while making for rich ethnographic narrative, were deadly serious because, notwithstanding the many dangers to the welfare of the expedition, the more successful encounters enabled explorers to gaze ahead at the contour of coastlines yet to be "discovered." A few native informants like Omai, brought back to London from Tahiti by Captain Furneaux in 1774 (Smith 1985, 80–84), were fawned over in fashionable society, but the majority never left their own shores and remained anonymous, known to European readers only through a particular description of their physical features or a general description of their ethnicity.

This display of the geographical gift implied a highly structured sense of time: longitude, the geometric division of time and space, regulated by the movement of the planets and the sun, or alternately on the other hand, the stages of human development, variations on a teleological account of progress by enlightenment philosophes like J. R. Forster, Rousseau, and Monboddo. For Forster (1729–98), who had traveled with Cook on his second voyage (1772–75), to voyage through space also meant to travel through time, to see civilized society as it once was. This latter geographical narrative of the history of humanity, in principle, embraced and incorporated all peoples, so that the incommensurability of all societies could and would eventually be overcome through contact, study, commerce, and the fruits of civilization. This ideal placed geography at the heart of enlightened conceptions of humanity's history (Forster 1996; Smith 1985).

The relationship between navigation by instruments and by following native informant is seldom questioned. Terms like 'guide' and 'native' implied just that: local knowledge was supplementary, contingent, and unessential to geographical truth. The inadequacy of these terms and their hierarchical assumptions, besides subordinating the native point of view, also served to make encounters safe as a topic of ethnographic writing.

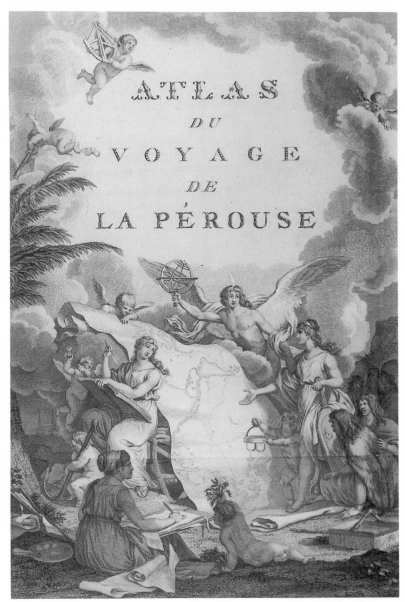

Fig. 7.1. Frontispiece to the *Atlas du Voyage de La Pérouse* (1797), drawn by J. M. Moreau le Jeune, engraved by Ph. Friere. Reproduced with the kind permission of the Cambridge Philosophical Society.

Yet the role of language in encounters ran much deeper than mere labels. For centuries language had remained central in European discourses about the problem of humanity and incommensurability, and whether the latter could be overcome in order to grasp the former. In fact, this process of appropriation was inherent in ethnographic description for all those observers who attempted to come to grips with radical cultural difference, among them, Columbus, Barchilon, Léry, Diderot, and Humboldt (Pagden 1993, 47–48). The act of possessing ethnographic knowledge, Pagden has shown, is, in general, preceded by a sequence of three actions: attachment, recognition, and naming. The initial stage, which Pagden calls the "principle of attachment," is to detach or dislocate the people's actions from the unfamiliar and to attach them instead to a familiar context. This prepares the ground for the act of recognition of their humanity, an act of dislocation which allows "the stark incommensurability of the two . . . to be, dissolved in the supposed common recognition" (Pagden 1993, 21). Pagden discusses this process of recognition and intelligence gathering as a journey in itself.

Pagden illustrates the sequence of steps which begins with the principle of attachment by borrowing from Bruno Latour's 'actor-network theory' (1987) the ideas of *material traces, immutable mobiles,* and *centers of calculation.* Latour presents a general model of the field sciences where the engine of capitalism situated at metropolitan centers archives and combines the many different recorded observations gathered by explorers and ethnographers at different times in different places. The power of the sciences, he argues, comes from this process of accumulation and combination. That being the case, there is a tremendous onus upon scientific travelers who visit far-flung places to ensure that their records survive the voyage and are intelligible to those at home. They must, therefore, be as immutable, portable, and combinable as possible: hence the term *immutable mobile.*

Any model of scientific travel must come to terms with both the journey into the field and the experience in the field itself. By borrowing Latour's model of cycles of capital accumulation, Pagden shows that ethnographic observations can be treated as resources possessing some unspecified currency or economic value. This allows him to change scale between local ethnographic dialogues and global systems of trade and colonization. The journey then becomes a matter of moving traces between the field and the center, between the culture of the observed and that of the observer. It allows Pagden to show the highly varied and subtle intellectual problems faced by

MICHAEL T. BRAVO

individual ethnographers attempting to overcome incommensurability, especially with regard to philosophical and practical problems of interpreting language, while emphasizing how much this enterprise shares in common with other collecting enterprises: minerals, plants, birds, fishes, and other living kinds. Conversely, Latour, who has a model of science in which the metaphors of journeys, exploration, and cartography figure prominently, and who is fundamentally concerned with materially objectified *inscriptions* or *traces,* needs a mechanism of appropriation, in the form of a gesture not unlike Omai's, which will give a sense of place and origin to the traces. In his discussion of exploration, Latour's mechanism is the native map maker, a "Chinese savage," who on the beach at De Langle Bay at Sakhalin in 1787 draws a map in the sand to explain to Lapérouse that Sakhalin is, indeed, an island (Latour 1987, 215–257). It is an enlightened gift where nature offers up her secret, a trace drawn in the sand. The reciprocal obligation of the explorer is to register wonder or astonishment and to offer tribute to his own patrons and savants by naming and returning with evidence of some kind such as a description or a specimen.

This relationship between navigation (space) and ethnography (time) or between civilized nations and the people of nature was a general foundation for much of what is taken as enlightenment (Fabian 1983; Outram 1995; Smith 1995). Nevertheless the relationship demands closer scrutiny. Was the dichotomy between navigation and surveying on the one hand, and the history of the nations encountered on the other hand, actually well founded in Lapérouse's discovery practices? Were they fundamentally different kinds of knowledge, the former realist, making reference to the substance of the natural world, and the latter socially and culturally relativist, ordering the worlds of diverse societies? It is a canon of enlightenment that the scientific imagination constructed a global time and space for the peoples of the New World: "never again would they be able to call either space or time their own" (Pagden 1993, 30). Notwithstanding local capabilities to resist colonial frameworks of domination, I wish to pose the question, how did enlightened navigators impose or distill time and space in the course of their encounters? Were the encounters essentially distinct, highly local, unrelated events, united principally by their common geographical substrate, as the frontispiece suggests?

I propose to elaborate on the concepts of the *geographical gift* and of *ethnographic navigation.* Enlightened explorers choosing to solicit local geographical knowledge, allowed for their navigation routines to be interrupted by these encounters. As far as possible, explorers adopted strategies to maximize the value of their geographical dialogues, while minimizing the risk. Central to these strategies was the offering of various kinds of gifts, with the hope of inspiring a sense of obligation or debt, as well as good will, on the part of those who received them (an idea I borrow loosely from Marcel Mauss's essay on *The Gift* [1925]). The role of acquiring elements of native languages proved crucial in these strategies and the ensuing encounters. Language played a key role in initiating relationships based on exchange. Navigators could use ethnographic descriptions and labels to interpret the meaning of native sketches, a process I refer to as the *geographical gift*. They could sometimes also orient themselves by mapping spaces in terms of ethnicities, a process I refer to as *ethnographic navigation*. This strategic use of gifts to solicit local time and assistance, and to interrupt one's own established instrument-based systems of navigation, added a complexity to the process of appropriation, and demonstrates how ethnographic inquiry organized spaces along the lines of ethnic groups, physical features, theories of climatic determinism, and degeneration.

Given the scope of this paper, I illustrate my argument with one case study, that of Lapérouse's ethnographic navigation of the Bay of Tartary in 1787 and the geographical question regarding whether or not Sakhalin is an island. I suggest that the argument carries some general force on the grounds that the institutional framework of Lapérouse's encounter was typical of an enduring tradition which began with the mid-century discovery voyages of Bougainville in France, Cook in Britain, and Malaspina in Spain, and carried on throughout the nineteenth century. However, the encounter strategies deployed varied between nations and over time and therefore, deserve further study. In order to demonstrate the key features of the *geographical gift* and *ethnographic navigation,* I discuss in some detail his string of encounters with the inhabitants of Sakhalin or "Tchoka" as its Ainu inhabitants knew it.

My discussion of Lapérouse's encounters in the Bay of Tartary, and of his staging of linguistic performances, also takes the form of a critique of Bruno Latour's analysis of the encounter at De Langle Bay. In historical terms, his account offers a different and competing reading to my own. Latour's interest in Lapérouse, to be fair, is not historical. Nevertheless, there is more at stake for each of us than the mere mechanics of interpreting drawings in the sand. Latour uses this encounter to illustrate his actor-network theory of science and technology, to emphasize the importance of the map as a metaphor

MICHAEL T. BRAVO

for the processes of capitalism inherent in the sciences: making stable, portable inscriptions. Given the importance of Latour's work and its widespread import into geography and other disciplines, a critique of its weakness as a vehicle for understanding encounters is warranted.

While subscribing to the importance of Latour's metropolitan perspectives on science and engineering, my theory is oriented from the perspective of the field encounter. My interpretation of Lapérouse's visit to Sakhalin shows how enlightened scientific practitioners established cross-cultural commensurability in their field sites, and how ethnographic inquiry could be a central feature of conducting research in the field sciences. I identify the field strategies adopted to produce the asymmetric relations that are commonly used to distinguish science from other local knowledge traditions. The journals of Lapérouse and the native map makers of Sakhalin permit us an examination of these two quite distinct theories, each of which signals the importance of cartography as a central metaphor for the sciences: Latour's, a theory of capitalization seen from the metropolis, and mine, a theory of linguistic commensurability and exchange seen from the perspective of the field. The study of Lapérouse at Sakhalin allows us to weigh their relative merits where they intersect.

7.2. The Ethnic Background of the Lower Amur River and Sakhalin

When Jean-François de Galaup Comte de Lapérouse (1741–c. 1788) embarked on the voyage in which he planned to sail round the world to the Bering Strait, where he would seek the entrance to a Northwest Passage to return to the Atlantic Ocean, he knew very little indeed about Sakhalin. The subject of much conversation among armchair geographers, the location and identity of Sakhalin was a geographical conundrum, a matter of dispute among cartographers and navigators, not only in England and France, but also in Russia and Japan. Was Sakhalin synonymous with the islands known by the Japanese as *Jeso* and *Oku-Jeso?* Which, if either, did the Japanese refer to as *Karafuto?* Was Sakhalin adjacent to either of these? Were they separated from the mainland by a strait named Tessoy? And if so, was Tessoy a navigable trade route, or merely a deep gulf sealed off at its northern extremity?

Cartographic sources supplied contradictory evidence on this question. In D'Anville's first edition chart (1735), "Saghalien," the most northerly of three islands, was placed at the mouth of a great continental river. A later (1752) edition of D'Anville's chart placed *Sahalien-Wa Hata* as the more northerly of a pair of islands, the third apparently vanishing (Stephan 1971, 34–35). Furthermore, his 1770 map, compared with his 1761 map, was judged by Milet-Mureau, the editor of Lapérouse's journals, to have given "geography a retrograde course" (Lapérouse 1799, II: 57). Other cartographers, like Robert (1767), Robert de Vaugondy (1775), Brion (1784), and W. de Lisle and Buache (1784), adopted one or another, and compounded the confusion. Lapérouse's own preliminary instructions remarked that the Japanese confused Jesso with the Kuriles (Lapérouse 1798, I: lxxxviii).

Lapérouse's instructions from the Academy of Sciences demanded that he attempt to solve a whole range of conundrums in a section called "Notes Geographical and Historic" (Lapérouse 1798, I: lxxx). Many of the geographical questions, including that of Sakhalin and the Kuriles, explicitly referred to previous sources and their limitations (though some references are excluded from this English translation). For example, one of the Kuriles, Siaskutan, is described as "very thinly peopled, but the inhabitants of neighbouring islands meet in it to trade" (Lapérouse 1797, I: 147; 1799, I: 57). They also speculated that Japanese cartographers mistakenly believed Jesso to be the name for all the Kuriles (whereas in fact it derived from the Japanese word *ezo* meaning 'foreigner' and referred to the frontier of Ainu lands to the North, which the Japanese named *Ezochi*). By the late eighteenth century Ezochi included East and West Ezochi on Hokkaido, Oku-Ezochi, and North Ezochi in the south of Sakhalin.

The identity of Sakhalin was in part an intellectual geographical and historical puzzle, a cartographic curiosity, but it was also much more than that. Situated at the frontiers of China, Russia, and Japan, it became the focus of political challenges to emerging state boundaries, and feudal trading interests at the peripheries (Siddle 1996; Stephan 1971). Although its strategic importance as a buffer zone waxed and waned in the eighteenth century, the (Bakufu) Japanese imperial government annexed the territory of Ezochi in the early nineteenth century, which included both the south of Sakhalin as well as Hokkaido (Siddle 1996, 39–40). This process was in part provoked by the extension of the Russian fur trade to the neighboring Kuriles, together with a number of surveys and raids. Russian (1783, 1805, 1806), French (1787), British (1797, 1804), and Japanese (1806, 1808–09) explorers or surveyors visited Sakhalin, mapping its rugged coast, but made little impact in its dense interior forests. Strong nationalist sentiment in Japan was perhaps triggered by a maverick Russian raid-gone-wrong, on Cape Aniwa in 1806. The level of tension increased with a Russian invasion in 1853, which led to joint occupation. One century after Lapérouse's survey, Paul Labbé dubbed Sakhalin "the Orient's 'Alsace-Lorraine'" (Stephan 1971, 42).

When Lapérouse arrived in the region, no state was in possession of either the lower Amur River or Sakhalin. An array of ethnic groups, practicing either hunting, fishing, or some combination of limited agriculture, populated this region. Today they call themselves "indigenous" peoples, though this should not lead one to ignore the importance of travel, trade, and migration to their ways of life. Certainly in political and economic terms, they were forced into different systems of tribute-based trade with their powerful Chinese, Russian, and Japanese neighbors (see Siddle 1996, 26–48; Stephan 1971, 19–48; and Forsyth 1992, 136–153, 201–219, respectively). The Chinese military presence in this northeastern periphery of Manchuria (just as many Russians in Moscow today conceive of Siberia as a gulag frontier) extended to the mouth of the Amur River and the adjacent northern area of Sakhalin, but not significantly further. As far north on the mainland as the upper Amur, they incorporated various Tungus, Dahur, and Nanai groups into their military ranks. Beyond that, Chi'ing dynasty authorities operated a system of tribute which demanded an annual payment of furs from an appointed elite of clan rulers (Forsyth 1992, 204).

The indigenous complex ethnic composition and movement in this region was complex and can only be described with limited ethnohistorical accuracy. The peoples of the densely populated lower Amur and the North of Sakhalin, in contrast to the more homogeneous groups who lived upstream, were ethnically varied and mixed. Most of these groups—the Udeghes, Nanais, the Ulchis, the Oroches, and the Oroks—belonged to the same Tungus family of tribes as the Evenks (Forsyth 1992, 103, 209) which collectively spans much of Siberia from the Pacific Ocean in the east to the Yenisey River in the west. In terms of kinship systems, these groups were exogamous, marrying outside their clans (ibid. 208). This together with the economic prosperity of the lower Amur, resulted in a great deal of intermarriage, mobility, and ethnic mixing. From an ethnohistorical perspective, the overlapping ethnicities mean that attaching eighteenth-century ethnographic descriptions to more recent ethnic labels is highly speculative. In fact, one of the principal aims of this paper is to show how ethnic labels were attached, and yet were tentative and fluid, to accommodate subsequent ethnographic observations.

The group of Oroches, who were to meet Lapérouse in the course of his survey on the coast at their summer residence, were hunters of wild reindeer, inhabiting a relatively small area between the lower Amur and the coast of the Bay of Tartary. They, like the Oroks who had migrated over the shallows to Sakhalin, were small subgroups of the Orochon or Reindeer-Tungus family. (The term *Orochon* is in fact the Evenk term for 'reindeer people,' which may also refer to the domesticated, saddled reindeer that many Orochons would ride for hunting.) Oroches made planned moves according to the seasons and movements of the wild reindeer which provided them with food and clothing (Forsyth 1992, 206; Kwon 1998, 116–117).

Sakhalin, the large island at the mouth of the Amur River, was home to indigenous groups—Ainus, Oroks (Forsyth 1992), and Nivkhs (Munro and Seligman 1996; Siddle 1996). Ainus inhabited the Kurile archipelago of islands, stretching from Kamchatka to the North of Japan, as well as southern parts of Sakhalin. Those whom Lapérouse encountered on the East coast of the Bay of Tartary (he neither knew of nor met Oroks) lived both around the Amur estuary (to the northwest) and the north of Sakhalin. They spoke an entirely distinct language from their Tungus neighbors and physically appeared quite distinct in spite of the relative permeability of their society. They lived a seasonal life, based almost exclusively on hunting marine mammals and fishing, traveling in small boats and canoes (Forsyth 1992, 207–211).

Most visitors to Sakhalin, including Lapérouse, were surprised by the distinct physical appearance of Ainu, most notably their ample body hair, though this met with varied responses. At the hands of Japanese merchants, the Ainu suffered a long legacy of extreme racism, which has continued into the present day. The Japanese generally treated them as a dehumanized other, and labeled them as demonic, barbarous, 'dogs' (Siddle 1996, 42–48). Japanese attitudes towards Ainu in the late eighteenth century fell into two categories. The Bakufu (government), who periodically sent expeditions to inspect their territory, envisaged Ainu as potential allies in their service, and adopted a 'civilizing mission' ideology in the belief that Ainu could acquire Japanese manners, language, and education, and thus be made a part of Japanese sovereignty. However, the matsumae merchants who had been exacting tribute from Ainus since

MICHAEL T. BRAVO

the beginning of the seventeenth century and had been granted a monopoly over *basho* (trading territories) became reliant upon harnessing Ainu labor for fishing and the large-scale production of fish fertilizer. The matsumae set out to control the Ainu's *Santan* trade to the Amur River, as well as their own in the south. To what extent the Ainu's violent resistance to matsumae penetrated as far north as Sakhalin is unclear. Nevertheless, the invocation of demonic stereotypes and the extreme ill regard for Ainu, is well established, especially by illiterate *dekasegi* seasonal migrant workers who worked alongside them as laborers or overseers (Siddle 1996, 32–44).

From the perspective of European powers, access to Japan in the eighteenth century was highly regulated and restricted to Nagasaki. The location of good harbors and natural resources in neighboring areas, near but outside the sovereign boundaries of Japan, was an incentive to the French, who lacked even a foothold in the region, to explore. Hence Lapérouse's arrival at the Bay of Tartary in 1787, while encountering no other national presence, was not without a wider imperial significance. Lapérouse like the bakufu, wishing to uphold ideals of his enlightenment, also envisaged indigenous peoples like the Ainus as 'other,' but according to different aesthetic codes than those adopted by the Japanese.

In relation to ethnographic description and images of natural man and the noble savage, Lapérouse's expedition needs to be placed thematically as well as chronologically alongside Cook's last voyage and his death (1779) at the hands of the Hawaiians. The encounters between the French Navy vessels and the Oroches and Ainus were staged by Lapérouse according to the conventions established by previous Pacific encounters, especially those of Bougainville and Cook. Lapérouse and Cook shared much in common. These expeditions were government sponsored, navy led, and hosted a range of supernumerary gentlemen of science, including astronomers, naturalists, and their artists; they were issued official instructions with copious directions to survey, to collect, to observe, and to record; they belonged to the same mold of scientific discovery.

The former expedition has been singled out as marking a transition between European perceptions of Pacific peoples from what Bernard Smith has termed the aesthetics of 'soft primitivism' to 'hard primitivism' (Smith 1985, 5–6, 49–50, 88–89). Lapérouse's instructions made greater, more explicit, ethnographic demands than those of Cook (Smith 1985, 138–139). Lapérouse himself described the objective of ethnographic observation by modern navigators as being to complete the history of man (Wolfzettel 1996). Establishing peaceful relations was of the highest priority. Although they were aware of the dangers of a violent death such as that suffered by Cook, events of the expedition only took a dramatic turn for the worse at Samoa, where a party of twelve including de Langle and La-monon, the naturalist, were ambushed and killed (Smith 1985, 140–145). Although this marked a key moment of disillusion among Europeans towards Pacific peoples, it occurred after the encounters at the Bay of Tartary. With the prolonged absence of the expedition, and the information provided by de Lesseps who had returned from Asia overland, the revolutionary government sent out search expeditions. However, owing to a certain lack of enthusiasm for a voyage initiated by a monarch as patron, eight years elapsed before the diaries, records, and notes, were finally published in 1797. Published ten years after his death, the narrative was assembled from correspondence and copies of Lapérouse's journal sent overland with de Lesseps during the expedition. Lapérouse explicitly stated that he wished his journals to be edited by a gentleman of science rather than a man of letters so as to avoid sacrificing veracity for style. The expeditions were made from the same Enlightenment mold, national differences notwithstanding. In the following year, 1798, two translations into English were published, followed by another edition in 1799.

Lapérouse, like the self-consciously 'enlightened' explorers who had preceded him, enacted prescribed arrival rituals to promote the likelihood of nonviolent exchanges, as a prelude to revealing, recording, and classifying nature's undisclosed truths, and to be the eyes of Europe's *philosophes*. In the last three decades of the eighteenth century, the Pacific was, to borrow Dening's term, a *theatrum mundi*, where a European and American audience could, from a distance, witness national expeditions enacting their encounters on foreign shores. It was a theater of mirrors: "a time of intensive theatre of the civilised to the native, but of even more intensive theatre of the civilised to one another. The civilised jostled to see what the Pacific said to them of their relations of dominance." It made "that unreal metaphor . . . of the 'expansion' of Europe . . . real at home and abroad" (Dening 1992, 373). With respect to Sakhalin, the case and setting of the *theatrum mundi* should be reoriented to take account of the Bakufu civilizing mission, a nativist discourse about the "other" *(Kokugaku)*, which rejected the long-standing tradition of Confucianism and its notion of barbarism, and instead extolled the superiority of Japanese culture (Siddle 1996, 38–40) as a vehicle of sovereign incorporation. The strength of the Kokugaku, in spite of the failure to convert Ainu to Japanese ideals any more than to those of the French enlightenment, nevertheless exerted a much stronger political impact upon the region.

The surveying encounters of explorers, and the ethnocartographic exchanges in particular, were significant stage settings in the *theatrum mundi*. Enlightened explorers self-consciously performed and narrated these exchanges for their reading publics. They entertained and they edified, reinforcing for the reader the categories of "explorer," "native," "map," "progress." Often performed hurriedly, they were nevertheless methodical and sometimes subtle. As strategic constructions, they aimed to maximize peaceful relations and informative exchanges.

My aim is to demonstrate how and why the solicitation of native sketches belonged to an ethnographic strategy of navigation, and hence to illuminate the relationships between enlightenment, ethnography, and geography. The term "ethnography" is preferable here to "anthropology" because the former term implies stronger linguistic connotations than the latter and draws attention to the history of nations and the movement of peoples. The episodes of sketching by local people, anxiously observed by the European navigators, were, I contend, language performances in which listening, speaking, writing, gesturing, and drawing were each important ingredients in soliciting a geographical gift. These aspects of communication and exchange, taken collectively over time, constituted a larger linguistic exercise, the construction of a regional ethnographic portrait or ethnic map.

The process of defining ethnic labels like Kamschatdale, Koriac, Mantcheoux, or in the case of the Ainu, simply "fishermen" and "islanders" was more flexible and powerful than simply compiling lists of ethnographic descriptions. These ethnic labels functioned as signs in an open, symbolic language. Their meanings were mutually defined in terms of comparisons (e.g., Oroches are in most respects like Oroks; both have many Chinese manners). Meaning was also ascribed by drawing on a wide range of descriptive characteristics (predicates), which included physical appearance, language, manners, customs, habits, religious ideas, and so on (e.g., Oroches are physically frail, short in height, wear their hair shaven at the front, display gratitude when receiving gifts and are superstitious about writing). The push and pull of these two different levels of ethnic description could precipitate more elaborate, or even different, meanings and hence change the basis of ethnic locations and comparisons. This will become clearer in the course of the chapter.

At the very opening of many enlightenment encounters, there was, in general, an agnostic attitude about the degree of commensurability or incommensurability between the ethnicity of the local people and other known ethnic groups. Establishing just how similar or different, and in which respects, depended on the particular people concerned, the number of related symbols and predicates with which to compare, and the relative ease or difficulty of ascribing affinities in the time available. In other words, the process of ethnic mapping could be linked to many contingencies. Enlightened explorers nevertheless, did not start with a tabula rasa. They possessed a readily available symbolic language of ethnicities. From the start then, they could describe a 'new people' in terms of similarities and contrasts with cultural features of other peoples. Ethnic labels provided the basis for drawing lines of linguistic commensurability and incommensurability. These lines are statements of relations between ethnic labels. These statements either affirm comparability, or alternately, declare a lack of it.

Statements of commensurability and incommensurability are generated by staged observation. The ritual of displaying peaceful intent and leaving presents is meant to 'provoke' the recipients, in the literal sense of calling forth a voice. The explorers anticipate some form of reciprocity, where their presents are answered by gratitude and a cultural auto-exposition through performances of some kind. The term 'auto-exposition' is appropriate here because the performance, including the gestural medium, is an integral part of the ideas, exclamations, emotions, or knowledge communicated. Only later is the performance distilled, leaving a residue of explicitly geographical knowledge.

For example, when the *Astrolabe* and *Boussole* were visited by men in canoes off the coast of Japan on the 5th of May, 1787, Lapérouse hazarded a guess that they might be *Kumi,* describing them as "neither Japanese nor Chinese, but, situated between the two empires, they seem to partake of both people. Their covering was a shirt, and a pair of cotton drawers. Their hair, tucked up on the crown of a head, was rolled round a needle, which seemed to us to be gold: each of them had a dagger, the handle of which was gold also." Pressed for time, the ships sailed on, and that is all the reader learns of the *Kumi.* Nevertheless, already Lapérouse is exercising the use of ethnic distinctions, putting to work his ideas about Chinese and Japanese people, and creating some uncertainty in his own mind, as

MICHAEL T. BRAVO

he observes people who appear to be a hybrid of the two.

By situating these visitors between the two, Lapérouse was effectively claiming a measure of ethnic commensurability between them. That is to say that descriptions of Chinese and Japanese people were deemed sufficiently relevant that a third ethnicity could be described in terms of similarities and differences between the other two. How does the idea of ethnic commensurability compare with what Anthony Pagden has called "the principle of attachment," whereby eighteenth-century travelers identified similarities and differences between themselves and the people they encountered? My concept of ethnic commensurability differs in two ways. The comparison and contrast of the *Kumi*, for example, involves 'triangulating' or placing the ethnicity in question in a wider frame of reference which draws upon a whole vocabulary of ethnicities and associated descriptions. And whereas ethnic commensurability, like the principle of attachment, is concerned with similarity relations, I place much more importance than Pagden on language, and in particular, the use of a vocabulary of ethnic categories, which carry with them associated ethnic characteristics. This will become clearer further on in my analysis of Lapérouse's encounter with the ethnic groups around the Bay of Tartary.

My contention is that in the course of a sequence of encounters around the Bay of Tartary, Lapérouse, by leaving presents, soliciting gifts, and asking local people to draw sketches, was drawing and reorganizing the lines of commensurability (and incommensurability) between neighboring ethnic labels. As the sequence of encounters progressed, his placement of the ethnic characteristics shifted. Some become more important as the basis of distinct comparisons and others less so. As a web of relations that define the ethnic labels becomes more complex and fixed, a map of ethnic portraits of the region around the bay coalesces. Moreover, as this portrait of ethnic commensurability became more incisive and clearly defined, the capacity of Lapérouse and his officers to read the sketches changes accordingly to the point where he is able to use them as a vehicle of navigation: hence the term, *ethnographic navigation.*

How does a linguistic, ethnographic portrait hold the key to interpreting traces drawn in the sand? A process of classifying language terms accompanies the search for stable meanings. As the ethnographic portrait becomes more clear, the navigators use the sketches to distinguish or distill geographical knowledge from the rest of the ethnographic portrait. As they try to discern which linguistic terms provide geographical knowledge, they are, simultaneously, attempting to assess the accuracy and validity of that knowledge. The accuracy is to some extent a measure of the degree to which the ethnographic portrait resonates with the ethnographic vocabulary as a whole, as well as the measure of confirmation through surveying. The spatial distances between terms in the ethnic vocabulary help to explain how knowledge travels by situating centers and peripheries: the ethnic terms provide a measure of both social and geographical distances. Lapérouse's imputation of astonishment to himself as the witness of these proceedings adds an important rhetorical twist to the credibility of the sketches, their apparent capacity to speak for themselves, and the imputation of accuracy to the native authors.

7.3. Navigating the Bay of Tartary

The argument in this paper about Lapérouse's solicitation and meaning of geographical gifts requires some familiarity and a basic outline of the weather conditions and the tracks of the ships in the course of the six stops made while tracing the coastlines around the Bay of Tartary.

The *Boussole* and the *Astrolabe* passed through the strait separating Japan from Manchuria in June 1787, sailed north, and began the exploration of the Bay of Tartary (Manche de Tartarie) later the same month (fig. 7.2). The chronology of this section of the voyage spanned approximately seven weeks. The ships stopped first on the Tartar mainland at Terney (Baie de Ternai) on June 23rd for four days, and then briefly at Suffren Bay, two hundred miles up the same coast on June 27th. At the fiftieth parallel, they sailed east across Tartary Bay and sighted a new coastline. A brief break in the persistently foggy weather gave them a glimpse of land to the north, possibly sealing off the channel. Anticipating the time that would be lost having to sail north to confirm this, and beat back against the southerly wind, and given that it was already late in the season, Lapérouse decided to seek shelter and to send a landing party to seek local advice. The ships anchored on July 12th at De Langle Bay, where they spent the evening and the following day. Encouraged by what their indigenous informants told them, they sailed northwards through six more days of fog. With a break in the clouds, they decided to anchor for the night of the 19th in D'Estaing Bay, where they sent out another landing party. They then continued northwards as far as where the channel narrowed and the sea became too shallow to proceed, which, together with the strong southerly wind, had built up a dangerously heavy sea. On July 27th, two small boat crews were sent to explore the shallow waters to the north, but the difficult sea conditions prevailed. Crossing over to the Tartary (west) coast, they anchored at Kastri Bay (Baie de Castries) on July 29th, where they stayed for four days. After learning as much as possible about their environs, they beat back to windward, surveying the mainland coast, then crossed over, back to the island. Squalls unexpectedly brought about a reversal of wind direction, which dramatically increased their progress, with the result that they were able to anchor off Cape Crillon, the southern tip of the island, on August 11th. The following day, they rounded Cape Aniwa to the east and headed for open sea (Lapérouse 1798, II: 39–116).

The crux of my argument is that, collectively, the encounters on the beaches of the mainland and the island held the key to Lapérouse orienting himself. Moreover, the episodes of natives drawing sketches in the sand, however important, were never, individually, decisive. Instead the meaning and the medium of cross-cultural communication were spread over the course of seven weeks of conversations with locals, highly gestured utterances, and discussions on board the ships. However, the argument that there was a decisive moment and medium has been put forward by Latour (1983, 203–236; 1987, 215–223) in his classic disposition of actor-network theory. Specifically, he claimed that the map-making episodes at De Langle Bay on July 17, 1787, were crucial and decisive transmissions of local knowledge from "Chinese savages" to Lapérouse and De Langle, his junior captain, commander of the *Astrolabe*. Lapérouse's diaries demonstrate that the idea of a decisive moment is false. Latour is also mistaken in the date of Lapérouse's arrival at De Langle Bay. He made a preliminary visit the evening of July 12th, carried out extensive ethnographic enquiries the following day, and weighed anchor on the 14th. However, the actual date does not affect his argument.

Lapérouse's geographical confusion, expressed in his journal as he sailed north from Suffren Bay, in fact persisted after his one-day map-making episode at the Baie de Langle. When plotting his course to sail eastwards from the coast of Tartary along the 50th parallel, he had hoped to reach Jeso or Oku-Jeso, rather than Sakhalin, or failing that, to arrive at the Kuriles en route to Kamschatka (Lapérouse 1798, II: 40). After sighting land on July 7th, Lapérouse expressed considerable doubt as to what he had seen: "I at first believed this land to be the island of Ségalien [Sakhalin], of which the southernmost part had been placed by geographers two degrees too much to the northward: and I judged that if I steered my course up the channel, I

Fig. 7.2.   Chart of discoveries made in 1787 in the seas of China and Tartary by the French frigates *Boussole* and *Astrolabe*. Plate 46 in *Atlas du voyage de La Pérouse*. Reproduced with the kind permission of the Cambridge Philosophical Society.

should be forced to pursue it till it opened into the Sea of Okhotsk, on account of the obstinacy of the south winds, which at this season incessantly prevail in these seas (Lapérouse 1798, II: 41). When three days later (still prior to landing on the island), Lapérouse could see the coastline stretching out in both directions, he mused,

MICHAEL T. BRAVO

"these new reflections . . . almost induced me to believe, we were not in the channel of the island of Ségalien, to which no geographer has assigned so southerly a situation, but rather to the westward of Jesso . . . and that we had, without perceiving it, entered the gulph which Jesso forms with this part of Asia. It only remained for us to discover whether Jesso is an island or a peninsula, forming with Chinese Tartary the same figure as Kamtschatka with Russian Tartary" (Lapérouse 1798, II: 42). As they prepared to leave, after the native map-sketching episode (recounted by Latour) at the Baie de Langle, named in honor of his fellow captain's success in soliciting local geographical knowledge, Lapérouse was confident that he had located a large island, but was no longer clear that it was Jesso: "It was of the greatest importance to discover whether the island we were ranging along was that to which geographers have given the name of Sakhalin" (*ibid.* 49). Evidently, Lapérouse suspected that it was not, because following his next stop at D'Estaing Bay, he recorded in his diary a change of heart: "[I had] . . . entirely recurred to my former opinion, and could no longer doubt that the island we were ranging along from 47°, and which, according to the information of its inhabitants, must extend much farther to the southward, was the island of Sakhalin . . ." (Lapérouse 1798, II: 59–60).

This is sufficient to make the point that these cross-cultural map-making episodes, alone, were not immediately decisive, in the sense of providing conclusive evidence to resolve the international cartographic dispute or to confirm the identity of the newly sighted land. The journal shows that it was not simply a matter of first identifying whether or not the newly sighted land was an island, and then determining its true name. No such instantaneously conclusive transfer of knowledge from a network of locals to the French Navy took place in the sand on the beach at De Langle Bay.

7.4. Ethnonavigation

How did the navigators resolve the meaning of such ethnocartographic sketches? The local people at De Langle Bay, D'Estaing Bay, and Kastri Bay all spoke of the land to the east side as *Tchoka*. To make matters more confusing, they called the major continental river draining into the channel (Amur) *Sagaleen*. A point to emphasize is that they did not locate themselves by starting with the names *Sakhalin* and *Oku-Jesso* and then finding them on the native sketches. It was, rather, the other way around. By constructing an ethnographic portrait of the people, their customs, their trade routes, their boats, and so on, the explorers were able to orient themselves; the labels came last. Only late in July (around the 23rd) did Lapérouse refer to the island as Sakhalin. Likewise, the ethnic labels were fixed last. The officers compiled a great deal of ethnographic information about the people of De Langle Bay and Cape Crillon, labeling them as 'fishermen,' and later as 'islanders,' though never actually giving them a proper name (Ainu). Given the absence of these key geographical and ethnographic terms, why did the sketching episodes in fact prove so convincing that Lapérouse was willing to base his conclusions on them without having circumnavigated the supposed island for himself?

Although Lapérouse was not a geographical *savant,* he was an expert navigator, his brother-officer, De Langle, was a fine observer and negotiator, and Lavaux, the *Astrolabe*'s surgeon, was an able linguist. The corps of officers constructed an ethnographic portrait of *seasonal* trade, travel, and ethnicity of the region around Tartary Bay. The ethnographic process took the form of rapid, systematic encounters, which the officers keenly observed and recorded in their journals according to the typical categories of travel literature (manners, customs, habits, language, ideas of religion, ideas of government, etc.). Information relating to navigation, like systems of exchange, technologies of transportation, evidence of literacy, sources of food, patterns of migration, were all grist for the mill. Recall that for the officers of the *Boussole* and *Astrolabe* it was already late in the season of exploration. They were still on their way to Bering's Strait to search for the entrance to a Northwest Passage. Local traders on the other hand, were not exploring. The officers could conjecture that these various native peoples were in the summer phase of a different way of life, different travel routes, different size boats. By constructing an ethnographic portrait of their seasonal navigation— call it *ethnonavigation*—they were able to create commensurability, a basis for translation, between the two societies of navigators from world's apart.

The *geographical gift* is an apt name for this ethnographic process of exchange, performance, and translation because the surrender of the ethnogeographical knowledge (more than other ethnosciences) draws attention to its own contour as the source of geographical knowledge. The explorers' possession and interpretation of the sketch testifies to the success of creating commensurability between cultures in the form of a visual language of the highest importance. Its legend, the key to interpreting it, lies in the cumulative ethnography. The entire performance of the native sketching in the sand,

MICHAEL T. BRAVO

and the officers observing, working out the artist's intent, is a display of the officers' enlightened leadership. The subsequent use of the sketch by the officers as a vehicle for settling a key geographical question demonstrates its utility. Its eventual displacement as the source of truth by a survey made with precise astronomical instruments positions the native map as an ethnographic artifact and the survey as geographical knowledge. This entire process, which I call the *geographical gift*, strategically reorders local relations of navigation, exchange, dialogue, and performance into categories of space and time, scientific knowledge and local knowledge, geography and history; it occupies the epicenter of enlightenment. In order to illuminate the framework of the geographical gift, I will examine the mechanics of the encounters in greater detail.

Approaching the Baie de Langle, Lapérouse was explicit about his desire "to procure information from the natives." His description of the encounter conforms to the tradition of the enlightened arrival ritual. He did not immediately approach the fleeing local inhabitants to draw maps. First, Captain Langle was sent ashore to announce the arrival of the ships and to carry out reconnaissance, taking the utmost precaution to avoid danger and to allay local fears by whatever means possible. Discovering signs that the huts had been vacated moments earlier, Langle followed the procedure of leaving different kinds of gifts: some useful pieces of iron, some pretty beads, and so on. Before returning, a canoe with seven men arrived. Of course, Lapérouse knew absolutely nothing about them: their social organization, population, history, beliefs, or even their principal place of residence! He soon began to assemble the outline of an ethnic description. After observing and describing their clothes, bodily appearance, and conduct (e.g., showing no inclination towards theft), he judged their manners to be "grave, noble and affectionate" (Lapérouse 1798, II: 45). Believing them to belong to a society of merchant fishermen, Langle proceeded to give them more presents, gesturing that he would return. Offering gifts, encouraging trust, collecting local specimens of plants and fauna, and studying the customs, manners, and habits of the local people was the first step in the civilizing mission.

The following morning, the officers took up the next stage of the encounter. Equipped with pen and paper, they set about creating a vocabulary list. Instead of either hiding or fleeing as they had previously done, the fishermen by this stage appeared only too happy to assist their visitors. We are told that having built trust, the two parties proceeded to discuss vocabulary, point to objects and name

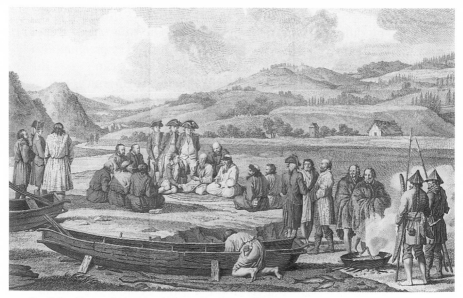

Fig. 7.3.   Dress of (Ainu) inhabitants of De Langle Bay, drawn by Duché de Vancy, engraved by Cathelin. Plate 50 in *Atlas du voyage de La Pérouse*. Reproduced with the kind permission of the Cambridge Philosophical Society.

places; they identified their base or place of origin as *Tchoka*. Their ease in assisting the officers, together with their facility for handling a pen in the manner of the Chinese, persuaded Lapérouse that "the art of writing is probably known to them" (Lapérouse 1798, II: 47) (fig. 7.3).

We are told that when the officers expressed their wish, presumably through gestures, that the locals draw maps of their country, as well as that of the Mantchéoux (from the mainland), an elderly man proceeded to make a sketch in the sand. This was followed by another more detailed map drawn on paper by a younger man. Both maps showed a narrow strait at the north end of the channel separating an island from the mainland.

Lapérouse's narrative then relates how a young man drew the river flowing into De Langle Bay, thereby indicating the relative distance to the northern and southern extremities of the island. Distance was laid down in terms of a number of strokes representing the number of days necessary for traders traveling by boat to reach a given place (Lapérouse 1798, II: 47). A translation of strokes into standardized units of time was not possible, a classic dilemma in enlightened encounters, the fundamental reason being that navigation was a culturally specific phenomenon; Ainu seasonal movements depended upon the habits and rituals of traveling, eating, sleeping,

MICHAEL T. BRAVO

fishing, and trading. Lapérouse's strategy was to estimate a maximum distance taking into consideration local technologies of travel (fig. 7.4). Given the fact that pirogues, unlike ships, "never go above a pistol-shot from the shore," the officers reasoned that "they did not advance in a strait [sic] line above nine leagues per day; for as the coast admits of landing every where, they go ashore to dress their provisions and take their meals, and it is probable they make frequent rests. We therefore estimated our distance from the extremity of the island at sixty-three leagues at most" (Lapérouse 1798, II: 48).

As the ethnic group ('islanders') acquires an increasing number of characteristics or predicates, moments where translation can take place become possible. Establishing distances and scales is the aspect of the encounter which, when successful, resembles translation as much as ethnography. To show the distance between the island and the mainland required establishing commensurability in terms of using body gestures to make comparisons. The younger man placed "two hands perpendicular and parallel, at the distance of two or three inches from each other, gave us to understand, that he meant thus to describe the width of the small river where we got our water. By widening them, he signified the width of the river Sakhalin, and in the same manner the much greater width of the strait which separated his country from Tartary." To ascertain the depth of the river, Lapérouse plunged a pike into the water: "He seemed to understand us, and placed one hand over the other, at a distance of five or six inches. We thought he meant thus to signify the depth of the river Ségalien. He then extended his arms to their full length, as if to communicate the depth of the strait. It now remained to ascertain whether he had been describing absolute or relative depths" (Lapérouse 1798, II: 49).

After this encounter, Lapérouse could be confident that he had positioned or calibrated these people with respect to the Chinese. The success of the effort, time, and expense spent on this encounter is demonstrated by the fact that in spite of acknowledging the contrast in race and the incommensurability of language between the islanders and the Chinese, he has established partial commensurability of navigation by recognising their links through trade, customs, and manners. This piece of ethnography answers the question, "Is this an island?".

At the north end of the channel the expedition found its way blocked by sand banks, navigable only by shallow-draught canoes but not by the larger discovery vessels. Nearby to the west on the

PIROGUE DE TCHOKA.

PIROGUE DE L'ÎLE DE PÀQUE.

Fig. 7.4.   Pirogues from Tchoka and L'Île de Pâque. Drawn by Blondela, engraved by Masquelier. Plate 61 in *Atlas du voyage de La Pérouse*. Reproduced with the kind permission of the Cambridge Philosophical Society.

Fig. 7.5.   Dress of (Orochon) inhabitants of De Castries Bay, drawn by Duché de Vancy, engraved by Dennel. Plate 54 in *Atlas du voyage de La Pérouse*. Reproduced with the kind permission of the Cambridge Philosophical Society.

mainland, at the place he named Kastri Bay (Baie de Castries), Lapérouse put in for five days, where he took the time to renew his encounter strategy, except this time, with the Oroches, a minority in this region whose language belongs to the Tungus-Manchu family of languages (fig. 7.5). Their economy was based primarily on hunting wild reindeer and, to a lesser extent, fishing. According to a recent anthropological study of Orochon hunting, they travel on domesticated reindeer, moving as silently as possible to mask their human presence from the animals. The act of hunting is reciprocated. Immediately after the animals are killed, the hunters are vulnerable to a counterattack by the animals' spirits. The hunters return home by a different route, ensuring that they do not cross the path of their outgoing journey. For the Oroches, hunting is lived as a complex narrative; "the Orochon hunter hunts with words" (Kwon 1998, 124). While this has long been commonly understood by Oroches, Lapérouse, who met them at their summer camps, down near the coast, did not have access to this knowledge of their culture. He saw little beyond the coastal contour, or surface, of the land.

Nevertheless, Lapérouse and his officers also acquired their knowledge of this land through words, though in a quite different sense than the Oroches, and constructed their enlightened journey as an ethnographic narrative. They crisscrossed their own tracks in an attempt to cover the bay and to link the peoples and sites of their visits as comprehensively as possible. Lapérouse's Chinese interpreters had much more success in communicating with the Oroches than with the Ainu fishermen at De Langle Bay. Lapérouse, encouraged also by evidence of Chinese customs, such as kow-towing, supposed (crudely) their language to be a dialect of Chinese. He understood that their individuality, "inviolable fidelity . . . and their almost religious respect for private property" (Lapérouse 1798, II: 85) occupied a significant place in their social organization.

Lapérouse, confident of linguistic commensurability, built on this trust through reciprocal exchanges of respect. For example, he praised their generosity with food, but also noticed their suspicion of writing. They wondered what purpose these inscriptions served, and reminiscent of the hide-and-seek in hunting, which is characterized by a perilous interchangeability between the human hunter and the human prey (Kwon 1998, 119), in their discomfort, they fell silent. Lapérouse recorded in his journal: "they appeared to consider the motion of the hand that wrote as magic signs, refusing to answer our enquiries, because as they gave us to understand, they considered it an evil" (Lapérouse 1798, II: 74). This aversion, far from hindering Lapérouse, became a piece in the larger puzzle of the regional ethnography contours, the wider framework for judging the credibility of the sketches.

As the ships sailed to the south end of the island, they discovered to their delight the existence of a large channel to the east. Seeking to confirm whether it is actually a channel or just a large gulf, the expedition put in to consult the locals. Lapérouse's description of their encounter at Cape Crillon was, once again, rich in ethnographic detail. This time he was able to draw the lines of ethnic commensurability more easily. The local people shared strong linguistic similarities to the fishermen from De Langle Bay, while standing in marked contrast to the Oroches from Kastri Bay on the mainland. At this settlement, ethnographic navigation proved both easier and more coherent than previously. By using words from the vocabulary list constructed with the Ainu fishermen at De Langle Bay, their initial suspicions were quickly overcome. The officers managed this by changing the scale of their encounter. Rather than presenting themselves as first-time arrivals to Cape Crillon, they were able to present themselves as more aware, second-time visitors to the fishermen. At this point in the narrative, Lapérouse had grown more confident that he had overcome the main hurdles of ethnic incommensurability.

His ability to use the ethnic labels to shift scale between local, regional, and global comparisons is very striking and in marked contrast to his tentative categorizations when he arrived in the strait earlier that summer. He can now position his ships' tracks with some precision by naming and describing the peoples and places he has visited.

Lapérouse, confident that he had discovered the geographical identity of Sakhalin, commented that he knew part of the island even better than the locals, and proceeded to solicit more sketches to compare with the previous ones in the hope of solving the question of the southern channel! Observing that these people seemed "to be in the habit of delineating countries, for, with a single stroke, they described the part we had just explored, as far as the Segalien, leaving a narrow passage for their canoes: and they marked each night's resting place and gave it a name" (Lapérouse 1798, II: 89). Where seven weeks earlier he had not been confident of understanding the sketches, he could at once see their imperfections at Cape Crillon. The channel projected as a series of lines on a map could be seen to divide fundamentally different races, bridged only by the commerce of the canoes traveling over the shallow sandy strait at the northernmost end. He surmised that whatever knowledge the people of Cape Crillon possessed derived from the importance of the Sakhalin river for regional trade between the Mantchéoux Tartars, the Chinese, the Bitchys, the Oroches, and the islanders. Without this, they would "know as little of the Chinese and their merchandise, as the inhabitants of America" (Lapérouse 1798, II: 89).

The sketching episodes on the beaches of Sakhalin in July 1787, by epitomizing the problematic concept of the "native map," open up wider issues about the relationships of enlightenment, knowledge, capitalism, and empire in the late eighteenth century. In this essay, I have examined how Lapérouse distilled relations of space and time, geography and ethnography, as characteristic axes of enlightenment geographical science. One feature of my study is that it is not limited to the periphery: on the contrary, it attempts to show how Lapérouse, deeply immersed in the navigation of the Bay of Tartary, was able to construct the region as the periphery using an ethnographic process of mapping to create distinctions between geography and ethnography, between local seasonality and the regulated astronomical time of his chronometers (a measurement of distance from the Royal Observatory in Paris). The beauty of ethnographic mapping for Lapérouse is that he is able to construct many kinds of distance. When asked, "where is Sakhalin?", doubtless he could give the latitude and longitude of some sections of the coast-line. He could also locate the island in terms of its ethnic topography in relation to the rest of the world. He contrasted the (Ainu) is-landers of Sakhalin with the Oroches of the Tartar mainland, who he believed to share a common history with their northern (Tungus) neighbors, the Koriacs and Kamschatdales. The relationship was not a total contrast because the ethnic distinction is qualified by the shallow body of water separating Sakhalin from the mainland, a well-traveled trade route, regularly negotiated by canoes. Moving to a larger regional scale, Lapérouse declared them "a very superior race to the Japanese, Chinese, and Mantchou Tartars." Then moving from a regional to a more global scale, Lapérouse judged the Ainu "coun-tenance more regular and more similar to those of the Europeans" and described the islanders as "particularly beautiful, and of a very regular proportion, . . . strong made, . . . and their arms, neck and back, were covered with hair. I make this observation merely be-cause it is a general characteristic, for it is easy to find individuals equally hairy in Europe. I think their middle stature about an inch less than that of the French. . . ." And then at a global scale, he classified Tartars alongside (Sàmi) Laplanders and Samoides, who are "to the human species what their stunted birches and firs are to the forest trees of the more southern climates" (Lapérouse 1798, II: 86–89) (fig. 7.6).

7.5. Latour's Account of Ethnogeography

Bruno Latour's account of Lapérouse's survey of Sakhalin illumi-nates relationships between cartography, his exemplary science *par excellence,* and capitalism. Eighteenth-century historians of science like Pagden (1992) and Miller and Reill (1996) have found Latour's work very attractive because it presents a model of knowledge in which the collecting of individual pieces of information on the pe-riphery can be seen as incremental steps in the accumulation of knowledge in centralized, metropolitan, scientific bureaucracies; it helps to show why a navigator or fieldworker, less powerful than the people he visits for the first time, is in a much stronger position when he or someone else arrives the second time with a chart. Knowledge, contends Latour, is like arriving somewhere for the second time.

Lapérouse's encounter with the Ainu of Sakhalin at De Langle Bay is an excellent choice of example for illustrating actor-network the-ory. In the late eighteenth century, the French empire, having suf-fered a number of major setbacks, is unequivocally weak in Eastern

MICHAEL T. BRAVO

NISKANI, AOUCANTOURI et EROUGANTOI.

Habitans de la Baie de LANGLE.

OROTCHIS.

Habitans de la Baie de CASTRIES.

Fig. 7.6. Dress of Niskani, Aoucantouri, and Erougantoi, inhabitants of De Langle Bay; Orotches, inhabitants of De Castries Bay. Both drawn by Duché de Vancy, engraved by Simonet. Plate 55 in *Atlas du voyage de La Pérouse*. Reproduced with the kind permission of the Cambridge Philosophical Society.

Asia, where the intersection of the Chinese, Russian, and Japanese interests collide. French economic interest in the regional fur trade is insignificant and Lapérouse's exploration would seem comparatively innocent insofar as his primary interest is cartography. When de Lesseps arrives in Europe with copies of his journals and maps, they assume their importance as elements *(immutable mobiles)* in a wider network of other maps, archivists, draftsmen, and savants. The particulars of the case study therefore seem to have little to distract Latour from the centrality of the cartographic traces.

Latour's interest in the material cartographic traces, and the contrast between sand and paper as media of inscription, effectively omits all the ethnographic and historical aspects of the geographical gift. He describes the Ainu informants at De Langle Bay as "Chinese savages," whereas Lapérouse drew precisely the opposite conclusion. While noting that they shared many customs with the Chinese (their greetings, the length of their nails, their way of holding a pen), he was more struck by the fact that the expedition's Chinese interpreters could not understand a single word of their language. As we have seen, Lapérouse constructed a geographical division between the mainland Tartars and the island fishermen. Latour focuses on the transfer of the *traces,* contrasting the portability, longevity, and immutability of the inscribed paper in contrast to the sketch in the sand which the tide will wash away. This is crucial for Latour because the material trace is the fundamental unit or element of actor-network theory. Traces are effectively material resources.

Latour (1987, 217) claims, though wrongly, I think, given the convergence of national interests in the region, that the relationship between Lapérouse and the islanders is much the same at the moment he weighs anchor to leave as when he first arrives: "Lapérouse crosses the path of the Chinese at right angles; they have never seen each other before and the huge ships are not here to settle." A different way to read Latour, here, is that the traces relinquish their connection with the islanders and take on the voice of Lapérouse. A different way to read the same passage is that the transfer of information takes place independently of the meanings each party ascribes to it. Latour (personal communication, 1996) claims that in his actor-network theory, traces have an ontological content, but no epistemological value. Traces or inscriptions have an existence, but no meaning until it is assigned or attributed to them. When Latour refers to the two parties crossing at right angles, I take him, actually, to be saying that the cartographic exchange itself is ontological and takes place at the level of the trace, whereas the interpretations take

MICHAEL T. BRAVO

place, more or less separately, within the confines of each respective culture. The "right angles" metaphor is another way of saying that ontology and epistemology can be clearly distinguished. Traces are ontological resources, which only acquire meaning when later placed in a wider context of network relations (Eisenstein 1983).

What is the price paid for a geographical theory which emphasizes ontology to such an extent? Curiously, Latour finds himself aligned with the mild cultural relativists. As he (Latour 1983, 228) puts it: "I should not be astonished by Lapérouse's astonishment in finding geographer-savages. Until recently, most people would not hesitate to say that the knowledge of Lapérouse is *superior* by nature to that of the savages . . . However we see very well from this dialogue with the savages that the difference in knowledge systems comes not from the content, intelligence, or even the graphical form—it is this which most astonishes Lapérouse—rather it comes from the displacement of the spokesman." However, the chief importance of my argument about the geographical gift is to show how the geographical gift creates this displacement between the cultures. The gift is the engine of the displacement of the spokesman from the Ainu fishermen to Lapérouse. Latour explains that Lapérouse's astonishment derives from the unexpected ease with which a map in the sand moves from the savages to Lapérouse's thought system—the map. But how does this displacement come about? Even during the key sketching episode on the beach at De Langle Bay, most of the communication takes the form of a sequence of spoken, verbal utterances, a broken conversation, intermittent references to a vocabulary list, the imaginative use of body gestures—rather than the detail of the drawing. In fact, the innocence associated with interpreting the meaning of native sketches on the beach is a complete illusion. Rather it is the result of building an ethnographic portrait around the Bay of Tartary (learning simple Ainu words and phrases, their interpreter conversing with continental traders visiting D'Estaing Bay) contextualized in terms of regional trade and navigation. In other words, the displacement on the beach is actually the result of an ethnographic contrast between the Oroches reindeer people of mainland Tartary and the Ainu fishermen of Sakhalin. It is already a great divide in Lapérouse's eyes, as he maps the contrast between the physical appearance and manners of the Ainu fishermen which differ so greatly from those of the Oroches. However, Lapérouse can measure this cultural divide more precisely in terms of the trade routes which link the island to the mainland: the distance across the strait at the Baie de Castries, the shallow draft of the canoes, the use of Chinese interpreters, and so on. These tie the Ainu to the economy of the Ségalien or Amur River system.

7.6. Conclusions

The terms 'enlightenment' and 'geography' are daunting in terms of everything they encompass in the eighteenth century. General pronouncements about the constitution of geography in terms of the mapping of time and space should be made cautiously. Although standardization and measurement became ideals of enlightened geography, they were scarcely all-pervasive. If imperial cartography was actually locally circumscribed and metropolitan, it is important to identify those practices and systems of production which maintained it as a global discourse. The geographical conundrum of Sakhalin posed such a question: could French savants resolve cartographic disputes in the most distant reaches of empire? This not only tested the abilities of scientific travelers to survey and chart inhospitable seas and coastlines. It also required them to engage with other peoples who might very well perceive the interlopers as aggressors and respond with acts of murder, as happened to De Lange, Lamonon, and other crew members of the *Astrolabe*. The *theatrum mundi* in which explorers enacted their ideals of enlightened encounter, threatened to be overwhelmed by a *theatrum belli,* where the economic and sovereign interests of nations converged.

The single gesture of the native mapmaker delineating geographical forms in the most convenient medium at hand was highly symbolic. It was the complement to the explorer's characteristic gestures of precision in using astronomical instruments for orientation. It indicated an ethnographic relativism about how peoples know the world: the explorers, like their ships' names, represented by compasses and astrolabes on the one hand, and the natives by their everyday local knowledge on the other hand. It also symbolized the civilizing mission ideology wherein the surrender of local knowledge took the form of enlightened homage in exchange for the gifts of civility and commerce. In other words, the gesture of the native cartographer was central to the whole theatrical syntax of exploration as a sequence of partly controlled, partly unpredictable, face-to-face encounters between disparate cultures.

The concept of the geographical gift does more than draw attention to the encounter as a theatrical vehicle or narrative device of imperial exploration. Focusing on the importance of a strategy of exchange set in a broader regional political framework displaces the symbolic gesture of the native mapmaker, destroys the presumed symmetry between cultures, and expresses the false innocence. The native map need no longer be read as a sign of the transfer of an object of local knowledge, or a trace, across a boundary from one culture to another. It comes into existence only as an ethnographic exercise in cross-cultural production of knowledge. The scale of this exchange is not as local as might first seem to be the case. The meaning of the mapping gesture is derived by a sequence of translations between strictly local, temporal dialogues and a regional, atemporal portrait of ethnic descriptions. However, the term "translation" is not sufficiently specific to give an accurate sense of this change of scale, particularly with respect to the role of language.

The initiation of exchange by offering gifts like beads, mirrors, and pieces of metal is the starter motor whose purpose is to inculcate some sense of peaceful obligation among the hosts. It is directed to establish a host-guest relationship in which local people are forthcoming with their language and statements about themselves. The native who draws even an elaborate map but who is otherwise mute is of little use to any explorer. Language was the key. Building lists of words with approximate translations (in the strict sense) gave explorers a vocabulary with which to probe. Typically these lists comprised terms which were useful for navigation: orientation, numeracy, social hierarchy, and trade. They enabled explorers to surmise, or more often to hazard guesses, about the way in which the peoples they encountered traveled, traded, communicated, and identified themselves. These tentative statements of commensurability were essential to establish the mutual positions of groups of people living in different kinds of ethnic organization. The language of ethnic distinction depended heavily on traditions of navigation and customs of exchange.

The relations of distance between social groups could not be reduced to distinctions based simply on different kinds of exchange, for example, one society being commodity driven, while another was organized around the gift. Enlightened encounters in the field were understood and judged with an emphatic attention to linguistic commensurability. In the case of native map making, commensurability as translation between systems of measurement and orientation is obvious. However, this is only a special case of the relationship between geography and ethnography. The privilege of translation is contingent in no small measure on many other assertions of both commensurability and incommensurability. This supports the more general thesis of the importance of language in enlightened encounters (Pagden 1993), not only in intellectual terms, but equally in the scientific practices of the explorers.

If geography epitomized the rationalism of space during the French enlightenment, its rule was hardly transcendent. The idea that time and space were made secular and uniform was only true in the fictions of metropolitan cartographers: that mythical reality was strictly local (Sahlins 1981, 1985). Lapérouse's journals demonstrate that he ordered the geographical spaces of encounter with indigenous groups, situated between nations, according to varied, subtle, distinctions. This ethnographic survey was one of many attempts to complete the survey of the world's coastlines and to map all of humanity, a characteristically enlightened project. In remaking the map of Sakhalin, Lapérouse composed a portrait of the island which was clear and succinct, though ethnographically incomplete (in contemporary terms). Given his rich ethnographic comparisons and contrasts, his consummate narration of the geographical gift, and his strong belief that a shallow channel separated the island from the mainland, which contemporary European reader could possibly doubt that Sakhalin was an island? Beneath the tide of competing national strategic interests which continued to flow and ebb in this region, the contour of the island remained controversial, but in the long run, was less important than the existence of a navigable strait. The British *Critical Review* (1798), trusting deeply in navigation, chastised Lapérouse for not persevering enough to see the channel with his own eyes. The geographer Ludwig Lindner had fewer doubts when he "celebrated the superiority of European science after recapitulating that the voyages had 'proved' Sakhalin to be a peninsula" (Stephan 1971, 35). When Arrowsmith's map of Asia (1822) showed Sakhalin as an island, the orientalist Klaproth dismissed him as "the most ignorant of those whose occupation is cartography" (Klaproth 1826, 3; c.f. Stephan 1971, 35).

Lapérouse's network was never at right angles to those of the Ainu, the Russians, or the Japanese. Savants in France and Japan were engaged in similar geographical battles. Japanese interests in Karafuto (Sakhalin) involved a balance of interests between those of the bakufu and matsumae. The unfortunate Ainu periodically put up fierce demonstrations of resistance. Their political freedom was tenuous. They were further away from their Japanese oppressors but closer to the Chinese. Their feudal relations were giving way to stronger

claims to sovereignty, annexation, and eventually colonization. Lapérouse's instructions demanded that he clarify the Japanese top-
onymy of Sakhalin and Hokkaido (Ezogashima). The apparent innocence of his encounter together with the Ainu and his pursuit of
pure geographical knowledge, like the relationship in the frontispiece of navigation illuminating Lapérouse's exploration for history,
was an ideal of French enlightenment.

Note

A special word of appreciation to Bruno Latour for his generosity and
openness in helping me develop my critique of his work. A discussion of an
earlier version of this paper by the students of the Summer Academy of the
Max Planck Institute for the History of Science, Berlin, helped me to clarify
the argument. I also wish to acknowledge the generous assistance and criticisms of Dan Brown, Marie-Nöelle Bourguet, Lorraine Daston, Francis Herbert, Everett Mendelsohn, Dorinda Outram, Simon Schaffer, Piers Vitebsky,
and Michael Wintroub, the editors of this volume, and two anonymous
referees.

References

Allen, E. W. 1959. *The Vanishing Frenchman: The Mysterious Disappearance of
Lapérouse*. Rutland, VT: Charles E. Tuttle Co.
Belyea, B. 1992. "Amerindian Maps: The Explorer as Translator." *Journal of
Historical Geography* 18(3): 267–277.
Bourguet, M.-N. 1997. "Voyage, Mer et Science au XVIIIe Siècle." *Le Bulletin
de la Supplément à la Revue d'Histoire Moderne et Contemporaine* 44(1–2):
38–55.
Bravo, M. T. 1996. *The Accuracy of Ethnoscience: a Study of Inuit Cartography
and Cross-Cultural Commensurability*. Manchester: Manchester University.
Chapin, S. 1990. The Men from Across La Manche in *Background to Discov-
ery: Pacific Exploration from Dampier to Cook*. Edited by D. Howse. Berkeley
and Oxford: University of California Press, pp. 81–127.
Czaplicka, M. A. 1914. *Aboriginal Siberia: A Study in Social Anthropology*. Ox-
ford: Clarendon Press.
Dening, G. 1992. *Mr. Bligh's Bad Language: Passion, Power and Theatre on the
Bounty*. Cambridge: Cambridge University Press.
———. 1996. *Presenting the Past: History's Theatre Performances*. Melbourne:
University of Melbourne Press, pp. 101–200.
Eisenstein, E. 1983. *The Printing Revolution in Early Modern Europe*. Cam-
bridge: Cambridge University Press.
Fabian, J. 1983. *Time and the Other: How Anthropology Makes Its Object*. New
York: Columbia University Press.
Forster, J. R. 1996. *Observations Made During a Voyage Round the World*. With
introductory essays by N. Thomas, H. Guest, and M. Dettelbach. Hon-
olulu, University of Hawai'i Press.

Forsyth, J. 1992. *A History of the Peoples of Siberia: Russia's North Asian Colony, 1581–1990*. Cambridge and New York: Cambridge University Press.

Gaziello, C. 1984. *L'Expédition de Lapérouse, 1785–1788: Réplique Française aux Voyages de Cook*. Paris: Comité des Travaux Historiques et Scientifiques.

Greenblatt, S. 1992. *Marvellous Possessions: The Wonder of the New World*. Oxford: Clarendon Press.

Harley, J. B. 1989. "Deconstructing the Map." *Cartographica* 26(2): 1–20.

Klaproth, H. J. 1826. *Observations sur la Carte de l'Asie Publiee en 1822 par M. Arrowsmith*. Paris: n.p., p. 3.

Kwon, H. 1998. "The Saddle and the Sledge: Hunting as Comparative Narrative in Siberia and Beyond." *The Journal of the Royal Anthropological Institute* N.S. 4(1): 115–127.

Lapérouse, J.-F. d. G. 1797. *Voyage Autour du Monde, 1785–1788*. Edited by M. L. A. Milet-Mureau. 4 vols. with Atlas. Paris.

———. 1798. *Voyage around the World, 1785–1788 . . . Annexed, Travels over the Continent [of Asia] with the Despatches of La Pérouse in 1787 and 1788, by M. de Lesseps*. Translated by F. A. Maurelle. 2 vols. London: J. Stockdale.

———. 1799. *Voyage around the World, 1785–1788 . . . Annexed, Travels over the Continent [of Asia] with the Despatches of La Pérouse in 1787 and 1788, by M. de Lesseps*. Translated by F. A. Maurelle. 2 vols. London: G. G. and J. Robinson.

Latour, B. 1983. "Comment Redistribuer le Grand Partage?" *Revue de Synthèse* 110 (Avril–Juin): 203–236.

———. 1987. *Science in Action*. Milton Keynes: Open University Press.

Law, J. 1987. "On the Social Explanation of Technical Change: The Case of the Portuguese Maritime Expansion." *Technology and Culture* 28: 227–253.

Lewis, G. M. 1998. *Cartographic Encounters: Perspectives on Native American Mapmaking and Map Use*. Chicago: University of Chicago Press.

Livingstone, D. N. 1993. *The Geographical Tradition: Episodes in the History of a Contested Enterprise*. Oxford, UK; Cambridge, NY: Blackwell.

Mauss, M. 1924. "Essai sur le Don. Forme et Raison de l'échange dans les sociétés archaïques." *L'Année Sociologique* n.s. 1: 30–186.

Miller, D. P., and Reill, H. P., eds. 1996. *Visions of Empire: Voyages, Botany, and Representations of Nature*. Cambridge: Cambridge University Press.

Mundy, B. E. 1996. *The Mapping of New Spain: Indigenous Cartography and the Maps of the Relaciones Geográficas*. Chicago: University of Chicago Press.

Munro, N. G., and B. Z. Seligman. 1996. *Ainu Creed and Cult*. London and New York: K. Paul International. Distributed by Columbia University Press.

Outram, D. 1995. *The Enlightenment*. Cambridge; New York: Cambridge University Press.

Pagden, A. 1993. *European Encounters with the New World*. New Haven and London: Yale University Press.

Sahlins, M. 1981. *Historical Metaphors and Mythical Realities: Structure in the Early History of the Sandwich Island Kingdom*. Ann Arbor: University of Michigan Press.

———. 1985. *Islands of History*. Chicago: University of Chicago Press.

Siddle, R. 1996. *Race, Resistance, and the Ainu of Japan*. New York: Routledge.

Smith, B. 1985. *European Vision and the South Pacific*. New Haven and London: Yale University Press.

MICHAEL T. BRAVO

Sorrenson, R. 1996. The Ship as a Scientific Instrument in the Eighteenth Century, in *Science in the Field, Osiris,* edited by H. Kuklick and R. E. Kohler, 2d series, 11:221–237.

Stephan, J. J. 1971. *Sakhalin: A History.* Oxford: Clarendon Press.

Thomas, N. 1991. *Entangled Objects: Exchange, Material Culture, and Colonialism in the Pacific.* Cambridge, MA, and London: Harvard University Press.

Turnbull, D. 1993. *Maps Are Territories: Science Is an Atlas.* Chicago: University of Chicago Press.

Wolfzettel, F. 1996. *Le Discours du Voyageur: Pour une Histoire Littéraire du Récit de Voyage en France, du Moyen Age au XVIIIe siècle.* Paris: Presses universitaires de France.

Chapter Eight

From Enlightenment Vision
to Modern Science?
Humboldt's Visual Thinking

ANNE MARIE CLAIRE GODLEWSKA

Méditation, le mot n'est pas trop fort pour rendre compte de l'intensité et de la
continuité de la réflexion humboldtienne.
—Charles Minguet, *Alexandre de Humboldt historien et géographe de l'Amérique*

8.1. Introduction

The early nineteenth century has been described by Foucault as the
approximate moment of a shift from the Classical to the modern
episteme. It was, as Foucault has argued, in this period that the nature of the life, earth, and human sciences shifted from description
to explanation and from a focus on surficial phenomena to the theoretical exploration of interior structures (Foucault 1966).[1] Not at all
coincidentally, it is also generally regarded as a foundational period
for a variety of modern disciplines revolving around the earth and
life sciences, including geology, botany, zoology, physical geography, and biology. It is impossible to identify Humboldt exclusively
with any one of these sciences. Rather, he had a commitment to understanding the natural laws that governed all of these realms, and
the links between the natural world and human society, human perception, and human creativity, whether scientific or artistic. It is also
very difficult to assign him comfortably to either Foucault's classical
or modern episteme. Indeed, he is a transitional figure.

Today Humboldt is remembered in the life and earth sciences less
for substantive research than for a synthetic way of thinking. Modern geographers tend to look at him as the closest thing to a geographer to be found in the early nineteenth century. About that, they
are certainly correct. The ideas he developed in both his "Plant geography" and his *Cosmos* are far closer to a modern geographer's
concerns than any of the work produced by the composers of universal geographies or the map makers who still defined the discipline

in the early nineteenth century. Yet the relationship between Humboldt and the geography of the late eighteenth and early nineteenth centuries is problematic but important. As I have argued elsewhere, geography, certainly in France, but probably in Europe in general, was a very different field than its modern namesake. Further, by the early nineteenth century it had moved into a period of crisis. For centuries geography had been the science that described the earth (including not only the natural world but the human structures upon the earth such as borders and nations), either textually or graphically. By the early nineteenth century, it was beginning to see its purview seriously eroded by the newly forming life and earth sciences. Geography's methods were increasingly deemed primitive and inadequate to the exploration of both the natural and the human world. And its huge scope, which sought to include everything, seemed to consign it to the role of "vestibule" to the more serious sciences.

In early nineteenth-century parlance, Humboldt was not a geographer. His identity and role were not tied to the ideas and understanding of men such as Jean-Denis Barbie du Bocage, Edme Jomard, Adriano Balbi, Jean-Baptiste Bourguignon d'Anville, or even the illustrious Conrad Malte-Brun. Largely self-educated, Humboldt adopted no field but paid attention to any research in a wide array of fields that might allow him to seize the fundamental unity of the Cosmos.[2] Nevertheless, there is a consonance (if not some identity) between his ideas and the struggle in which geographers were engaged to keep their field whole while entering into a new era of modern science. Compared to the writing and thought of geographers of the early nineteenth century around the cosmos, the natural world, description, and graphic representation, Humboldt was an original and innovative thinker. The key elements of his work which, no matter how commonplace today, were truly innovative and even in some cases controversial, included: a focus on theory, explanation, and cause, which arguably stretched his ability to maintain the unity of the sciences to the limit; attention especially to what is dynamic and changing rather than what is static; an interest in distributions not just location; a concern to enhance interdisciplinary analysis through the use of mathematics and statistics; a heightened sense of the importance of observation and especially theory-driven observation; awareness of the importance of scale and its impact on the focus of study, with a chosen focus on geographic-scale forms; the development of crucial concepts such as the natural region and landscape; focus on the "way of life" of plants, animals and humans;

and an awareness that the key to appreciating this "way of life" is an understanding of the internal functioning of phenomena from geologic processes, to plant communities to human societies. In all of this innovation, Humboldt was trying to retain a holistic vision of the cosmos while taking full advantage of the insight to be gained from empirical, experimental, and theoretical science. In a sense, he was trying to reconcile a nineteenth- and twentieth-century science shaped by disciplinary structures with the less-defined and delimited Enlightenment science.

Enlightenment science is not easy to characterize. In scientific circles, the Enlightenment in France was characterized by a rational enthusiasm or the sense that, with the right questions, most puzzles could be elucidated; by a belief in progress, both human and natural; by a passion for exploration, observation, and the collection of information (and specimens) and the sense that the collection and classification of observations would in itself lead to enlightenment; and by an Aristotelian cosmological legacy that suggested that all in the Cosmos was linked by a universal chain of being (with or without angels and God at its apex). Bernard Smith has discussed the artistic vision that accompanied Enlightenment science as an artistic neoclassicism which regarded unity of mood and expression as more important than fidelity and detail of depiction. Smith argues that Humboldt sought a new unity—an ecological unity—which could reconcile the increasing development of an empirical science tending toward the disruption of unity and the holism of an artistic vision and expression. This artistic unity was not neoclassical. This art did not entail the sacrificing of nature's detail, roughness, and particularity to some higher sense of unity. It was artistry—and unity—derived from, and based on, an analytical and penetrating approach to the depiction of landforms and landscapes (Smith 1988, 4, 18, 27–29, 203–212).

My argument runs parallel to that of Smith. He argues that Humboldt provided the "typical landscape" (or a landscape designed to capture the look of an ecological system), then evolving among landscape artists such as Hodges and Webber, with "theoretical justification." Smith based his argument largely on Humboldt's writings about the look of nature, landscape, landscape painting, and even his few but emphatic comments about photography. In this chapter, I want to explore that "theoretical justification," or the hard thinking behind Humboldt's attitude to representation. Mine is perhaps closer to the focus of Michael Dettelbach's concern (Dettelbach 1996). Instead of concentrating on landscape painting, I have chosen to explore the innovative ideas and approaches developed by Humboldt and their link to representation. Otherwise put, I am interested in the intellectual structure of the unity Humboldt was trying to discern and its relationship, less to landscape painting than to his experimentation with maps, protothematic mapping, graphs, and diagrams. It is my contention that in his scientific graphics he was trying to develop or adapt from the work of others a language— or a way of seeing—that would encourage both conceptual depth and rigor and holistic vision. I am not saying that Humboldt was successful in this endeavor, but that the endeavor was coherent and powerful.

I consider Humboldt's endeavor relevant to the geography of his era because, although he was on the margins of what was then geography, he was grappling with essentially the same problem geographers were grappling with: how to combine the holism of the cosmos with modern empirical research without abandoning what was then geography's central methodology: description and particularly graphic description, or cartography.

Although he did not focus on Alexander von Humboldt, in *Les mots et les choses*, Michel Foucault predicted much of the innovation I describe in this chapter. However, despite his focus on the representational basis of science in the eighteenth century, Foucault was relatively insensitive to the different modes of representation employed in science. Thus, apart from his somewhat disconnected analysis of the Meninas, he rarely mentions and does not take into account the wide variety of pictorial media such as maps, graphs, trees, diagrams, and their changing function in science through and across the epistemes. As a result, he misses an important consequence of the changes taking place in the shift from Enlightenment to modern thought: while representation was being abandoned as the guiding ideal of scientific thought, many of the sciences moved toward a far greater use of scientific illustration. This is neither paradoxical nor contradictory of Foucault's argument that the sciences were becoming more focused on invisible interior structures and their functional roles. Indeed, the graphics and cartographics employed by disciplines such as geology, and later climatology, botany, zoology, epidemiology, etc., were, as Cambrosio, Jacobi, and Keating have pointed out for immunology, as Jane Camerini has pointed out for plant biology and as Martin Rudwick has pointed out for geology, theory-loaded (Cambrosio et al. 1993; Camerini 1993; Rudwick, 1976, 1985). Together with sections, sketches, and landscapes, maps were used extensively in the development, elaboration, and testing of theory. Indeed, as LeGrand has shown for a later period, such graphics and cartographics could even replace experimentation in the testing of theory (LeGrand 1990). Thematic cartography, then, came to be because the use of maps in representing what could not be seen but could be hypothesized required the development of a new and extensive symbolic vocabulary.[3] It is not surprising, then, that there emerged in this period and especially in the work of Alexander von Humboldt a very different role for the map—indeed, a different type of map altogether, requiring a different kind of map-maker (Latour 1987, 228–41).[4] This new map ceased to aspire to the universal flat description typical of the topographic map and began to move toward what some people call thematic mapping; that is, the mapping of interactions and change in which the map functions as an analytical device which embodies a theoretical and scientific argument about the nature of the world.

This paper explores the innovative nature of Alexander von Humboldt's thought and the new graphic expression consequent to it. It is argued that in his thought, and that of a number of colleagues, a language was in formation: a language capable of expressing and exploring the new kinds of questions being asked in the life, earth, and human sciences in the early nineteenth century.

8.2. Humboldt's Innovation in the Geographic Realm: Theory, Explanation, and Cause

At a time when many—though by no means all—geographers, repulsed by geology's excesses and Buache's stunningly misguided generalizations, readily condemned "theory," Humboldt embraced theory, explanation, and cause wholeheartedly. Equally disturbed by some of geology's more ridiculous pronouncements, he eschewed the investigation of origin, or more particularly, the investigation of ultimate origin (Humboldt 1823, 1, 5, 34, 85; Rudwick 1985, 20).[5] Nor did he in any way accept the Buachian claim to be able to predict the mountainous structure of the globe from the hydrographic system (Beck and Bonacker 1969, xlix–l).[6] Theory in scientific hydrography, as opposed to the sort of divinatory hydrography advocated by Buache and his followers, had to be built from the ground up:

> It is from the intimate knowledge of the influence exerted by the inequalities of the surface, the melting of snow, periodical rains and tides, on the swiftness, the sinuosities, the contradictions, the bifurcations and the form of the mouths of the Danube, of the Nile, of the Ganges, and of the Amazon, that we form a general theory

ANNE MARIE CLAIRE GODLEWSKA

of rivers, or rather a *system of empirical laws,* that includes all that is common and analogous, in local and partial phenomena. (Humboldt 1823, 78–80)

In contrast to the focus on identity and analogy of form that characterized traditional natural history, theory, for Humboldt, entailed the search for analogies, parallels, and equivalents (Humboldt 1823, 18). Empirical laws, or extraction of what was analogous in physical phenomena, and the combination of those laws or analogies was theory. And the elaboration of theory was the point of science.

Identification of analogies and empirical laws, however, demanded understanding of cause. In his introduction to the concept of the use of contours as isothermal lines, Humboldt cautioned that if the various causes of local temperature were not understood, then attempts at generalization and averaging might well eliminate the most important factors determining the distribution and development of organic life "as outside and disruptive circumstances" (Humboldt 1813, 469). One must, Humboldt argued, be very careful in averaging out phenomena which "depend by their nature on locality, the constitution of the soil, the disposition vis-à-vis the sun's rays of the surface of the globe" (Humboldt 1813, 469). Above all, when one generalizes temperature, one must bear in mind its multiple causes and their relative weight (Humboldt 1813, 470–71).[7] Attention to cause required a careful balance between, on the one side, excessive empiricism, and, on the other, neglect of the facts. Thus, while explaining the theoretical importance of his work, Humboldt wrote against pure empiricism:

> In this *geognostical essay,* as well as in my researches on the *isothermal lines,* on the *geography of plants,* and on the laws which have been observed in the *distribution of organic bodies,* I have endeavoured, at the same time that I presented the detail of the phenomena, to generalize the ideas respecting them, and to connect them with the great questions in natural philosophy. I have dwelt chiefly on the phenomena of *alternation,* of *oscillation,* and of *local suppression,* and in those which result from the *passage* of one formation to another in consequence of *interior development.* These subjects are not mere theoretical speculations; far from being useless, they lead us to the knowledge of the laws of nature. It would denigrate the sciences to make their progress depend solely on the accumulation and study of particular phenomena. (Humboldt 1823, vi–vii)

While convinced that the sciences could not be profitably structured or driven by empiricism, he was equally certain that ". . . it would harm the advancement of science to try to rise to the level of general ideas while neglecting the knowledge of the facts themselves" (Humboldt 1816, 1:6).

Humboldt's work was well laced with theory, analogy, generalization, and exploration of cause. It is difficult to privilege any one of the many theories he developed over others. Perhaps, however, the most pervasive theory in his work, which was, in fact, common to all of his theories, was the argument that location, conceived of in three dimensions (latitude, longitude, and altitude; or in geognosic terms, position, and superposition) was the key to understanding the natural world. Location, as he demonstrated in his famous thematic diagram showing plant distribution as related to altitude, embodies the particularities of temperature, the chemical composition of the air, the presence or absence of particular rock formations (which themselves are a function of the historical action of similar forces), intensity of light, humidity, refraction, but also location in terms of position relative to bodies of water and continents . . . etc. (Humboldt 1973 [1805]). Thus, Humboldtian location was a concept far more complex and rich than location in the traditional geographic sense. Other scientists found his approach to location sufficiently convincing and powerful to build on. Certainly, Decandolle's theoretical argument that plant geographers should expect to find analogous species in similar habitats can be seen as an elaboration of some of Humboldt's arguments about the role of location in the natural world (Decandolle 1820, 383–4, 413).

8.3. Movement, Change, and Distribution

One of the most pronounced differences between Humboldt's research and most of the work carried out by both geographers and most natural historians in France in his day was his attention to the dynamic and changing elements of the landscape. Geography's attachment to description, and particularly to the map, made movement and change precisely the dimension excluded from depiction and thus from possible analysis. In natural history, classification was still something of a craze in the late eighteenth and early nineteenth centuries, and this directed the field's focus to the static and, as already discussed, to identity. Humboldt was dissatisfied with this approach to the natural world. He expressed his frustration with this perspective eloquently in a letter to Schiller in 1794:

> To date, the history of nature has been studied in such a way as to retain only differences of form in the study of the physionomy of

plants. This study of the physionomy of plants and animals has made the teaching of characteristics and identification such a sacred science that our botanical science can only seem an object of meditation for men given to speculation. But you sense, as do I, that something a little higher must be sought; that there is something that we must rediscover. We must follow Aristotle and Pline who included in their description of Nature both the aesthetic sense and the artistic education of man. Those ancients certainly had a larger vision than our miserable archivists [Registratoren] of Nature. These are, to my mind, the objects which seem worthy of attention and which are almost never taken up: the general harmony of form; the problem of whether there was an original plant form, which is now to be found in thousands of gradations; the spreading of these forms over the surface of the earth; the diverse impressions of joy and melancholy that the world of plants evokes in sensitive men; the contrast between the massive rocks, dead, immobile (and even the trunks of trees which seem inorganic) and the living carpet of vegetation which, in a sense, delicately covers the skeleton with a more tender flesh; the history and the geography of plants, that is, the historical description of the general extension of plants over the surface of the earth which is an unstudied part of the general history of the world; the investigation of the oldest primitive vegetation to be found in those funerary monuments (petrifications, fossils, carbon minerals, coals); the progressive habitability of the surface of the earth; the migrations and journey of plants—social plants and isolated plants—with the use of maps in this; which are the plants that have followed certain peoples?; a general history of agriculture; a comparison of cultivated plants and domesticated animals, the origin of two degenerations; which plants more or less strictly or more or less liberally abide by the law of symmetrical form; the return of domesticated plants to a wild state (American plants as well as Persian ones—wild plants from the Tage to the Oby); the general confusion in plant geography caused by colonisations. I worry about these questions constantly but the noise being made around me on this subject makes it impossible for me to abandon myself to it in a systematic fashion. I find that I have expressed myself as one demented. Still, I hope that you fully share my sentiments. (Minguet 1969, 76–7)

In their study of plant geography, both Humboldt and Decandolle, whose work on plant geography was a refinement of Humboldt's, had to resist strong tendencies to static classification. Decandolle's emphasis on the importance of the "station," or the special conditions under which each species would grow, entailed far more than the addition of a line of description at the bottom of a classification. In fact, he argued against simple-minded or any attempt at a *final* classification of stations. What he was trying to get at were the interactions between the exterior forces acting on plants, between competing plants, and the influence of these and other factors (such as differing reproduction modes, and thresholds of tolerance to sub-ideal conditions) on both plant structure and distribution. A static picture of plant distribution was not the final aim of this research: understanding the interaction of forces was.

Arguably, however, Humboldt's interest in the migration of plants (and animals and humans), their colonization of particular territories, and their tendency to form relatively uniform masses of vegetation, or not, was part of a larger interest in movement, change and indeed exchange. This extended beyond the plant world to the movement or tilt of rocks (whose real extent and significance he could not quite believe) and well beyond the natural world to movement and exchange in precious metals. In the *The Fluctuations of Gold*, Humboldt studied the dynamic forces behind the distribution of precious metals and their impact on prices with his accustomed sensitivity to the complexity of interactions (Humboldt 1900, 30).[8]

There was another significant difference between Humboldt and Decandolle's curiosity and that of most geographers and natural historians of the early nineteenth century. Humboldt, especially, was as interested in distribution as in location. His plant geography, and that of Decandolle, necessarily focused on the distribution of organic life in relation to the distribution of influencing factors. Humboldt's "géognosie" had a primary focus on the location, distribution, and comparison of rocks. His isoline work sought to provide the means to depict the distribution of temperature around the globe. He even extended his interest in plant distribution to fish distribution (under water at different levels of oxygen) and to parasites within the fish or animal (to different levels of oxygen in different organs) (Humboldt and Bonpland 1811, 298–304). Finally, his work on the fluctuation of prices was all about the global distribution of precious metals. In developing these ideas, he frequently demonstrated a strong sense of the spatial and dynamic or fluid nature (at least over time) of phenomena (Humboldt 1900, 43).[9]

8.4. Multidisciplinarity, Mathematization, and Observation

Accuracy, numeracy, and measurement were, as Edney shows here, major preoccupations of late eighteenth century geographical and, more generally, scientific thought. On first glance, Humboldt seems to have participated in this. If, however, we compare the overwhelming preoccupation with measuring and mapping "anything and everything" displayed by geographers on the expedition to

ANNE MARIE CLAIRE GODLEWSKA

Egypt with Humboldt's own description of his mapping and astronomical operations in the Americas, there is a decided difference in flavor (Godlewska 1995, 1999). Humboldt's concern was not so much accuracy and somehow capturing on paper and in number the true America. Instead, his aim was to provide information in a form which would enhance data comparability and data analysis from multiple points of view. He was attempting an abstraction. The purpose of this abstraction was not, however, possession and recreation but enhanced analysis. Accuracy, then, did not have the same importance. Thus, Humboldt explained that the careful reader of his geographic "Map of New Spain . . ." would see that it did not quite agree with his "Map of the Route from Mexico to Durango." This, he explained, was simply the result of a combination of the different sources used in the two maps and the process of reconciliation employed in compiling each (Beck and Bonacker 1969, plate 6 and discussion). So, for Humboldt, if the discrepancy were known and its source understood, then there was nothing further to worry about. The absolute truth of the picture was not the issue.

What measurement, number, and statistics did allow was a certain transferability of data and analysis across fields: a true interdisciplinarity. For Humboldt, ever concerned with the unity of nature and the sciences, this was power indeed. His memoir on isolines was designed precisely to bring together and compare temperature measurements so that "theory can draw on the corrections provided by the diverse elements considered" (Humboldt 1813, 462). The quality of the data, especially the number of observations and the quality of instrumentation used was important, but some of these problems could be minimized through the use of probability theory (Beck and Bonacker 1969, lxxx; Humboldt 1813, 462–3, 542, 602, 496). What was more important was having information on temperature in a form which could equally well be used to elaborate and test theory in plant geography, in economics, and even in geognosy.

For Humboldt, the isoline and more generally a numerically based graphic scale was analogous to and perhaps a continuation of the revolution brought about by the thermometer in the seventeenth century (Beck and Bonacker 1969, lxxii). The thermometer for the first time allowed tracking exactly what was happening on the surface of the earth, comparing yesterday with today, comparing here with there and recording information. Data in comparable numerical form was the basis from which the study of phenomena as complex as, say, fossil remains (historic plant and animal geography) could begin to be useful to the elucidation of geologic chronology.

> To compare formations with relation to fossils, is to compare the *Floras* and *Faunas* of various countries at various periods; it is to solve a problem so much the more complicated, as it is modified at once by space and time. (Humboldt 1823, 58)

Without number and comparable statistics, how could the processes behind geologic formations be studied at all?

Some areas of inquiry were more open to mathematization. "The details of natural history are foreign to statistics," Humboldt observed. But natural geography, by virtue of its ability to convert natural history's data to number and statistic, could substantially contribute to forming an exact idea of the territorial wealth of a state (Humboldt 1811b, 46). In this sense, through comparable numerical data, it could be a truly unifying discipline capable of linking nature and society.

Of course, for Humboldt, behind number lay observation. His voyage through South and Central America was all about observation and even occasionally experimentation (Humboldt 1811c, 49–92, 253–59). In the account of his travels to South America—which features prominently in Outram's account of travel knowledge—he pointed out that while the eighteenth century had been characterized by extraordinary developments in instrumentation, there had not been a growth in observations sufficient to take advantage of those instrumental improvements (Humboldt 1816, 7). He was prepared to use the observations of others, as indeed, given the scope and scale of his focus, he had to (Humboldt 1823, 76).[10] Nevertheless, it was his own observations which gave him the ability to sort through and judge that great mass of facts.

8.5. Scale, Landscape, and the Natural Region

One of the most fascinating features of Humboldt's work was his sensitivity to scale. He had a strong tendency, particularly in his work on the geography of plants, to move up and down the scale from a microscale discussion of a particular plant, to the analysis of local climate and conditions, to observations and generalizations on a continental scale. He was, however, acutely aware of the dangers of this sort of elision in scale, particularly in the realms of analysis and evidence (Humboldt 1811d, 1–8).[11] Successful comparison across large areas demanded careful attention to local conditions and mechanisms.

Humboldt did have a favorite scale of analysis. This could be described as geographic, ecological, or landscape-based. It was this

ANNE MARIE CLAIRE GODLEWSKA

scale of analysis which he felt had the power not just to interest but to move the human spirit. Certainly, he argued, it was as valid as the exploration of any other scale of existence.

> The configuration of great mountain masses, the great diversity of the contours of the high summits, situated like the lower lands in the midst of the agitations of the atmospheric ocean, are amongst the elements that constitute what we might call *the physionomy of nature*. The look of the mountains contributes no less than their form, their size and the grouping of plants, nor less than the different species of animals, the nuance of the celestial vault and the intensity of reflected light in determining the character of a landscape and the general impression made upon man by the different zones of the earth. (Humboldt 1854, 5)

Perhaps more than anything else, Humboldt was interested in landscapes and their characteristic look. What, for example, was responsible for the very different look of the landscape in temperate and tropical zones? In answering this question, and in his attention to landscape in general, he never lost sight of cause and process or of the action of process over time. Thus, for example, he described the topography of the Valley of Mexico, which he considered ideal for trigonometric mapping, in the following terms: "the vast plains of Zelaya and of Salamanca, united like the surface of the waters which seem to have covered the soil for many centuries . . ." (Beck and Bonacker 1969, 6). But it was the beauty of the landscape— of the human-scale landscape—of the view of the volcanoes from Mexico City—to which Humboldt was most susceptible and only a geographic-scale focus could bring that beauty home (Beck and Bonacker 1969, lxxviii).

Humboldt's focus on landscape closely approximates another concept that became important in the course of the nineteenth and twentieth centuries: the natural region. This idea has been traced by Michel Chevalier and Paul Claval to Girault de Soulavie, the provincial prelate who, during the first half of his highly checkered career, wrote *Histoire naturelle de la France méridionale* (1780–84) and the unfinished *Histoire du Vivarais* (1779–88) (Chevalier 1986; Claval 1968). Soulavie and Humboldt, separated by age, politics, nationality, and religion, nevertheless shared the same preoccupation with the vegetation characteristic of particular altitudes, climates, soil types . . . in short, natural regions. In his 1820 encyclopedia article on plant geography, in which he provided something of a historiography of the field, Decandolle linked the work of Girault de Soulavie and Humboldt (Decandolle 1820). Soulavie had developed some of

the key basic ideas behind plant geography, albeit in an impressionistic almost literary way, while Humboldt had laid down the essential elements of the science. However, both Decandolle and Humboldt, more concerned with the processes behind plant distribution
than with the delimitation of regions, can only be said to have given
the concept a nudge. The natural region, as an idea, was instead
picked up, developed and given a more geological emphasis in the
1820s and 30s by both Omalius d'Halloys and Coquebert de Montbret as the "physical region" (Omalius d'Halloy 1823).

8.6. Interior Structure and Function

If, rather than describing Humboldt's interests in our own disciplinary terms, dividing the natural from the human sciences, we see
them with his unifying vision, we could say that Humboldt had a
primary interest in the "way of life" of plants, animals, humans, and
even, in a sense, of rocks. Today, we would distinguish these as an
interest in the way of life of plants, animals, and humans and a focus
on the interactions between physical phenomena. Yet, the curiosity
is fundamentally the same in both cases. Humboldt wanted to understand how things worked and how they worked together to create what they are to the casual or admiring human gaze. Thus the
differences in "morals or customs" noted by travel writers were worthy of attention but only if the narrative writer could remove him-
or herself from the center of the stage to focus on the relations between the people under consideration. Humboldt did not doubt that
it was of such interactions between individuals that the larger society was composed (Humboldt 1816).[12] The failure of most travel writers to enter into this sort of social depth frustrated Humboldt. Indeed, he had so much trouble with the overbearing flow of the travel
narrative that he found himself, in the end, unable to write one.[13]

Humboldt's interest in the interactions behind social characteristics is paralleled by his attention to the aspects of animal physiology that might influence animal behavior. This sort of inquiry required dissection and Humboldt averred that as animal anatomy
was not his principal focus, his occasional work in dissection might
lack refinement. Nevertheless, he saw the dissections he carried out
in South America as a minor contribution to Cuvier's larger efforts
to get beneath the surface in understanding the relationship between physiological structure and animal behavior and to relate
these to classification (Cuvier 1811; Humboldt 1811e; Humboldt
and Bonpland 1811). But here again, Humboldt found his dual aims

of systematic anatomical study and his effort to acquire a sort of a cartographic overview of the territory through which he was moving (perhaps analogous to the travel narrative) in conflict (Humboldt 1811f, 306).[14]

As Humboldt sought to understand how human societies worked and animals behaved, he strove to understand plant geography. Thus, for both Decandolle and Humboldt, understanding the external forces acting on plants provided only part of the explanation. Humboldt was primarily concerned to refocus the work of natural historians on the geographic forces creating plant communities. However, he was aware that plant communities and landscapes were the result of the interaction of external forces and internal structures and constraints. Whether the external layers of a particular plant were composed of carboniferous or resinous material might have everything to do with its survival in cold conditions. Survival in different levels of light might be determined by the plant's particular root or leaf structure. And survival in different conditions of wetness would have everything to do with the plant's relative sponginess, its number of pores, etc. Indeed, Decandolle was convinced that the precise conditions under which a species would germinate could only be understood at the level of the excitability of its fiber or tissue (Decandolle 1820, 365–6, 369, 371–2). Moving beyond the surface layer—which had been the principal preoccupation of natural historians for generations—was, once again, essential to understanding the way of life of plants.

Humboldt was similarly concerned with exploring complex internal interactions in his geognostical studies. His aim was to move beyond the classification of rock types to acquire understanding of the geognostical structure of the globe. Indeed, he equated the kinds of questions he was asking in terms of geology, stratigraphy, and landform to the types of questions asked by Cuvier of the interior structures of animals (Humboldt 1823, 54).[15] Rocks, of course, do not have a way of life. Still, the course of life or nature, over time, had helped to create the earth's geognostical structure. In Humboldt's view, then, his study of geognosy went beyond Dolomieu and Saussure's surficial topography to Werner's preoccupation with the interior structure of particular regions (Humboldt 1823, 81). Humboldt's aim, as in his studies of society, animal behavior, and plant geography, was not to focus on the particularity of phenomena, but to understand the complex interactions between particular phenomena that had created, in the case of geognosy, the physical region (Humboldt 1823, 83–84).

8.7. A New Kind of Map

Humboldt did not "invent" the thematic map. A small number of such works predate Humboldt by hundreds of years.[16] He did, however, develop and explain some of the basic forms of expression used henceforth in such maps. He also thought through and demonstrated the relevance of these maps to the new science he was practicing (Robinson and Wallis 1967, 119).[17] Denis Wood has argued that there is no such thing as a thematic map per se; that all maps have themes and, he might have added, that all maps embody a theory (perhaps scientific) about the nature of the world (Wood 1992, 25). Nevertheless, even those relatively little informed about the history of mapping would find most thematic maps fundamentally different in nature from, say, a topographic map. What, then, is the difference, and why it is important?

There are as many assumptions about the nature of reality and what is important to depict in it behind the topographic map as behind the thematic map. Nevertheless, there is a different *sense of* certainty and reality surrounding the two maps. The stated aim of the topographic map is mimesis. It is designed to replace reality with a more useful and simplified picture of commonly-accepted salient aspects of reality. In that sense, grammatically speaking, it is a simple declarative sentence: one very extended descriptive statement. Recognition of the importance of this sort of statement, almost classification, to state administration was, as Edney has also remarked here, one of the great products of the eighteenth century. It was in the eighteenth century that codification of the language of this description took place and began to spread throughout Europe.

The thematic map has a logical structure more akin to an hypothetical argument: if we take this (the basic topographic information on the map) to be more or less true, and we use it to give a spatial structure to this (hierarchically structured data not generally found on topographic maps) and perhaps also this (other hierarchically structured data), then we have an interesting pattern not generally visible either on the topographic map or in the landscape. The thematic map, then, embodies an argument about invisible phenomena, or phenomena undergoing change, or about interactions between phenomena. It does not function as a record but as an analytical device, set aside when the question at hand is resolved. Arguably, then, although everything is a mental construct, there are different kinds of mental constructs.

Another difference between the topographic map and the the-

ANNE MARIE CLAIRE GODLEWSKA

matic map is its place in the division of labor. The topographic map was, and is still, made by map specialists; people primarily concerned with the correct, consistent, and complete expression of an elaborate descriptive topographic statement: people who came to be called cartographers. The thematic map was, and is, more commonly designed and often made by the scholars and scientists who are specialists in the hierarchically structured data: social scientists, earth scientists, etc. Their concern is less with the language of description than with the logic of the argument and any innovations in expression flow from the nature of the argument.

All this might be little more than an interesting definitional aside except that the development and elaboration of the thematic map correspond to a shift in the nature of social and natural science. It is not coincidental that the thematic map came into its own in the early nineteenth century. It is also not coincidental that it received special attention from Alexander von Humboldt, whose work and preoccupations exemplified the shift from descriptive science to theoretically driven explanatory science focused on the observation and analysis of change, distribution, interactions, and the interior functioning of natural and social phenomena.

By 1807 Humboldt had a knowledge of maps which could rival that of any topographical engineer, commercial cartographer, or academic geographer. Not only did he collect astronomic, geodetic, and itinerary data for many of his own maps and plans; he understood the critical decisions behind the compilation of that data. Thus, in his *Atlas géographique et physique du Royaume de la Nouvelle-Espagne*, he was able to explain his maps, his sources, and his decisions in the way that few before or since d'Anville managed. Indeed, Humboldt had a compilation cartographer's sense of what parts of the world had been mapped to what scales and what degree of accuracy. Certainly, he was aware of just how few places, even in the heart of Europe, had been well determined (Beck and Bonacker 1969, lxxxii).[18] It was his knowledge of far more than maps that gave his understanding of maps themselves even greater depth than is to be found in any of d'Anville's works. For example, his sensitivity to landscape form and his sense of its relatedness to hydrology, geology, and history gave him a more critical attitude to relief depiction. Thus, he complained about the contemporary maps of the Valley of Mexico:

> In spite of how interesting this country is on three levels, in terms of its history, its geology and its hydraulic architecture, there is not a single map in existence whose contemplation will give birth to a real sense of the valley's form. (Beck and Bonacker 1969, lv)

How, then, should such form be depicted? In practice, he generally chose hachures. However, his understanding of just how little was known about landscape and topography in most parts of the world made him cautious in advocating the use of either hachures or contours. Thus, at a time when many cartographers were arguing that contours by virtue of their greater accuracy ought to replace the more traditional hachure method of relief depiction in topographic mapping, Humboldt argued that pictorial hill symbols might more accurately reflect the state of current knowledge (Beck and Bonacker 1969, lxvii).[19] Humboldt was far from naive about traditional topographic maps and their limitations for the interpretation of the phenomena he sought to study.

It was from this base of knowledge, both about maps and about physical and social processes, that Humboldt began to experiment with graphic representation. He was not alone in his experimentation but was conscious of developing something analogous to some of the explanatory graphic tools being developed by people like Girault-de Soulavie, August Crome, Guettard, J. L. Dupain-Triel, and, most important, William Playfair. While his graphic experimentation was intellectually playful, it was also purposive. Humboldt was seeking a more analytical spatial language that would allow the almost intuitive transfer of understanding from one graphic genre to another and from one specialist body of knowledge to another. Humboldt was trying to find a form of expression capable of revealing his vision of the unity of nature with the newly found rigor of the systematic sciences. To that end, he experimented with isolines, distribution maps, flow maps, a map of error, proportional squares, something he called "pasigraphy," and a multidimensional pictorial graph. These eloquent graphic arguments about dynamic relationships in space, distributions, and interactions which often revealed patterns not visible to the naked eye and created a new systematic, rather than geographic, time/space, seemed to Humboldt to suggest that his intuition about the unity of nature was sound.

Isolines (or lines drawn between measures of equal values of a particular phenomena, such as height, temperature, degrees of cloud cover) were not a recent innovation. Nevertheless, although they had been developed by Halley over a century earlier to show patterns of magnetic declination, Humboldt pointed out that their true potential had not been realized. Humboldt rescued isolines from relative obscurity precisely for their ability to show patterns not visible to the eye or patterns so complex as to be obscure to the senses. Isolines were a method by which disparate numerical data could be

252   ANNE MARIE CLAIRE GODLEWSKA

brought together and rendered accessible to theoretical interpretation (Humboldt 1813, 462).[20] With such a rigorous yet simplifying
mode of graphic representation great volumes of data could be managed, rendering it possible to even consider "the influence of local
causes of perturbations" (Humboldt 1813, 463). Indeed, the use of
isolines, by revealing new patterns and enhancing comparability of
data, would cause scientists to reflect back on the data they were collecting and to reexamine its rigor and value. Once this sort of care
had been employed in the collection and interpretation of the data,
Humboldt predicted that it might even be possible to adequately
theorize solar action upon the earth and to calculate the distribution of heat received from the sun around the globe.

The greatest value of graphic representation, and of isolines more
particularly, was its remarkable ability to reveal relationships between and the relative importance of, for example, latitude and con-
tinentality; altitude and wind patterns; or humidity, heat, light, and
pressure (Humboldt 1813, 510–11).[21] Geographic maps and their
power to reveal geographic patterns in locational data were analogous to thematic maps and their ability to reveal heretofore invisible patterns in systematic data. The space and especially the time
displayed by these maps was a little different than geographic space
and the relatively thin temporal code embodied in most maps. The
space was systematic, structured so as to reveal the relationships between physical phenomena.[22] The time instead of expressing, for
example, France in 1756, might describe temperature variation over
months, decades, or even hundreds of years. Given all this, one
would expect Humboldt to have demonstrated the cartographic
power of his isolines with a world map. As Robinson and Wallis have
explained, we do not know precisely why Humboldt did not include
the map that was ultimately published in the *Annales de chimie et de
physique* in his article on isothermal lines. It may have had more to
do with the publication trials and tribulations of the *Mémoires d'Arcueil* (Robinson and Wallis 1967, 122). Certainly, Humboldt knew
that at times the space of the map can be problematic and intellectually constraining. Thus, he may have considered that his 1813 essay on "Des lignes isothermes et la distribution de la chaleur sur le
globe" was as well illustrated with a table structured by "isothermal
bands" of 5 degrees celsius. In fact, this was entirely in keeping with
one of the principal aims of his essay, which was to unshackle the
concept of temperature from latitude. He created a table, entirely
relatable to a map, but whose space was defined not by latitude and
longitude but by bands of average annual temperature (Humboldt

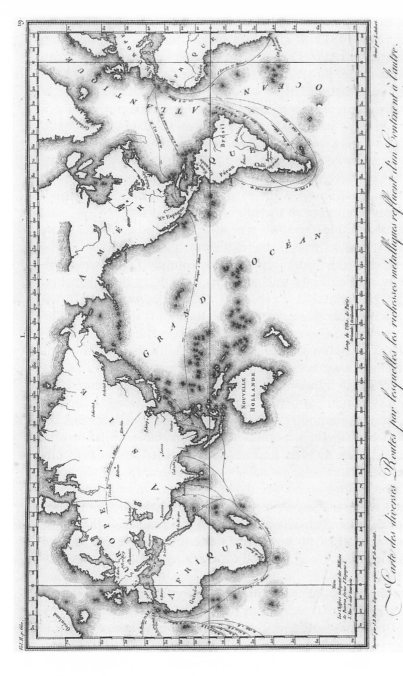

Fig. 8.1. A map by Alexander von Humboldt, 1812. "Cartes des diverse routes par lesquelles les richesses métalliques refluent d'un continent à l'autre." From Alexander von Humboldt, *Atlas géographique et physique du royaume de la Nouvelle-Espagne, fondé sur des observations astronomiques, des mesures trigonometriques et des nivellemens barometriques.* Paris: G. Dufour.

1813, 602, "Bandes isothermes, et distribution de la chaleur sur le globe"). This amounted to giving "systematic space" priority of expression over "geographic space."

Maps, tables, and graphs were all very well, but intelligent use of isotherms had to be informed by knowledge of physical and natural systems. Relationships might only be apparent, or unimportant, if unsupported by data and theory from, for example, plant geography, physics, chemistry, etc. Isolines, then, offered the possibility of truly capitalizing on numerical data while also increasing the likelihood of more intuitive insight (Humboldt 1813, 545).[23]

In some cases, Humboldt's experimentation with thematic mapping was more speculative than real, often because the data were not available to undertake or even mock up the cartography. Such was the case with a proposed botanical map of the regions inhabited by single species (or social plants). Humboldt speculated that this map would almost certainly reveal such plants to have posed significant obstacles to human settlement and the movement of armies. They would, he thought, be seen to have formed barriers to human movement as significant as the mountains and the seas.

Standard topographic or geographic maps have limited means of depicting flow and movement. Humboldt produced at least one graphic designed to show the flow of metals between Europe and North America (fig. 8.1). The map, really a pseudothematic map, carries relatively little information. It is a map of the world centered on the Pacific. It shows only, as its title suggests, the routes by which precious metals were moved between the continents. Humboldt, however, placed the map alongside four graphs (fig. 8.2) showing the volume of precious metals extracted from mines in the Americas since 1500; the amount of gold and silver extracted from Mexican mines since 1700; the proportion of gold and silver mined in different parts of Central and South America; and the proportion of gold produced by America, Europe, and Asia. It is but a tiny step from this map and its associated graphs to a thematic map, such as those produced by C. J. Minard, showing volume by thickness of arrow. It is important to note, however, that in separating out map and graph, Humboldt was able to fully develop both the dimensions of space and time. He, thus, suggested the depth and importance of the relationship between the Americas and Europe (Beck and Bonacker 1969, lxxxiii).[24] Although his graphic is relatively informationally thin, Humboldt was fully aware of the interesting insight that might result from the mapping of human activities of all sorts and even from the comparison of these results with the results of investigations into the spatial patterns of the physical world.

On a somewhat more sardonic note, Humboldt explored the alternative spaces created by contemporary and early geographers in the south-central region of North America in a map of error. On this map, entitled "Map of the False Positions for Mexico, Acapulco, Veracruz and the peak of Orizaba," Humboldt drew the outline of Mexico and placed these locations according to Arrowsmith, d'Anville, Covens, Harris, and the Connaissance du temps of 1804, among others. Many of these authorities effectively placed Acapulco well out in the Pacific, Mexico City almost anywhere in Mexico either in the Gulf of Mexico or close to the Pacific coast. This map, based on Tobias Mayer's *Mapa critica Germaniae,* was designed to show "just how imperfect the published maps of Mexico had been" (Beck and Bonacker 1969, lxvi), but it also nicely portrays the imaginative geography practiced by many of Humboldt's contemporaries.

In one of Humboldt's more playful graphics he compared the relative territorial extent of Spain and the Spanish colonial possessions by means of proportional squares (fig. 8.3).[25] Alongside this graphic is another which compares four dimensions: the comparative population size of European versus colonial territories and the comparative territorial extent of these same areas. This representation is highly suggestive of questions about the nature of colonialism that would only be explored in the twentieth century. Humboldt commented about this depiction:

> The figures combined on this plate demonstrate what is said below on the extraordinary disproportion observable between the extent of the colonies and the area of the European metropoles. The inequality of the territorial division of New Spain has been rendered apparent through a depiction of the indendancies by concentric squares. This graphic method is analogous to that first ingeniously used by Mr. Playfair in his commercial and political atlas and in his statistical maps of Europe. *Without attributing too much importance to these sketches, I cannot either regard them as mere intellectual games unrelated to science.* [emphasis added] It is true that Playfair showed the growth of the English national debt which had a strong resemblance to the peak of Teneriffe. But physicists have long used similar figures to show the rise and fall of the barometer and the average monthly temperature. It would be ridiculous to try to express by curved lines moral ideas, the prosperity of peoples, or the decadence of their literature. But anything that has to do with extent or quantity can be represented geometrically. Statistical projections, which speak to the senses without tiring the intellect have the

ANNE MARIE CLAIRE GODLEWSKA

advantage of bringing attention to a large number of important facts. (Beck and Bonacker 1969, lxxxiii–lxxxiv)

Humboldt was not to know that it was precisely the uses that he deemed absurd for such representations that would become the most persuasive in both social science and the public realm.

Humboldt was more interested in natural than social phenomena and far more interested in landforms and landscapes than he was in geological structures. He was, however, aware of research and innovation in all of these areas. He had noticed the stratigraphic diagrams being developed in geological studies and became convinced that

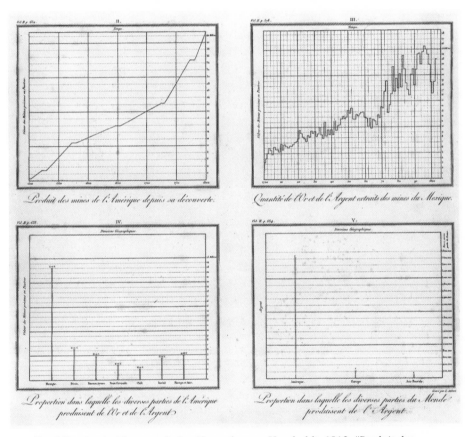

Fig. 8.2. A set of four graphs by Alexander von Humboldt, 1812. "Produit des mines de l'Amérique depuis sa découverte; Quantité de l'or et de l'argent extraits des mines du Mexique; Proportion dans laquelle les diverses parties de l'Amérique produisent de l'or et de l'argent; Proportion dans laquelle les diverses parties du Monde produisent et de l'argent." From Alexander von Humboldt, *Atlas géographique et physique du royaume de la Nouvelle-Espagne, fondé sur des observations astronomiques, des mesures trigonometriques et des nivellemens barometriques.* Paris: G. Dufour.

I. *Tableau comparatif de l'étendue territoriale des Intendances de la Nouvelle-Espagne.*

Fig. 8.3a.   A graph by Alexander von Humboldt, 1812. "Tableau comparatif de l'é-tendue territoriale des Intendances de la Nouvelle-Espagne." From Alexander von Humboldt, *Atlas géographique et physique du royaume de la Nouvelle-Espagne, fondé sur des observations astronomiques, des mesures trigonometriques et des nivellemens barometriques.* Paris: G. Dufour.

they could be modified to enhance the study of geognosy. By geog-nosy Humboldt meant the comparative study of the superposition of rocks and rock types around the globe. The purpose of this study was not the detailed mapping of bed stratigraphy but the acquisition of a sense of common form, common development, and the ways in which landscapes had altered as a result of local conditions. Thus,

ANNE MARIE CLAIRE GODLEWSKA

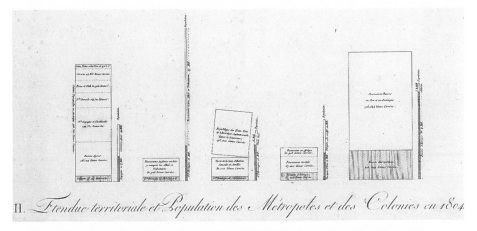

II. Etendue territoriale et Population des Métropoles et des Colonies en 1804.

Fig. 8.3b. A graph by Alexander von Humboldt. Etendue territoriale et Population des Métropoles et des Colonies en 1804. From Alexander von Humboldt, *Atlas géographique et physique du royaume de la Nouvelle-Espagne fondé sur des observations astronomiques, des mesures trigonometriques et des nivellemens barometriques*. Paris: G. Dufour.

he was not interested in the sort of chemical and mineral detail pursued by geologists but sought a more generalized picture which would allow him to immediately apprehend similar or identical patterns of superposition in widely separated areas. To this end he advocated two techniques of data representation, one graphic ("imitative or figured") and the other "algorithmic." The graphic technique Humboldt proposed looked very much like the geologist's stratigraphic diagram except that it covered a far larger area, paid more attention to the landforms in question, and generalized the formations into parallelograms with perhaps some additional stippling to suggest "the relations of composition and structure" which so preoccupied geologists (Humboldt 1823, 476–7). Humboldt's "Esquisse géognostique des formations entre la Vallée de Mexico, Moran et Totonilco" drawn in 1803 and engraved in 1833 seems to be the only example of this graphic pasigraphy (fig. 8.4). It bears ample sign of inadequate information on the structures beneath the surface topography for much of the area depicted. Humboldt also produced block diagrams designed to reveal and accentuate the topography of Central Mexico (fig. 8.5).[26] He was looking for a particular type of information, about the succession and relative age of rocks. He believed he could decipher these by "fixing the attention on the most general relations of *relative position, alternation,* and the *suppression* of certain terms of the series."

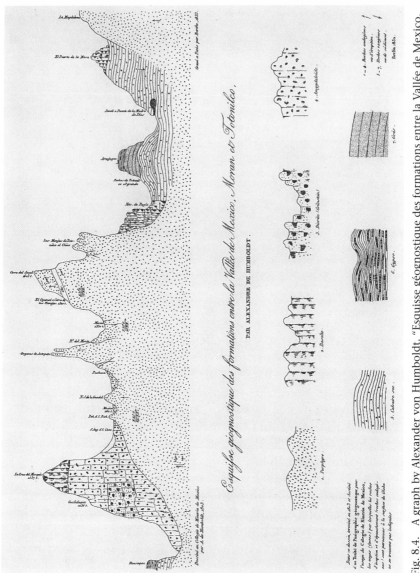

Fig. 8.4. A graph by Alexander von Humboldt. "Esquisse géognostique des formations entre la Vallée de Mexico, Moran et Totonilco." From Beck and Bonacker, 1969. Facsimile reproduction of: Alexander von Humboldt, *Atlas géographique et physique du Royaume de la Nouvelle-Espagne.* Stuttgart: A. Brockhaus, plate 21.

The whole geognosy of positions being a problem of *series,* or the simple or periodical succession of *certain terms,* the various super-imposed formations may be expressed by general characters, for instance, by the letters of the alphabet. . . . The more we make abstraction of the value of signs (of the composition and structure of the rocks), the better we seize, by the conciseness of a language in some degree algebraic, the most complicated relations of position, and the periodical return of formations. The signs "α," "β," and "γ" will no longer represent granite, gneiss, and mica-slate; red sandstone, zechstein, and variegated sandstone; chalk, tertiary sandstone with lignite, and Parisian limestone; they will only be the terms of a series, simple abstractions of the mind. (Humboldt 1823, 466–67)

Beyond looking for patterns in the structures beneath the earth and between continents, Humboldt was looking for modes of representation that could generate ideas capable of bridging the gap between different realms:

In this *geognostical essay,* as well as in my researches on the *isothermal lines,* on the *geography of plants,* and on the laws which have been observed in the *distribution of organic bodies,* I have endeavoured, at the same time that I presented the detail of the phenomena, to generalize the ideas respecting them, and to connect them with the great questions in natural philosophy. I have dwelt chiefly on the phenomena of *alternation,* of *oscillation,* and of *local suppression,* and in those which result from the passage of one formation to another in consequence of *interior development.* These subjects are not mere theoretical speculations; far from being useless, they lead us to the knowledge of the laws of nature. (Humboldt 1823, vi)

If the final aim was the decipherment of the laws of nature and his method was generalization to emphasize a few key concepts, he also sought to make the link between his geognostical and botanical work on a more banal level. Thus, his landform diagrams included information on temperature, the snow line, etc., that was designed to provide measurement-based scalar links to "the large picture [depiction] attached to my *Géographie des plantes*" (Beck and Bonacker 1969, lxxii).

The most influential of Humboldt's experiments into graphic expression was without question the remarkable multidimensional pictorial graph reproduced in his *Essai sur la géographie des plantes* and entitled "Géographie des plantes équinoxiales (fig. 8.6). Tableau physique des Andes et Pays voisins." Humboldt's overriding concern in this essay, and in his Tableau, was both to introduce a new way of looking at the world of plants to his natural history colleagues

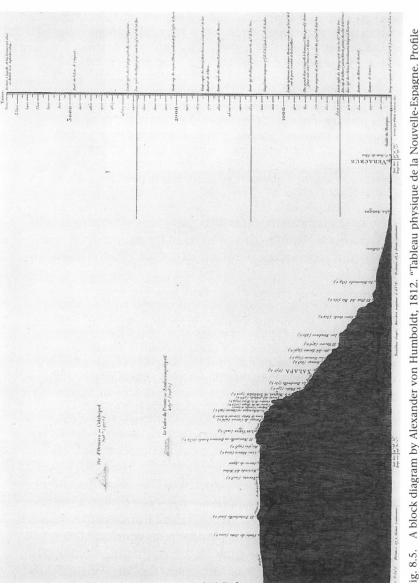

Fig. 8.5. A block diagram by Alexander von Humboldt, 1812. "Tableau physique de la Nouvelle-Espagne. Profile du Chemin d'Acapulco à Mexico, et de Mexico à Veracruz." From Alexander von Humboldt, *Atlas géographique et physique du royaume de la Nouvelle-Espagne, fondé sur des observations astronomiques, des mesures trigonometriques et des nivellemens barometriques.* Paris: G. Dufour.

and to the larger public and to demonstrate how interconnected all of nature's forces and aspects really were. In a sense, then, his *Essai sur la Géographie des plantes,* more than any of his other works, was a concise version of the *Cosmos.* Arguably, too, his "Tableau" was an even more concise version of the *Essai.* In spite of the length of the five-volume *Cosmos,* Humboldt believed in the power of concise and precise expression which might convince skeptics of the importance of looking for interconnections between different realms and different lines of enquiry (Humboldt 1805, v–vi).[27] The "Tableau" presented a view of three peaks of a mountain chain, one of which was colored with vegetation while the other two carried the names of the plants to be found at that altitude. On the sides of the illustration there was a scale marked in toises (1 toise = 1.949 meters) and in meters. Columns parallel to the meters/toise columns indicated: minimum and maximum temperatures at given heights; the chemical composition of the air at different heights (oxygen, hydrogen, carbon); the lower limits of perpetual snow at different latitudes; the typical animals to be found at each elevation; the temperature at which water boils (at different elevations); "Vues géologiques" in which he observed that rock type was independent of elevation but that in any given area there is an order of superposition, inclination, and direction of beds which is determined by a "système de forces particulier." In that same column he then described the standard superpositioning and pointed out what was geologically particular about the equatorial regions: depth of the beds, the elevation of the postgranitic formations, etc. Subsequent parallel columns described: the intensity of light at different elevations; the humidity level at different elevations; how blue the sky looked measured in cyanometers at different elevations; the gravitational force at different elevations; the kinds of agriculture engaged in at various elevations; the incidence of electrical phenomena at various elevations; measured heights in different parts of the world; the distance at which the mountains would be visible from the sea, making abstraction of refraction; and the refraction at a given height and at 0 degrees of temperature.

In the text of his *Essai,* Humboldt then tried to draw the reader to his own conclusions by discussing linkages between phenomena that he had been unable to bring out in the graphic. These included: the link between plant geography and geology to be found in fossil remains of plants (which added the dimension of geologic time) (Humboldt 1805, 19–23); the link between plant geography and the political and moral history of "man" forged by the human need for

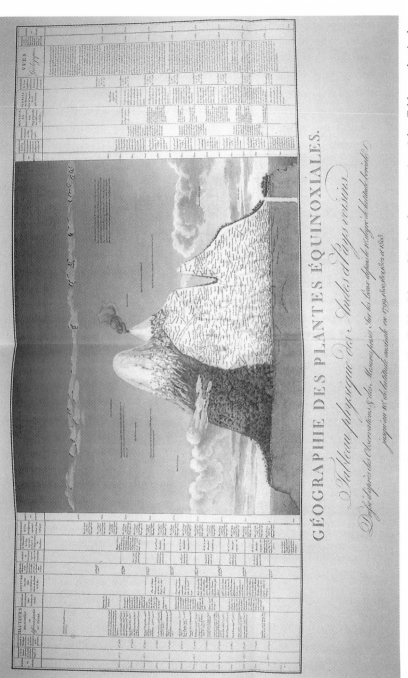

Fig. 8.6. A multidimensional pictorial graph by Alexander von Humboldt. Géographie des plantes equinoxiales: Tableaux des Andes et Pays Voisines. From *Voyage de Humboldt et Bonpland: voyage aux régions equinoxiales du nouveau continent, cinquième partie: essai sur la géographie des plantes.* Paris: Theatrum orbis terrarum, 1805. (Facsimile integral de l'édition Paris 1805–1834).

food (which added the dimension of human material history) (Humboldt 1805, 27–30); and the link between nature and "man's" spiritual being to be found in the imitative arts (which added the dimension of intellectual history) (Humboldt 1805, 30). These linkages raised fundamental scientific questions about both the history of the earth and of human history on the earth—precisely the sorts of questions Humboldt wanted his readers to ask.

At the end of this remarkable essay, which illustrated his even more remarkable "Tableau," Humboldt, one of the greatest early nineteenth-century exponents of field research, spoke of the power of study through books and art. Broaching a theme strikingly reminiscent of those later developed by Foucault in his "Fantasia of the Library" (Foucault 1977), he commented:

> That is how, without a doubt, enlightenment and civilisation most influence our individual happiness. They [books and art (or text and graphic)] make us live at one and the same time in the present and in the past. They gather around us all that nature has produced in the diverse climates and put us into contact with all of the peoples of the earth. Sustained by the discoveries already made, we can leap into the future, and with a premonition of the consequences of phenomena, we can establish the laws of nature. It is through this research that we open the way to intellectual delight, to a moral liberty which fortifies us against the blows of destiny and which no exterior power can undermine. (Humboldt 1805, 35)

This, then, was the purpose of his oeuvre: to place "man" in a position from which he could observe nature, himself, and the workings of the entire cosmos.

8.8. Conclusion

This chapter has sought to capture the essence of Humboldt's "meditation" on the cosmos over a lifetime of exploration, research, experimentation, discussion, and thought. When we examine his work and contrast it with the natural history or geography of the Enlightenment, its innovative nature jumps out at us. Certainly the contrast with the type of questions being asked by contemporary geographers—still focused on the flat and static surface of the map, or on literary description derived from second-, third-, or fourth-hand results—is stark. Standing astride a major shift in the nature of the modern earth and life sciences away from description and toward explanation, he embraced theory, its explanatory power, and its exploration of immediate cause. Not satisfied with the study of the unchanging and the static, his curiosity turned to movement, change, and distribution. Stretching beneath the surface, he strove to understand the interior structures and functions of phenomena, whether organic, inorganic, or social. He advocated and pointed the way to a new multidisciplinarity through the universal language of mathematics. He played with concepts of scale to try to arrive at a unit of analysis that would allow him to both analyze and synthesize.

Yet, Humboldt was also trying to retain something of the holism of traditional geography and, for that matter, of Enlightenment science which saw the cosmos as a whole that could be elucidated by description. In geography, that description took two forms, textual and cartographic, which sought to place "everything" within an x,y coordinate system. In natural history, the evolving descriptive system was classification. Both implied a whole which could be gradually filled in according to a usefully rigid conception of nature. Both required critical observation and data collection. Humboldt's work was innovative in his attempt to break through and go beyond the classification systems of the "miserable archivists of Nature" to see and combine what these cataloging systems could not perceive. In plant geography, that meant sacrificing classification based on superficial characteristics to a combined analysis of the interaction of interior structures and location-based exterior influences, a physics of nature. In mapping, it meant challenging and testing the veracity of maps and their classic forms of expression, going beneath and above the surface of the map to explore locationally based cross sections and elevations and reformulating the map from a descriptive statement to an argument infused with theory. All of these were part of a larger effort to see nature at once in great detail, using the insights from experimental physics, anatomy, geodetic, and astronomical observation, and, at the same time, with a holism capable of integrating organic, inorganic, and human nature in all of its diversity and complexity.

The rise of thematic mapping coincides with the development of the new technologies of graphic reproduction beginning with lithography in the 1820s. Undoubtedly thematic mapping was well served by these technologies, which permitted the use of color, were cheaper, easier to draw, and easier to integrate into texts. Likewise, the establishment of regular statistical survey led to an accumulation of data that began to facilitate the graphic display of geographic information by mid-century. It was also the kinds of questions that scientists were beginning to ask of the natural and human world

ANNE MARIE CLAIRE GODLEWSKA

that stimulated the development of this new graphic form of expression. Thematic mapping came to be in precisely the period which witnessed a shift from a more descriptive to a more explanatory mode of investigation. That is far from coincidental, as the thematic map enhanced analysis while permitting synthesis, both geographic and not. In his experimentation with maps, graphs, and profiles, Humboldt was looking for ways to express what was seminal in his work. He was, to borrow a phrase from Bravo in chapter 7, looking for commensurability between the richness of nature and a language capable of representing and ordering that richness. He was looking for a language at once highly descriptive but also analytical: capable of moving beyond geographic space to assume the spaces of scientific theory; capable of comparing like things in very different places; and one that could reveal new linkages. Humboldt's experimentation with graphic expression is interesting because it is suggestive of a shift in both the nature and language of science.

Notes

1. First developed in Michel Foucault *Les mots et les choses* (Paris: Gallimard, 1966) but qualified and elaborated upon in his *L'Archéologie du savoir* (Paris: Gallimard, 1969). An excellent guide to the writings of Michel Foucault is Alan Sheridan, *Michel Foucault: The Will to Truth* (London and New York: Tavistock Publications, 1980).

2. Michael Dettelbach, in his superb essay in *Visions of Empire,* sees this commitment to the unity of the cosmos as part of a larger Humboldtian project investigating global physics. He argues that Humboldt was deeply influenced by both Laplacian physics which sought to understand the world through the multiplication of measurement-based observations and the German aesthetic response (in particular the aesthetic writings of Freidrich Schiller) to the Napoleonic imposition of modern centralised (anti-local) state control in the Germanies. See Michael Dettelbach, "Global Physics and Aesthetic Empire: Humboldt's Physical Portrait of the Tropics," in David Philip Miller and Peter Hanns Reill, eds., *Visions of Empire. Voyages, Botany, and Representations of Nature* (Cambridge: Cambridge University Press, 1996), 258–292.

3. This is something of a simplification. As Arthur Robinson explains, thematic mapping can be seen as emerging from a whole series of confluent events and innovations from new concepts to new technologies to broader social change. In science, he focuses on the development of the law of universal gravitation, the refinement of quantitative description of observable facts, the development of mathematical forms of expression including differential calculus and probability, the development of universal standards of measurement, the rise of social statistics, and the growth in interest in the links between environment and disease. There were, in his view, equally important preconditions in industry, trade, and its supporting technology,

including transportation. See Arthur Robinson, *Early Thematic Mapping in the History of Cartography* (Chicago: University of Chicago Press, 1982), 26–43.

4. The rise to prominence of natural history and its evolution into biology and geology in the late eighteenth and nineteenth centuries entailed a shift from a focus on the mostly static depiction of form and location to an increasing appreciation of the mostly dynamic behaviors, interactions, and movements of living and nonliving phenomena. It also entailed a shift from the domination by the distant past (as laid down by God or long-past physical forces) to the immediate, interactive present moved by forces very evidently still active. Both of these shifts reduced the value of the static geographic map. Instead, it became necessary to move from the possibility of building one kind of model at different scales to that of building many different kinds of models each with a logic dictated by the dynamic, immediate, interactive phenomena under study (thematic maps and scientific models). As Bruno Latour has pointed out, each of these maps or models, from Mendeleev's periodic table to Reynolds's turbulence coefficient, created a new and distinct space and time. This space and time was the product of an intellectual construct or classification, the logic of which was related to the reshuffling of connections between elements (such as chemical elements or plant genera) which was made possible by the separation of these from their geographic context and their representation by mobile, stable, and combinable inscriptions. The space and time created by this reshuffling was and is accessible only to those trained to read the inscriptions and able to understand the logic behind the reshuffling. The space and time of chemistry, geology, hydraulic engineering, and sociology is not necessarily closely related to that of geography. Consequently, geographers were and are no more able to describe and map these space/time relations than any other noninitiate.

5. As did many nineteenth-century geographers.

6. One of his clearest statements of the nature of theory, its origins in empirical research and its great distance from speculation came from a gentle but unequivocal refutation of Buachian theory. "The waters certainly give some idea of the layout of the country. But the course of the rivers only indicates the difference in level of the terrain through which they flow. Knowledge of the large valleys or basins and the study of the lines of division [between basins] is of greatest interest to the hydrographic engineer. It is, however, by a false application of the principles of hydrography that, from the depths of their armchairs, geographers have sought to determine the direction of mountain chains in countries where they think they know the exact course of the rivers. They imagined that two large hydrographic basins could only be separated by high elevations or that a large river could not change direction unless obstructed by a group of mountains. They forgot that often, either as a result of the nature of the rocks, or as a result of the slope of the beds, the highest plateaus give birth to no river while the sources of the largest rivers are often far from mountain chains. Also, the attempts that have been made to date to make maps according to theoretical ideas have not been very successful. Indeed, it is that much harder to thus *guess* the configuration of the terrain as the pelagic currents and most of the rivers which actually shaped the earth, have completely disappeared.

Complete knowledge of those that existed and of those that still exist today could tell us something about the *slope of the valleys,* but nothing about the absolute height of the mountains, or about the location of their chains!"

7. "... The winds; the nearness of the seas which are immense reservoirs of a heat that varies only slightly; the inclination, chemical nature, colour, albedo, and evaporation characteristics of the soil; the direction of the mountain chains which act, either by favouring the play of descending currents or by providing shelter from certain winds; the forms of the land masses and their prolongation toward the poles; the quantity of snow that covers them in the winter; their temperature highs and reverberation in the summer; finally that ice, variable in its extent, which forms circumpolar continents and whose detached parts, drawn by currents, sensibly modify the climate of the temperate zone."

8. Discussing prices, he commented: "These numerical data are undoubtedly of great value; still, considerable caution is required in the interpretation of them, since the problem of the prices of corn, as well as of prices in general, is a very complicated one; and varying theoretical views, the influence of land-owners, as well as the unequal local accumulation of money and commodities, all produce their effect on legislation. Besides this, atmospheric changes (the mean warmth of the spring and summer months) affecting the cultivation of cereals, do not embrace at the same moment the entire area of agricultural Europe. A disproportionate increase in population, and consequently increasing intercourse, multiplies the demand for the precious metals. In the standard which we think to find in the fluctuating prices of grain, we have to deal with two contemporaneously fluctuating quantities. The increased price of grain, even for a particular country, no more indicates the relative increase of money, than it informs us of the state of weather and the quantity of sunbeams. Data which should embrace a considerable portion of Europe contemporaneously are nowhere to be found ..." This work was first published as "Ueber die Schwankungen der Goldproduktion mit Rücksicht auf staatswirthschaftliche Probleme," in *Deutsche Vierteljahrs Schrift* 1 (1838) IV, pp. 1–40. I am indebted to Dr. Ingo Schwarz for this reference.

9. "When we reflect how considerable must be the amount of spices in the present consumption of Europe, compared with what it was at the conclusion of the fifteenth century, though constituting the greater part of the then existing commerce, we shall discover another remarkable example of the potency of precious metals, when exercising their concentrated force in a narrow space; at that time the shores of the Mediterranean and Western Europe."

10. "Although in the course of a laborious life, I have had the pleasure of seeing a greater extent of mountains than any other geognost, the little I observed is lost in the great mass of facts which I have undertaken to display."

11. In his essay *Sur deux nouvelles espèces de Crotales* [Rattlesnakes], he demonstrated at once the rigor and subtlety of comparisons and the critical role of sensitivity to the scale of evidence available to support such comparisons. Speaking of the different incidence of snakes between different parts of the world, he commented: "The reasons for these differences is clear when one considers the greater or lesser humidity of the soil, the density of

the forests, the extent of the savannas that replace these forests, the variation in summer temperatures, and other meteorological phenomena which render countries at the same latitude more or less favourable for the multiplication of Ophidia. Many naturalists, more concerned with contrasts than with the relationships between the two worlds, have described all of America as recently emerged from the bosom of the ocean, covered with swamps, inhabited by reptiles of a greater variety of forms and more numerous than Africa and the Great-Islands. These ideas have been long accepted because they were conceived of by a man of genius, and consigned to works, many of which owe their fame less to the exactitude of facts than to the charm of their style and the grandeur of their conceptions. As a healthy criticism brought naturalists back to the path of observation, and they studied the uniform laws of nature, and the modifications of these laws under the influence of local circumstances, the two continents seemed less contrasted in terms of physical phenomena. They found amidst the rocks of both hemispheres that series of formations which attest to a succession of the same catastrophes on the surface of the globe. Then they recognised that the portrait of America cannot be drawn from that of the inundated lands between the Orinoco and the Amazon rivers. It is not from the large number of reptiles in Egypt's delta region or in the 'Ouanagarah' in 'Nigritte' that one can precisely judge the number of species indigenous to all of Africa."

12. Humboldt periodically expressed his disgust with travel accounts as on pp. 6–7. On pp. 48–51 Humboldt discussed the contradiction between the scientific study of peoples and their customs and the focus and approach of the travel narrative.

13. This is clear from his failure to complete his *Voyage aux régions équinoxiales* and from the choppy nature of its read. Indeed, throughout this work, Humboldt seems unable to choose between a systematic focus and narrative, making abrupt transitions from approach to the other. Systematic analysis was clearly his preference as he tells the reader that he reluctantly chose to write a narrative due to the delay and difficulty in publishing much of the systematic work.

14. ". . . Our voyage would have been much more useful to the progress of descriptive zoology if, instead of being perpetually on the move, our situation had allowed us to stay for several months in the missions . . ."

15. "Geognosy is not confined to the research for diagnostic characters; it comprehends the whole of these relations in which we may consider every formation: 1st, its position, 2ndly, its oryctognostic constitution (that is, its chemical composition, and the particular mode of aggregation, more or less crystalline, of its molecules); 3dly, the association of different organised bodies that are found imbedded in it. If the superposition of different heterogeneous rocky masses exhibits to us the successive order of their formation, why should we be interested in knowing the state of organic nature at the different epochs when those deposits were formed? . . . The study of formations is similar to that of organised beings. Botany and zoology, considered at present in a more elevated point of view, are no longer confined to the examination of some external characters, and distinctions of species; those sciences enter more profoundly into the study of the whole vegetable

and animal organisation. The characters drawn from the forms of shells suffice to distinguish the different species of testaceous acephalae. Shall we, on that account, regard as superfluous the knowledge of the animals which inhabit these shells? Such is the connection among phenomena, and their natural relations (those of life as well as those displayed by the stony deposits formed at different periods), that if we neglect any of them we shall form not only an incomplete, but most frequently, an erroneous picture of the world."

16. For an excellent discussion of these protothematic maps which Gilles Palsky describes as "cartes spéciales" and "cartes singuliéres," see Palsky (1996a, 1996b).

17. Arthur Robinson and Helen Wallis argue that Humboldt's use of the isotherm and his presentation of a paper on isolines to the French Academy of Sciences in 1817 was "the catalyst for many similar uses of the isarithm in thematic cartography. . . ."

18. Writing about the mapping of the Port of Acapulco, he commented: "We should not be surprised by how uncertain we still are about the position of a port in the southern sea. After all, the longitude of Amsterdam was uncertain, just a few years ago, and not by two or three minutes but by a third of a degree."

19. Speaking of these hill symbols, he commented: "This last, the most imperfect and the oldest of all [types of relief depiction], creates a mixture of two kinds of very heterogeneous projections. I won't try to conceal, however, that this disadvantage is almost balanced by a real advantage. The old method provides symbols that announce vaguely 'the terrain is mountainous,' or that 'there are mountains in this or that province.' The more vague this hieroglyphic language, the less exposed it is to error. By contrast, the hachure method forces the draftsman to say more than he knows, more even than it is possible to know about the geologic constitution of a huge expanse of terrain . . ."

20. "The distribution of heat on the globe is one of those phenomena about which we have long known in general terms, but which we have been unable to establish with rigour or to submit to calculation, so that experience and observation can provide data by which theory can be refined. The aim of this memoir is to facilitate the combination of data. Results drawn from a large number of unpublished observations will be gathered together according to a method which has never been tried before, in spite of the fact that its advantages, in the demonstration of magnetic declination and inclination, have been known for a century."

21. "If, instead of geographic maps, all we had were tables of coordinates of latitude, longitude and height, a large number of interesting relationships, clear in the configuration of continents and the inequalities of their surface, would have remained unknown." (Humboldt 1813).

22. In time, the space to be found in some thematic maps would become increasingly removed from geographic space, as in the case of genetic maps in which space is chemically rather than topologically defined. For a broad and popular sense of mapping in contemporary science, see Stephen S. Hall, *Mapping the Next Millenium. How Computer Cartography Is Revolutionizing the Face of Science* (New York: Vintage Books, 1992).

23. "Most natural phenomena have two distinct parts: one of which can be submitted to rigorous calculation; the other can only be attained by induction and analogy."

24. In his comments on this graphic he remarked: "This map shows the ebb and flow of metallic wealth. Here we can see the general west to east movement, which is in opposition to the flow of the oceans, the atmosphere and the civilisation of our species!"

25. This graphic was adapted from one developed by August Crome. For this link I am grateful to Sybilla Nikolow, "Rendering the Strength of the State Visible: August Crome's Statistical Maps Around 1800," conference paper presented at New Perspectives on Alexander von Humboldt. An International Symposium, Georg-August-Universität Göttingen, 29–31 May 1997.

26. In his description of those diagrams in which he exaggerated the height of the formations to give a clearer sense of their appearance, he suggested why he ultimately abandoned his graphic pasigraphy: ". . . It was after careful thought that I decided to separate the geological profiles which show the superposition of rocks, from the physical depictions which show differences in height of terrain. It is very difficult, I would almost say impossible, to sketch a geological cross-section of an extensive region if that cross-section must also be correct in terms of its height. . . . Those who have thought about graphic methods, and who have tried to perfect them, will feel, as I do, that these methods can never combine all advantages. Similarly, a map too crowded with symbols becomes confusing and loses its principal advantage which is to allow one to grasp a large number of relationships at once. . . ."

27. ". . . I prefer here to draw the attention of physicists to the great phenomena of nature in the lands through which I have travelled. It is their unity that I have considered in this essay. Here, I am presenting the general observations which I develop in detail in other works being prepared for the public. Here I consider all physical phenomena observable either on the surface of the earth or in the atmosphere which surrounds it. The physicist familiar with the present state of science, and especially of meteorology, will not be surprised to see such a large number of phenomena dealt with in such a small number of pages. *Had I been able to devote more time to this writing, my work would have been even shorter; because a picture [tableau] should show only the large physical perspective and results which are certain and susceptible of numerical expression* [emphasis added]."

References

Beck, Hanno, and Wilhelm Bonacker. 1969. Facsimile reproduction of: Alexander von Humboldt, *Atlas géographique et physique du Royaume de la Nouvelle-Espagne*. Stuttgart: A. Brockhaus.

Cambrosio Alberto, Jacobi Daniel, and Peter Keating. 1993. "Ehrlich's 'Beautiful Pictures' and the Controversial Beginnings of Immunological Imagery." *Isis* 84: 662–699.

Camerini, Jane. 1993. "Evolution, Biogeography and Maps: An Early History of the Wallace Line." *Isis* 84: 700–727.

Chevalier, Michel. 1986. "L'abbé Soulavie, précurseur ardéchois de la géographie moderne (1752–1813)." *Revue du Vivarais* 40, 2: 81–100.

Claval, Paul. 1968. *Régions, nations, grands espaces. Géographie générale des ensembles territoriaux.* Paris: Marie-Thérèse Genin.

Cuvier, George Léopold. 1811. "Recherches anatomiques sur les reptiles regardés encore comme douteux par les naturalistes; faites à l'occasion de l'Axolotl, rapporté par M. De Humboldt du Mexique par M. Cuvier," in *Recueil d'observations de zoologie et d'anatomie comparée faites dans l'océan atlantique, dans l'intérieur du nouveau continent et dans la mer du sud pendant les années 1799, 1800, 1801, 1802 et 1803.* Paris: F. Schoell and Gel Dufour, 93–126.

Decandolle, Augustin Pyramus. 1820. "Géographie Botanique," in *Dictionnaire des sciences naturelles par plusieurs professeurs du Jardin du Roi.* Paris: Le Normant and Strasbourg, Levrault.

Dettelbach, Michael. 1996. "Global Physics and Aesthetic Empire: Humboldt's Physical Portrait of the Tropics," in David Philip Miller and Peter Hanns Reill, eds. *Visions of Empire. Voyages, Botany, and Representations of Nature.* Cambridge: Cambridge University Press, 258–292.

Foucault, Michel. 1966. *Les mots et les choses.* Paris: Gallimard.

———. 1969. *L'Archéologie du savior.* Paris: Gallimard.

———. 1977. "Fantasia of the Library," in *Language, Counter-Memory, Practice. Selected Essays and Interviews,* Donald F. Bouchard, ed. Ithaca, New York: Cornell University Press, 87–109.

Giraud-Soulavie, Louis Giraud M. L'abbé. 1780. *Histoire naturelle de la France méridionale, ou recherches sur la minéralogie du Vivarais, du Velay, du Viennois, du Valentinois, du Forez, de l'Auvergne, de L'Uségois, du Comtat Venaissin, de la Provence, des Diocèses de Nismes, Montpellier, Agde, etc. Sur la Physique de la Mer Méditerranée, sur les Météores, les Arbres, les Animaux, l'Homme et la Femme de ces contrées.* Vol. 1. Nismes: C. Belle.

———. 1780. *Histoire naturelle du Vivarais.* Paris: J.-F. Quillan.

———. 1781. *Histoire naturelle de la France méridionale, ou recherches sur la minéralogie du Vivarais, du Velay, du Viennois, du Valentinois, du Forez, de l'Auvergne, de l'Uségois, du Comtat Venaissin, de la Provence, des Diocèses de Nismes, Montpellier, Agde, etc. Sur la Physique de la Mer Méditerranée, sur les Météores, les Arbres, les Animaux, l'Homme et la Femme de ces contrées.* Vol. 3. Paris: J.-F. Quillan.

Godlewska, Anne. 1995. "Map, Text and Image. The Mentality of Enlightened Conquerors: A New Look at the *Description de l'Egypte.*" *Transactions of the Institute of British Geographers,* New series, 20, 1: 5–28.

———. Forthcoming. 1999. "The Powerful Mapping Metaphor," in *Geography Unbound. French Geographic Science from Cassini to Humboldt.* Chicago: University of Chicago Press.

Humboldt, Alexander von. 1805. *Essai sur le géographie des plantes; accompagné d'un tableau physique des régions équinoxiales, fondé sur les mesures exécutées, depuis le dixième degré de latitude boréale jusqu'au dixième degré de latitude australe, pendant les années 1799, 1800, 1801, 1802 et 1803 par Al. De Humboldt et A. Bonpland.* Paris: Chez Levrault, Schoell et Compagnie, libraries, 1805, v–vi.

———. 1811a. *Sur un ver intestin trouvé dans les poumons du serpent a son-*

nettes, de Cumana; par A de Humboldt, in Humboldt, Al. De et A. Bonpland. *Recueil d'observations de zoologie et d'anatomie comparée faites dans l'océan atlantique, dans l'intérieur du nouveau continent et dans la mer du sud pendant les années 1799, 1800, 1801, 1802 et 1803.* Paris: F. Schoell and Gel Dufour, 298–304.

———. 1811b. *Political Essay on the Kingdom of New Spain.* . . . London: Longman, Hurst, Rees, Orme, and Brown, Paternoster-Row.

———. 1811c. *Sur la respiration des Crocodiles par A. de Humboldt* and *Observations sur l'anguille électrique . . . du Nouveau continent,* in Humboldt, Al. De et A. Bonpland. *Recueil d'observations de zoologie et d'anatomie comparée faites dans l'océan atlantique, dans l'intérieur du nouveau continent et dans la mer du sud pendant les années 1799, 1800, 1801, 1802 et 1803.* Paris: F. Schoell and Gel Dufour.

———. 1811d. *Sur deux nouvelles espèces de Crotales* in Humboldt, Al. De et A. Bonpland. *Recueil d'observations de zoologie et d'anatomie comparée faites dans l'océan atlantique, dans l'intérieur du nouveau continent et dans la mer du sud pendant les années 1799, 1800, 1801, 1802 et 1803.* Paris: F. Schoell and Gel Dufour.

———. 1811e. "Mémoire sur l'os Hyoïde et le Larynx des oiseaux, des singes et du crocodile," in *Recueil d'observations de zoologie et d'anatomie comparée faites dans l'océan atlantique, dans l'intérieur du nouveau continent et dans la mer du sud pendant les années 1799, 1800, 1801, 1802 et 1803.* Paris: F. Schoell and Gel Dufour, 1–13.

———. 1811f. *Sur les singes qui habitent les rives de l'Orénoque, du cassiquaire et du rio Negro; par A. De Humboldt,* in *Recueil d'observations de zoologie et d'anatomie comparée faites dans l'océan atlantique, dans l'intérieur du nouveau continent et dans la mer du sud pendant les années 1799, 1800, 1801, 1802 et 1803.* Paris: F. Schoell and Gel Dufour.

———. 1813. "Des lignes isothermes et de la distribution de la chaleur sur le globe," in *Mémoires de physique et de chimie de la Société d'Arcueil,* vol. 3. Paris: J. Klostermann fils.

———. 1816. *Voyage aux régions équinoxiales du nouveau continent, fait en 1799, 1800, 1801, 1802, 1803 et 1804, par Al. De Humboldt et A. Bonpland . . . avec un atlas géographique et physique.* Paris: À la Librarie Grecque.

———. 1823. *A Geognostical Essay on the Superposition of Rocks in Both Hemispheres.* London: Longman, Hurst, Rees, Orme, Brown, and Green, Paternoster-Row.

———. 1849–1852. *Cosmos. Sketch of a Physical Description of the Universe.* Translated by E. C. Otte. Vol. 1 (1849), 2 (1849), and 4 (1852). London: Henry G. Bohn.

———. 1854. *Volcans des Cordillères de Quito et du Mexique.* Paris: Gide et J. Baudry.

———. 1864. *Cosmos. Essai d'une description physique du monde.* Translated by Harvé Auguste Etienne Albans Faye and Louis Charles Galuski. Vol. 3. Paris: Théodore Morgand.

———. 1876. *Cosmos. Sketch of a Physical Description of the Universe.* Translated by E. C. Otte and W. S. Dallas. Vol. 5. London: George Bell and Sons.

———. 1900. *The Fluctuations of Gold.* New York: The Cambridge Encyclopedia Co.

————. 1973 (1805). "Géographie des plantes équinoxiales. Tableau physique des Andes et Pays voisins. Dressé d'après des observations et des mesures prises sur les lieux depuis le 10e degré de latitude boréale jusqu'au 10e de latitude australe en 1799, 1800, 1801, 1802 et 1803," in *Voyage de Humboldt et Bonpland. Voyage aux régions équinoxiales du nouveau continent* Cinquième partie *Essai sur la géographie des plantes* Paris, 1805. Facsimilé intégral de l'édition Paris 1805–1834 by Amsterdam: Theatrum Orbis Terrarum Ltd.

Humboldt, Alexander von, and Aimé Bonpland. 1811. *Recueil d'observations de zoologie et d'anatomie comparée faites dans l'océan atlantique, dans l'intérieur du nouveau continent et dans la mer du sud pendant les années 1799, 1800, 1801, 1802 et 1803.* Paris: F. Schoell and Gel Dufour.

Latour, Bruno. 1987. *Science in Action.* Cambridge, MA: Harvard University Press.

LeGrand, Homer E. 1990. "Is a Picture Worth a Thousand Experiments?" *Experimental Inquiries: Historical, Philosophical and Social Studies of Experimentation in Science,* 241–271. Dordrecht: Kluwer.

Minguet, Charles. 1969. *Alexandre de Humboldt historien et géographe de l'Amérique espagnole 1799–1804.* Paris: François Maspero.

Omalius d'Halloy, Jean-Baptiste-Julien d'. 1823. *Observations sur un essai de carte géologique de la France, des Pays-Bas et des contreées voisines. . . .* Paris: Imprimerie de Madame Huzard.

Palsky, Gilles. 1996a. "Aux origines de la cartographie thématique: Les cartes spéciales avant 1800," in *La Cartografia Francesca* Cicle de conferénces sobre Història de la Cartografia, 5è curs, 21, 22, 23, 24 i 25 de febrer de 1994, 129–145. Barcelona: Institut Cartogràfic de Catalunya.

————. 1996b. *Des Chiffres et des Cartes. La Cartographie Quantitative au XIXe siècle.* Paris: Comité des travaux historiques et scientifiques.

Robinson, Arthur. 1982. *Early Thematic Mapping in the History of Cartography.* Chicago: University of Chicago Press.

Robinson, Arthur, and Helen Wallis. 1967. "Humboldt's Map of Isothermal Lines: A Milestone in Thematic Cartography." *Cartographic Journal* 4, 2 (December), 119–122.

Rudwick, Martin J. S. 1976. "The Emergence of a Visual Language for Geological Science, 1760–1840." *History of Science* 14: 149–95.

————. 1985. *The Great Devonian Controversy. The Shaping of Scientific Knowledge among Gentlemanly Specialists.* Chicago and London: University of Chicago Press.

Smith, Bernard. 1988. *European Vision and the South Pacific.* 2d ed. New Haven and London: Yale University Press.

Wood, Denis, with John Fels. 1992. *The Power of Maps.* New York and London: The Guilford Press.

Travelings

It has become increasingly difficult to conceive of the period known as the European Enlightenment without recourse to scientific travel. Of crucial importance were the expeditionary adventures of a host of scientific explorers like Cook and Bougainville and the imagined geographies that such expeditions generated of the South Pacific. Soon the geographical variability and human diversity that such endeavors disclosed began to subvert long-held convictions about a common human nature and a cosmopolitan constitution. But it was not just scientific explorers and their associated instruments that moved around the globe; books and ideas also traveled and were appropriated differently in different places. Attending to travel in a variety of forms thus takes us to the heart of a number of key epistemological and moral dilemmas in Enlightenment philosophy.

Invoking the myth of Perseus as a trope for the duplicities of exploration and conquest, Dorinda Outram reflects on the moral and epistemological quandaries at the core of Enlightenment travel. For those journeys raised troubling doubts about the nature of human nature and universal rationality. At the same time, the knowledge transmitted by travelers to European centers of Enlightenment culture prompted questions about the reliability and status of explorers' reports. Accordingly, matters of testimony and trust occupied center-stage in the acquisition of distant knowledge. For acquiring dependable information about remote regions necessarily required trust in another's moral integrity and perceptual reliability— in short, in the mind and body of another person. But such matters, as Outram reminds us, only reveal with particular clarity the more general Enlightenment dilemma over how *anything* about the external world could be known. And yet field knowledge was frequently treated with suspicion by those schooled in laboratory science. The potential for what we might call ethical and empirical 'slack' between the knower and the known were just too great to allow a sedentary naturalist like Cuvier to take field investigators' reports at face value. Thus it is clear that questions of location and dislocation—so supremely exposed in the tradition of exploration—were fundamental to Enlightenment science. Indeed, such theoretical concerns crystallized in the very *body* of the explorer, which became the site where epistemic authority acquired its truth status.

The geographical discourse embedded in travel reports attracts Paul Carter's attention. Concerned with what he calls its mythopoeic origins, Carter points to geography's efforts to derive *deductive* certainty from *inductive* procedures. In consequence, Carter seeks to uncover the taken-for-granted frameworks of meaning and inference with which explorers operated. Too often, he believes, the physical setting of these encounters—a coast, an inlet, an island—are taken as 'givens,' as the unproblematic stage on which ethnographic encounters are played out. But such portrayals have been sustained only by suppressing the imagined geography associated with such spaces; in contrast, when those meanings are recognized, the data of geography are transformed into a "mythopoeically constructed space." By turning to the critical commentary of Vico, Carter hopes to unveil the Enlightenment origins of the myth that geographical knowledge is without mythic foundation. Indeed, it is here that the roots of the problematic nature of geographical knowledge lie revealed, for geography since the Enlightenment has found itself perennially poised between the universal and the particular, the logical and the speculative, the local and the general.

It was not of course just explorers who traveled the world during the period of Enlightenment. Reputations also traveled, and Nicolaas Rupke focuses attention on the different ways in which the work of that scientific traveler *par excellence,* Alexander von Humboldt, itself traveled. In so doing, Rupke uncovers a distinctive geography of reception. After reviewing the variety of ways in which the Humboldt corpus has been portrayed—as an exemplar of the Enlightenment tradition, as fundamentally Romantic, or as illustrative of German classicism—Rupke concentrates on critical responses to Humboldt's Mexican work in order to ascertain the relative significance of these writings. Rupke's careful scrutiny of a wide variety of reviews in a number of national settings issues in two significant conclusions which will necessitate a substantial revision of the conventional portrait of Humboldt contained in standard histories of geography.

First, Humboldt's rise to fame in the international arena should no longer be attributed to *Kosmos,* but to his account of Mexican political economy. Second, the Humboldtian moment in Enlightenment science was a complex phenomenon with its own characteristic geography.

These essays contribute admirably to a reassessment of the role of travel throughout the length and breadth of the European Enlightenment. Travel reports can no longer be seen as simply the source of imagined geographies; rather, they raised moral and epistemological questions fundamental to the Enlightenment ethos. At the same time, our conception of Enlightenment travel can profitably be expanded to take into account the movement of scientific reputations which helped produce the complex geography of Enlightenment itself.

Chapter Nine

On Being Perseus:
New Knowledge, Dislocation,
and Enlightenment Exploration

DORINDA OUTRAM

The eighteenth century believed perhaps more strongly than any other that travel makes truth. The problems raised by such a belief in an era which was both that of the Enlightenment and that of the exploration and mapping of that one-third of the earth's surface which is the Pacific ocean are the subject of this paper. The link between traveling and truth finding sits deep in the organizing mythologies of the West, whose origins in Greece and Rome were fully familiar to the Enlightenment and provided much of the basis for its self-fashioning. One of the most powerful of such myths is that of the Greek hero and wanderer, Perseus.

In Greek mythology, Perseus confronted the three-headed monster called the Gorgon. He cut off the monster's only mortal head, Medusa, whose head was covered with snakes and whose gaze turned to stone all who encountered it. In order to do this, Perseus made the monster confront her own reflection in his burnished shield. Once she was petrified by her own gaze, Perseus was able to lop off her head. We are told too that after this, Perseus returned to his own kingdom and found there his hall full of rival claimants to his throne. When he held up his terrible trophy on the threshold, the head of Medusa stared down the assembled throng and turned them to stone, frozen in mid-breath, in the very act of perceiving their danger. Perseus then presented Athena, goddess of clear sight, of rationality, with Medusa's head, which trophy this armed goddess hung henceforward on her own shield.

Even in this bare and incomplete outline, this is a myth that tells us much. It tells us about the power and the terror of sight. It tells us about the transformation of life into the frozen forms of death. It also tells us much about great journeys and their ending, for Perseus travels to the westernmost extremities of the earth to reach the

Gorgon. And it tells us much about the costs of such travel. For Perseus, in many versions of the tale, reaches the Gorgon's lair after many prior ordeals, in the depths of fatigue—reaches it not in a hero's poised certainty of victory, but with his courage already failing, already about to freeze in the held breath of his own risk and terror. This is why, at the last moment, one version of the story runs, his courage failed him. Turning his head away, he averted his eye from Medusa and relied on the goddess Athena to guide his arm in the fatal blow. Thus the myth of Perseus is also a myth about the relationship between heroism and vulnerability. It is also a myth about the interpolation of action, terror, and, that supremely Enlightenment category, rationality.[1]

This is a disturbing myth. There is so much of the terror, so much of the devastation of seeing, in such close companionship with such rationality and such power. By thinking about the myth of Perseus in the shape of this unease, we begin also to enter some of the unease with which explorers' actions and the specific forms of rationality associated with the knowledge they produced also affected the Enlightenment. But to talk about unease in this context may seem strange. The Enlightenment has often been presented precisely as an heroic age, one which saw the emergence, and triumphant success, of exploration conducted primarily for scientific ends. We can, after all, scarcely imagine the map of the world without the knowledge produced by the Pacific and Antarctic explorations of this period, or the Enlightenment itself without the knowledge produced through the journeys of such men as Cook, Pallas, Bougainville, La Condamine, or Alexander von Humboldt. The imagination of the Enlightenment is as incomplete without the new plants and animals, minerals, fossils, and geographies recorded and brought back by explorers. At an even deeper level, what explorers reported about the new human societies they encountered made the whole issue of human difference itself a central problem in the Enlightenment. For if there was one human nature, why then did human societies differ so profoundly? If there was one human history, and one human rationality, then how could the human social world be ordered with such absolute difference in its particular manifestations?

And yet, explorers' knowledge, an interpolation of rationality and action, was also a serious and disturbing problem, in ways specific to the Enlightenment. What sort of knowledge, after all, was explorers' knowledge? What secured its status as truth? The testimony of the explorer himself was, by itself, no guarantee of anything, when such testimony precisely concerned the distant, the new and the

DORINDA OUTRAM

previously unseen. Such knowledge could not be experimentally replicated and was not produced in front of an impartial and freely assembled audience, or in the context of any of the adjudicating institutions of European science. Explorers, it is true, often returned with artifacts and natural specimens from the distant lands they visited. But artifacts could bear witness only to themselves. They could by themselves prove nothing about the truth status of the narrative of discovery and exploration in which they were alleged to lie embedded.[2] Explorers' knowledge thus posed in stark form the problems of belief, trust, and facticity.

Trust has become a fashionable theme in recent writing in the history of science. The essence of the explorer's claim was to be trusted as an eyewitness to a world that few or no others had seen. Yet it was precisely here that the explorer ran straight into all the Enlightenment's problems with accepting such testimony as compelling belief. These were the problems which were discussed in David Hume's 1748 *Essay on Miracles,* or Voltaire's 1765 *Questions sur les Miracles* (see also Shapin 1994). Eyewitness testimony, as Voltaire and Hume demanded, became increasingly measured against the grid of a newly elaborated and powerful idea of probability. The linkage of the exotic, the hitherto unknown, and the marvelous, began to decline (on this point, see Daston 1988; Daston and Park 1998). But even here, explorers' knowledge, precisely insofar as it was knowledge of the new, resisted the grip even of an increasingly powerful probabilism. In the end, it could be accepted only on the basis of trust. Such trust could be built up by means of authorship, and it is not surprising that many explorers, most notoriously Alexander von Humboldt, invested perhaps as much in writing the narrative of travel as they did in that traveling itself. But the trust crucial to turn a traveler's tale into an exploration account, which makes much higher truth claims, could not be produced by authorship alone. The essential and prior step was to place trust in the explorer himself: trust in his moral integrity and trust in his perceptual accuracy. Both were important. The explorer had to be trusted both to see and to tell, had to be trusted to interpolate action with rationality. And such trust was very difficult to command precisely because the philosophy of the Enlightenment called into question this very link between the moral self and the perceiving self of the explorer. Exploration knowledge was a particularly acute form of this problem because so much knowledge produced in exploration depended on the sense impressions of the explorer. Even if he used instruments to extend and calibrate sense impressions, what the explorer himself saw was crucial to establishing the truth-status of his observations. This meant that exploration knowledge was profoundly at odds with attempts since the scientific revolution of the seventeenth century to find epistemological legitimacy for experiment-based science in a denigration of knowledge based on the senses (Dear 1987; Shapin 1988). Just as it was at odds with the Enlightenment's questioning of the necessary veracity of the eyewitness observer.

Enlightenment philosophy also went to some lengths to question the very possibility of the integrity of the knowing subject. A strong sense of the self that knows itself to be a self, and is thus capable of being a locus of moral identity, seemed increasingly for the Enlightenment to be placed in question by the very operations of the senses on which explorers relied to make contributions to knowledge. Hume and Condillac, for example, questioned whether there could be an epistemological unity of consciousness. Hume famously remarked in his 1739 *Treatise on Human Nature,* "I may venture to affirm of mankind, that they are nothing but a bundle or collection of different perceptions, which succeed each other with an inconceivable rapidity, in a perpetual flux and movement." (I, iv, 6) This was a theme pursued by the French philosopher Condillac in his *Traîté des Sensations* of 1754. In his work he uses the heuristic device of a living statue to ground his debate on human sense impressions. It is not long before the statue utters the painful words:

> . . . I know that the parts of my body belong to me, but I am unable to understand how. I see myself, I touch myself, in one word I experience myself, but I do not know what I am. And if I believed that I was sound, taste, colour, smell, now I know no longer what I should believe that I am (IV, 8, 6).

In other words, personal identity is constitutionally discontinuous. This means that the self which processes sense impressions may have no clear relationship with the moral self. The discontinuity of the self was not a new idea in the Enlightenment. But it was given a new twist by the increasing reluctance of the Enlightenment to find a saving unity in the idea of the soul. Instead, increasing attempts were made to ground identity in sense impressions at exactly the same time as the discontinuity of those impressions was admitted. This has the consequence that sense impressions in fact pull away from the possibility of moral authentication.

Explorers' knowledge might thus seem to be the exemplification of Enlightenment concerns about the difficulty of knowing anything securely about the external world, of making a secure relationship

DORINDA OUTRAM

between knowledge and the knower, or of persuading others of the validity of knowledge gathered through ineluctably individual sense impressions. Could the explorer, like Perseus, ever find a way of doing the work of Medusa, of freezing the world into the forms of communicable truth? If one reads what explorers themselves sometimes say about the actuality of their cognition, then it might appear that the philosophers were right, that to perceive the dazzle and the glitter of the world really did pose a threat to a unitary personality capable of moral discipline, capable of being trusted. Let us to this point read a famous letter by Alexander von Humboldt, written in July 1799, a few days after his first landfall in Venezuela with his traveling companion Aimé Bonpland. In a state close to ecstasy, he wrote home to his brother Wilhelm in Berlin:

> What a fantastic and extravagant country we're in! Fantastic plants, electric eels, armadillos, monkeys, parrots, and many real, half-savage Indians . . . and what trees! Cocoa-nut trees fifty or sixty feet high, *poinciana pulcherrima* with a foot-high bouquet of magnificent, bright red flowers . . . a host of trees with enormous leaves and scented flowers, as big as your hand, completely unknown to us. . . . And what colours in birds, fish, even cray-fish, sky-blue and yellow. We rush around like the demented. In the first three days we were quite unable to classify anything, we'd pick up one object only to throw it away for the next. Bonpland keeps telling me that he will go mad if the wonders do not cease soon.[3]

This is a description of the boundary between ecstasy and sensory disorganization. It is a description of falling suddenly into a world where everything was brightly colored and strangely shaped, and where nothing bore any relation to previous experience. Perception was thus totally unshaped by memory. The brilliance of the world, its hallucinatory dazzle, threatens to dissolve those two alleged entities called Humboldt and Bonpland into a cacophony of sense impressions bordering on insanity. And the power of the experience of the exotic is also powerful enough to shatter the aesthetic communities on which the possibility of its description rests. Not only might the explorer himself be dissolved by the intensity of what he saw, he might also be unable to describe what he saw to those who did not. This is why Alexander's letter from Venezuela to Wilhelm in Berlin walks a fine line between the possibilities and the limits of the description of the new. And this is exactly what exploration is about: establishing the new as a category of experience.

Everything I have said up to this point has emphasized the problems raised by exploration knowledge in the Enlightenment. It is not surprising that it aroused sustained attacks by contemporaries. To a savant of the stature of the anatomist and taxonomist Georges Cuvier, for example, exploration, like all field science, was highly suspect. In an 1807 review of Humboldt's *Tableaux de la Nature*, Cuvier forcefully argued against the claims of fieldwork to produce adequate knowledge of nature through direct observation. He did so through a direct attack on the cognitive and moral make-up of the field investigator. For Cuvier, the procedures of what he calls the *'naturaliste-voyageur'* were irreconcilable with those of the sedentary naturalist. The traveling observer was doomed to remain precisely that. Cuvier remarks that the field naturalist passes through ". . . at greater or lesser speed, a great number of different places, and is struck, one after the other, by a great number of interesting objects and living things," but ". . . he can only give to them a few instants of time and thus his observations are broken and fleeting." This sounds almost as though Cuvier had read Humboldt's letter of 1799. Cuvier then summons up the figure of the sedentary naturalist, the man who works in the collections and museums. He admits that such a man ". . . only knows living beings from distant countries through reported information subject to greater or lesser degrees of error . . . the great scenery of nature cannot be experienced by him with the same vivid intensity as it can by those who witness it at first hand." And yet, Cuvier argues, directly attacking Humboldt, ". . . if the sedentary naturalist does not see nature in action, he can yet survey all her products spread before him. He can compare them with each other as often as is necessary to reach reliable conclusions. He defines his own problems." Cuvier concludes with the remarks that ". . . the travelling observer can only travel one road. One can only roam freely through the universe, by staying in one's study. For that, a different sort of courage is needed . . . courage which does not allow its possessor to leave a subject, until by observation and . . . connected thought, he has illuminated it with every ray of light possible in a given epoch of knowledge."[4]

For Cuvier, the formation of new knowledge in continuous passage through space, the very essence of exploration, was deeply suspect. For him, travel does not make truth. On the contrary. For Cuvier, mastery over, and real comprehension of the order of nature, comes not from passage but from immobility. The apparently heroic field observer lies, in fact, under the tyranny of the immediate. Real courage, Cuvier argues, means to create an ordered structure out of the immediacy of experience, not to be tempted by its momentary vivid-

DORINDA OUTRAM

ness. Such an argument shows one way to discuss location and dislocation in relation to experience and epistemology. Cuvier argues that true knowledge of the order of nature comes from the very fact of the observer's distance. It depends exactly on not being there, on being anywhere which is an elsewhere. This was a way of avoiding all the problems which the Enlightenment had debated in relation to sense impressions, to eye-witnessing, and to the assessment of the moral integrity of the distant field observer in continuous transit. It is important that Cuvier does not refer to observation by means of scientific instruments to answer or avoid these questions. Just as much as Humboldt, he assumes that the central question is the assessment of the quality and location of the naturalist himself, not on his use of instruments. His version of 'objectivity' turns rather on the experiential categories of location and dislocation.

Was this an error? Surely instruments offered a golden opportunity to avoid all the problems raised by direct sense impressions of the new and the transitory? It is highly tempting to see instrumental readings as a way of freezing and transporting the alien reality with which the explorer was surrounded. James Cook took a battery of instruments with him to the observation station at Point Venus, on Tahiti, in 1769; Humboldt took a huge collection of the latest devices to Venezuela. The British Arctic expeditions of the 1820s were given specific mandates to test new forms of compasses and atmospheric gauges (Bravo 1992; Licoppe 1996; Schaffer 1988; Sorrenson 1996). We also know that the disciplines of creating a regular series of instrumental observations under difficult conditions were one of the guarantees of an explorer's probity. And from such series previously unsuspected global regularities were detected, such as Humboldt's ideas of the relationship of climatic and vegetation zones, or of the air temperature relations portrayed as isotherms.

However, knowledge gathered through instrument readings and knowledge gathered through sense impressions were not so distinct to the Enlightenment as they have sometimes appeared to be to us. Much use of instruments was designed precisely to return man to an Adamic state of sense perceptions perfect in their immediacy and accuracy. Instruments were thought of as enhancing human sense impressions rather than replacing them, were thought of as part of man's pilgrimage through the world on his way to regaining the perfect knowledge possessed by the first Adam in Eden. As the Swiss natural philosopher Jean Sennebier commented in his *L'Art d'Observer* of 1775, instruments, ". . . usually represent objects precisely;

or at least they show them in the way they would appear to our senses if they were keener than they are. Instruments thus become an essential part of our senses, which they perfect, and supports to their weakness, which they reduce" (I, 3). Such a view of instruments as extensions of the human body and senses was also bound to be at odds with an interpretation of instrumental readings as being connected with the production of standardized and hence potentially universal knowledge. This was hardly a possibility in eighteenth-century conditions. Before the true mass production of instruments, it was often very difficult to calibrate readings. The susceptibility of instruments to damage and distortion in field conditions was infinite. The blurring of the distinction between human and instrument was even intensified by the end of the century, by the focus on self-experimentation in the work of scientists such as Alexander von Humboldt himself, or Johann Wilhem Ritter.

Contemporary reactions to information gathered through instruments thus do not allow us to displace persons and their locations and dislocations, from the exploration science of the Enlightenment. Instrumental readings did not substitute for the narratives of which the explorer was also the author. They also do not remove the question of the person of the explorer. Nor do they dispense us from writing the history of the acts of cognition and the categories of experience through which explorers attempted to make sense of the glittering multiplicity of the world around them. By authorship, explorers transmit these acts and categories into the European world. Thus, the way in which explorers perceive the new terrain they discover is often the founding pattern for subsequent European perceptions of the area.

In this context, let us listen again to Alexander von Humboldt, a man who had already used his own body as a Leyden jar to measure the amount of electricity generated by electric eels. Humboldt tells the same story of the convergence of the instrumental and the human in his account of his ascent of Mount Chimborazo in January 1802, a climb which in a single day made him world famous. About halfway he commented that he and his companions

> Began to feel nausea and giddiness which was far more distressing than our difficulties in breathing. Blood exuded from our lips and gums, and our eyes became bloodshot. These symptoms . . . vary greatly from individual to individual according to age, experience, constitution, and redness of the skin. But in the same individual, they constitute a kind of gauge for the amount of rarefaction in the atmosphere, and the absolute height he has reached. . . . We fixed

DORINDA OUTRAM

up the barometer with great care found it stood at 13 inches, 11 and 2/20 lines. The air temperature was only three degrees below freezing, but after our long stay in the tropics even this amount of cold was quite benumbing. Our boots were wet through with snow-water . . . (Botting 1963, 154–155).

In this text, instruments and persons often substitute one for another. The individual body is "a kind of gauge," treated as valuable even in the presence of the barometer. Von Humboldt himself is highly present in the text, self-monitoring, recording instrumental readings, writing up his companions' responses and assessing their meaning. The supreme instrument present here is his witness.

The use of instrumentation is thus not the same as the history of experimentation. Emphasize the history of instrumentation, and we are unable to answer the question of what it is which makes exploration a specific form of human activity. Exploration is not merely metropolitan experimental and observational science carried out rather a long way from home. Overemphasis on instrumentation may also miss one obvious and important distinguishing feature of exploration, which is its close relationship to the definition of the new. It is this which decisively distinguishes exploration in the Enlightenment from all previous expressions in the West of the link between travel and truth, of which the myth of Perseus' search for the Gorgon is one of the most compelling examples. If the epitome of such a tradition in the Christian world is taken to be the practice of pilgrimage, then the difference becomes clear at once. The pilgrim is not traveling to find the new. He is traveling a road which will lead him but to a sacred site which is already known. An ample literature emphasized that it is the duty of the pilgrim precisely not to be curious about what he sees along the way, not to try if he can observe anything new, but to fix his thoughts exclusively on the already defined sacred goal of his journey.

The explorer in the Enlightenment is not a pilgrim. While his journey may have objectives, it cannot, by definition, have a known destination. The point of his journey is to see what has not been seen before, to pass over terrain which has not been previously traveled. It is the gaining of new knowledge in conditions of continuous transition. This distinguishes exploration from other forms of the production of the new which the Enlightenment produced. Here one thinks not only of experimental science, but also of the proliferation of taxonomic systems, whose abstraction, taken to its extreme in that of Linnaeus, was strikingly successful in acting as a diagnostic for the previously unseen. This search for the new can also be seen as a link between the experimental and the taxonomic sciences of the Enlightenment which are so often treated as separate enterprises. These few remarks clearly do not exhaust this potentially vast subject of the creation of the concepts and practices of the exemplification of the new in the Enlightenment. But this may at least be a more productive way to treat exploration in the Enlightenment, than to see it, as some postmodernists and feminists have done, as little more than an attempt to 'appropriate' the exotic natural world to European systems of understanding (for example, Pratt 1992).

Exploration knowledge is also unique in that it is produced in a situation of continuous transition. Even hostile contemporaries, such as Georges Cuvier, were in no doubt that location and dislocation were fundamental categories for understanding the nature of the knowledge so produced. For them it was a distinct category of experience, associated with a distinct sort of person. This is an important point. For quite unlike the witnessed and replicated forms of metropolitan experimental science, the authority of exploration science stubbornly came back to the person. On the sense impressions of the explorer, on his discipline, and above all on his success in welding together exactly the relation between the sensual and the moral which Enlightenment philosophers tried so hard to pull apart, was based the authority of the explorer's relationship with the new. In the end, the authority of the knowledge so produced was lodged in the authority of the person of the explorer and continued to be so at a time when other forms of scientific experience based on objectivity values were equally relentlessly pushing apart the observer and his observation. This was also at a time when the revolutions of 1776 and 1789 had equally demonstrated the possibility of the replacement of the personal authority of monarchy with the collective, representative authority of the republican systems which they founded. Against this background, far from being congruent with other forms of authority specifically developed in the Enlightenment, it is possible to see exploration as representing the survival and export into exotic worlds, of far older systems of authority.

The oldest locus of authority is the human body. And it was on this that the authority of the explorer was ultimately based. Contemporary accounts return repeatedly to the theme of the authentication of the explorer's travels by the trials of his body. For example, the hair of the Berlin-born explorer Peter-Simon Pallas turned white at the age of 37 after exhausting journeys in Siberia. A fall through

DORINDA OUTRAM

a treacherous ice sheet into the freezing waters of the river Ob permanently damaged his health. On returning to St. Petersburg, he found it impossible to make a permanent home anywhere, and by frequently moving house, endlessly replicated the permanent transition of the explorer's life in the field. Michel Adanson, explorer of Senegal, and author of its first flora, unflinchingly endured equivalent physical stresses and was unable to tolerate the pressure of human intercourse which greeted him on return to Paris. Thereafter a life-long recluse, his life prolonged the solitude of the field.[5] Alexander von Humboldt was permanently affected by rheumatism as a result of sleeping for months on the damp floor of the tropical rain forest. On his return to Europe, he kept his apartments in Paris and Berlin heated to tropical heights, so deeply had his metabolism internalized the demands of another continent. In another letter to his brother Wilhelm, he proudly proclaimed his physical adaptation to the tropical world into which he had chosen to journey, rather than to the Prussian plain from which he had come:

> I was made for the tropics. I have never felt so well as in the two years I have been here. I work hugely and sleep only a little. Often I make astronomical observations, hatless in the fierce sun, for five or six hours together. I have even been in towns where there has been an outbreak of yellow fever, and never had so much as a headache. (Quoted in Duviols and Minguet 1994, 15)

Men like Humboldt, Adanson, or Pallas thus did not merely bear the marks of their traveling upon the outside of their bodies; they also internalized the exotic environment mentally and metabolically. They lived in a mimesis of distance, of location in dislocation. Their bodies' suffering did not merely authenticate, in ways strikingly reminiscent of the Christian saints most of them would have repudiated as models, their claims to have traveled; they also made distant lands and oceans present to those who saw them. We all know that the body learns. And because it learns, it also teaches. What the explorer's body taught its witnesses allowed them to believe. That body was a living proof of the otherwise unseen and unseeable vastness of the world.

To say all this is not to collude with a representation of explorers as dominating, all-seeing masculine figures, which was once the darling of Victorian mythology, and is now subject to the repeated attack of the apparently equally vigorous mythologies of postmodernism. Far from it. Enlightenment explorers often manifest in their body, not domination but vulnerability. Of that vulnerability, the tattoos acquired by the sailors who accompanied James Cook into the Pacific were only the most decorative and benign reminders. Such vulnerability was usually a far more risky affair, more like walking on the thin ice of a Siberian river. But without that vulnerability the explorer could not manifest in his own person the moral economy which made his reporting acceptable as authentic knowledge. Such vulnerability, as we learned from the story of Perseus, is one of the places where action may be interrupted by reflection. It is the other side of the very love of ambitious risk that drove men onto the ice in the first place. This was acknowledged even by the taciturn James Cook, who described himself during a time of crisis in the Antarctic ice-edge search of January 1774 as a person who "had Ambition to reach furthest south . . . further than anyone had gone before . . . as far as it was possible for man to go" (Beaglehole 1969, II, 323). It was in this dance of ambition and vulnerability that the Enlightenment explorer lived. It was from this dance that knowledge of the new, such as the new continent of Antarctica, was created.

We learn from explorers that cognition has a history and politics inseparable from location and dislocation. And that the cognition of distant, new, and unseen things gains authority and reality from its incorporation in the bodies of particular men. This incorporation is not reducible to the many other ways in which exploration and explorers sponsored representations of the previously unseen. This is an important point to make not just in relation to our understanding of eighteenth-century explorers, but also to our understanding of ourselves. Today, we face very great problems about how we envision and represent the world. We seem to live in a world so awash with images of itself that the realities which these images relate to may seem less important than the games which these images of forever unwitnessed realities play among themselves. We worry about what Foucault called the "poverty of the real," and we do not replace it (how could we?) with anything which satisfies us more. At the same time, we increasingly denigrate knowledge based on direct sense impressions. We have not escaped from the dilemmas which haunted exploration knowledge in the Enlightenment. Principally, we have not yet come to terms with the cognitive and moral issues involved in making dislocated images of in fact located objects (Miller 1995; zur Lippe 1997). But in learning how to reconstruct the acts of cognition through which explorers came to know the world, in their obstinate reconciliation of the senses and the person, we also reconstruct the possibility of our own empathy with that world, and in

DORINDA OUTRAM

doing so become more convinced of its, and our, own reality. The head of Medusa does not have to be our weapon.

Notes

The subject of this paper was suggested by Lorraine Daston, whom I have also to thank for her comments on several versions. I also gratefully acknowledge the helpful responses of audiences at Harvard University, the Massachusetts Institute of Technology, the University of Edinburgh, the University of Cambridge, and the Max-Planck-Institut für Wissenschaftsgeschichte, Berlin. This paper was also fortunate enough to attract the commentaries of Irina Podgorny and Wolfgang Schaeffner at the Berlin Sommerakademie of 1995. Michael Bravo and Graham Burnett made essential contributions over a long period to this paper, as did discussions with Rudolf zur Lippe. I thank Stephan Müller-Wille for illuminating conversations on Linnaeus. I take this pleasant opportunity to thank the Max-Planck-Institut for its hospitable shelter in 1995–1996, which sustained the evolution of this paper and of much else besides.

1. For another modern reinterpretation, see Klaus Heinrich, *Floß der Medusa : Drei Studien zur Faszinationsgeschichte mit mehreren Beilagen und einem Anhang* (Basel & Frankfurt am Main, Stroemfeld, 1995). A portrait commissioned by Frederick II of Prussia from the artist Christian Bernhard Rode (1725–1797) represents the monarch as Perseus about to strike the fatal blow under the direction of Athena; see Michaelis (1989, 95–96). The portrait appeared as an engraving in 1759 and was exhibited, reworked as an oil painting, in 1789.

2. A contrary argument appears by implication in Marie-Noëlle Bourguet (1997).

3. *Briefe Alexander's von Humboldt an seiner Bruder Wilhelm, herausgegeben von der Familie von Humboldt in Ottmachau* (Stuttgart, 1880), 10–18; quoted passages 11, 12–13.

4. "Analyse d'un ouvrage, de M. Humboldt, intitulé Tableaux de la nature ou considérations sur les deserts, sur la physionomie des végétaux, et sur les cataractes de l'Orenoque" (Library of the Institut de France, Paris, Fonds Cuvier, MS 3159). This passage is discussed in Outram (1984, 62–63), and, differently, in Outram (1996). An opposing interpretation is offered by Dettelbach (1993). I would like to thank Michael Dettelbach for many discussions.

5. For example, Georges Cuvier, *Eloges Historiques, précédés de l'éloge de l'auteur par M. Flourens* (Paris, 1843), 234, 231, 238, 250, 136, 148. The eulogy of Adanson was written in 1807 and that of Pallas in 1811. For the making of representative individual bodies in this period, see, in more general context, Outram (1989).

References

Beaglehole, James C., ed. 1969. *The Journals of Captain Cook, 1772–1775.* Cambridge: Cambridge University Press. 3 volumes.

Botting, Douglas. 1963. *Humboldt and the Cosmos.* London: Sphere Books.

Bourguet, Marie-Noëlle. 1997. "La collecte du monde: voyage et histoire naturelle, (fin XVIIème siècle-début XIXème siècle)," in Claude Blankaert et al., eds., *Le Muséum au premier siècle de son histoire*. Paris: Editions du Muséum National d'Histoire Naturelle: 163–196.

Bravo, Michael T. 1992. "Science and Discovery in the British Search for a North-West Passage, 1815–1825." University of Cambridge Ph.D. thesis.

Cuvier, Georges. 1843. *Eloges Historiques, précédés de l'éloge de l'auteur par M. Flourens*. Paris: Paul Ducrocq.

Daston, Lorraine. 1988. *Classical Probability in the Enlightenment*. Princeton: Princeton University Press.

Daston, Lorraine, and Katherine Park. 1998. *Wonders and the Order of Nature, 1150–1750*. New York: Zone Books.

Dear, Peter. 1987. "Jesuit Mathematical Science and the Reconstruction of Experience in the Early Seventeenth Century." *Studies in the History and Philosophy of Science* 18: 133–175.

Dettelbach, Michael. 1993. "Romanticism and Administration: Mining, Galvanism and Oversight in Alexander von Humboldt's Global Physics." University of Cambridge Ph.D. thesis.

Duviols, J. P., and C. Minguet. 1994. *Humboldt: savant-citoyen du monde*. Paris: François Maspero.

Heinrich, Klaus. 1995. *Floß der Medusa: Drei Studien zur Faszinationsgeschichte mit mehreren Beilagen und einem Anhang*. Basel and Frankfurt am Main: Stroemfeld.

Licoppe, Christian. 1996. *La formation de la practique scientifique: le discours de l'expérience en France et en Angleterre, 1630–1820*. Paris: Editions la Découverte.

Lippe, Rudolf zur. 1997. *Neue Betrachtung der Wirklichkeit: Wahnsystem Realität*. Hamburg: Europäische Verlagsanstalt.

Michaelis, Rainer. 1989. *Deutsche Gemälde, 14.–18. Jahrhundert: Staatliche Museen zu Berlin, Gemäldegalerie*. Berlin: Staatliche Museen zu Berlin.

Miller, J. Hillis. 1995. *Topographies*. Stanford: Stanford University Press.

Outram, Dorinda. 1984. *Georges Cuvier: Vocation, Science and Authority in Post-Revolutionary France*. Manchester: Manchester University Press.

———. 1989. *The Body and the French Revolution: Sex, Class and Political Culture*. New Haven and London: Yale University Press.

———. 1996. "New Spaces in Natural History," in Nicholas Jardine, James Secord, and Emma Spary, eds., *Cultures of Natural History*. Cambridge: Cambridge University Press, 249–265.

Pratt, Mary Louise. 1992. *Imperial Eyes: Travel writing and transculturation*. London and New York: Routledge.

Schaffer, Simon. 1988. "Astronomers Mark Time: Discipline and the Personal Equation." *Science in Context* 2: 115–145.

Shapin, Steven. 1988. "Robert Boyle and Mathematics: Reality, Representation and Experimental Practice." *Science in Context* 2: 25–58.

———. 1994. *A Social History of Truth: Civility and Science in Seventeenth-Century England*. Chicago: University of Chicago Press.

Sorrenson, Richard. 1996. "The Ship as a Scientific Instrument in the Eighteenth Century." *Osiris* 11: 221–236.

DORINDA OUTRAM

Chapter Ten

Gaps in Knowledge:
The Geography of Human Reason

PAUL CARTER

It follows that the first science to be learned should be mythology or the interpretation of fables.

Verene 1991, 151; Vico 1984, 51

10.1. Introduction: In the Stride

"The history of mankind is the instant between two strides taken by a traveller" (Kafka 1954, 73). If, as the Neapolitan philosopher, Giambattista Vico thought, a figure of speech could be considered a compressed myth (Goetsch 1995, 32; Vico 1984, 129), then this gnomic remark of Franz Kafka's admirably introduces my theme in two ways. Enlightenment geographical discourse aspired to certainty: although its own procedures were irreducibly inductive, it aspired to the mathematical certainty of deductive logic. To this end—which was also the end of furnishing a global space paradigm which we are heir to—it necessarily repressed the process, the contingency of its own procedures. Emphasis was placed on the steady tread of intellectual (and imperial) advance. In contrast, the instant between strides was ignored. Yet the missing instant—in which, after all, progress originated—refused to disappear. This essay intends its recovery, and asks why it was viewed so suspiciously.

The notion that "the history of mankind" might be contained in a figure of speech brings me to the second term of my argument, and to the way in which the recovery of that missing instant might be undertaken. The fundamental premise of Vico's comprehensive attack on Cartesian logic contained in *La Scienza Nuova* (1725)—and on any science that attempted to ignore the human and historical origins of its own procedures—was that knowledge was *mythopoetic*, that is, the world it described was inseparable from the poetic techniques (the figures of speech) used in its production (Vico 1984, 14). The methods which human *ingegno* used to constitute the different fields of human knowledge were profoundly metaphorical—a metaphor, for example, was operationally a technique for bringing two ideas or *topoi* together (see Mali 1992, 179ff.)—a way, in other words, of linking Kafka's two strides, of naming the "instant."

My proposal, then, is that in denying the figurative nature of its knowledge, geographical discourse subscribed to the *mythos* of Enlightenment reason—precisely that idea that the new knowledge was not "poetic"—not produced mythopoeically—and therefore stood apart from the philosopher, the culture, and the nation that brought it into being. The results of this have been far-reaching and resistant. Even in our time of postcolonial revisionism, the imperium of geographical space remains intact, although it is true, too, that geographical theorists have been actively engaged in rethinking such fundamental tropes as space, place, and landscape and exposing their multilayered meanings (see, for example, Barnes and Duncan 1992; Barnes and Gregory 1997; Cosgrove 1984; Cosgrove and Daniels 1988; Entrikin 1991; Gregory 1994). It is a matter of urgency, I am suggesting, like others, that a newly critical (and self-critical) geography emerge. A first step towards this, it is argued here, is a recognition of geography's mythopoeic constitution. If it is difficult at first to accept the proposition that the conventional privileged sites of geographical discourse—the coasts, islands, and rivers—are poetic constructions, this merely reflects the nature of myths—that to bring them in doubt is, surprisingly, to find that our power to reason also collapses. Even so illuminating a scholar as Alain Corbin roots our appreciation of landscape in "coenaesthetic impressions (those which created a sense of existence on the basis of a collection of bodily sensations)" and seems to find it remarkable that "the sea, particularly for the population of the archipelagos, does not always constitute a barrier" (Corbin 1994; 1 and 292, note 71).

10.2. Geography's Myth

Among the arts and sciences that can trace their modern origins to the radical epistemic shift in the definition, kinds, scopes, and goals of knowledge associated with the European Enlightenment, geography seems to have survived relatively well the urgent, often openly hostile, critical, and metaphysical revisionism which, over the last fifty years, has subjected every aspect of the Western "project" to the intensest, and often most damning, scrutiny. Naturally, geographers have engaged critically with, say, the implications of poststructuralist and postcolonialist theory for their discipline. Some of this work, to be sure, might be considered as taking place on what might be described as already cleared and rationalized ground. But there is certainly also a lively concern with, say, the poetics of space and the metaphorics of landscape (for example, Buttimer 1993; Daniels 1993; Mayhew 1998; Sack 1997; Tuan 1974). Such work has concerns directly at variance with the kind of Enlightenment geography that, until recently, dominated the discourse. Even if it is acknowledged that, to adapt Vico, the world which geography describes is a projection of human interests, hence "it is within our ability to retrieve its principles from within the modifications of our own human mind" (Mali 1992, 60–61), this has not led many geographers to regard their world view as *mythic* in any constitutional sense (although see Lewis and Wigen 1997). Another place (that of the detached observer, however *engagé* he or she may be, and which is also the ideal place imagined by the map's projection of the world) remains intact as a conceptual vantage point.

So, to take an Australian case, I am not aware of any cross-disciplinary epistemological revisionism arising from cartography's encounter with the Yolngu Art of North-Eastern Arnhem Land. The Yirrkala Bark Petition presented to the Federal House of Representatives in 1963 used "a repertoire of esoteric signs and symbols . . . articulated according to a preexisting code or 'underlying template,' which is a map of the Dreaming associations of geographical features within clan lands" (Ryan 1990, 22–23). Yet Arnhem Land bark paintings, although they fold into their designs a spatiotemporal consciousness no Western map can begin to rival, remain toys, art objects, the mythic residue of collective dreams. Other work has begun to discuss, however, non-Western cartographies and to challenge notions of mapping as simply an Enlightenment enterprise (for example, Belyea 1992; Harley and Woodward 1994; Paulston 1996; Turnbull 1989).

This is not necessarily cause for self-congratulation. It may simply mean (what academic geographers have always feared) that geography is not a unified discipline, its net too wide-meshed to catch anything critical. Try to sift geographical knowledge from its matrix of disciplines—whose interests taken together embrace "the entire home of man"—and one is liable to end up with a conjectural residue as reasonless as Nicolaus Steno's tonguestones—he, it will be remembered, identified the *glossopetrae* of Malta as fossilized sharks' teeth by detaching them from one matrix (mythical) and attaching them to another (geological) (see Albritton 1989, pp. 10–12). On the other hand, and no less constrainingly, the robustness of geography in advancing its claim to supply an objective description of the globe (most dramatically evidenced in the near-universal acceptance of maps as faithful representations of physical space) may simply reflect geography's extraordinary success in championing a spatial paradigm of knowledge that the other human sciences now take for granted as the tacit dimension of their operations. Again, this has its cost: if geography's conquest of the globe now underwrites the metaphysical space of the arts and sciences, it is because geography, as a systematic curiosity about the local difference of the earth's surface, has been largely superceded. Eagerly providing the groundwork for "the best intellects and energies of expanding peoples" whose chief ambition has always been, according to Ellen Semple, "to annihilate space by improved means of communication" (Semple 1911, 200), geography has written itself out of a job; its earth-writing has come full circle rendering the earth a *tabula rasa* such as it was before the first map.

If this latter view has any merit, a critical revisiting of Enlightenment geographical discourse needs to do more than trace the genealogy of its links with the other sciences—more, too, than contextualize these critically within the larger territorial ambitions (both intellectual and commercial) of European imperialism: it needs to recognize the profoundly *mythic* nature of geographical knowledge. By mythic is meant here those beliefs or concepts that provide the unquestioned ground or frame of reference on or within which the operations of reason occur and build. They are the materials of civilization's collective dream; they keep us afloat in some sense. My contention is that geography furnishes the design of this mythic raft in a way, say, that history does not, and for exactly this reason geography's presuppositions remain largely invisible, unquestioned—even by the fiercest critics of European imperialism. The recent debate between Marshall Sahlins and Gananath Obeyesekere illustrates this point: Obeyesekere denied that Sahlins's account of the events leading up to Captain Cook's death offered an accurate cross-cultural perspective. Rather than being an anthropologically corrected history, it was, he said, another exercise in European myth-making, its narrative absurdly collapsing divergent, and conflicting, viewpoints into a vanishing point symmetrical with the historian's own ideologically informed interests. Sahlins strenuously rejected this, insisting on the reliability of the data on which he based his interpretations. But neither party—nor an interested public following the intellectual sparring match—had, I guess, difficulty comprehending the underlying proposition: that history, even of the most factual kind, complicit in legitimating the link between Enlightenment universalism and imperial expansion, is at best one-sided and ethnocentric, and at worst (as Obeyesekere charged) pure (because motivated) myth.

Compare the visibility of historical discourse with the invisibility of geographical discourse. In arguing their cases, Obeyesekere and Sahlins at least agreed on one thing: the physical setting, the coast, the island, the sea. In reconstructing what happened, these geographical facts are assumed to be given. The Hawaians ascribe to features of the local landscape and the sea cultural meanings of which the Europeans are aware—and these play their part in the unfolding drama of trespass. Yet the topographical *mise-en-scène* remains in both writers *theatrical*. This theatricalization of place, the evacuation of any mental geography associated with it, is an important elision. It is our first clue, in fact, that the historical act of encounter occurs within a mythopoeically constructed space—and not just "out there" somewhere, as the writers of heroic narrative would have us believe. It is no accident that playwright Oliver Goldsmith, writing in the same year that Cook first sighted the Australian coast, compared the theatrical confrontation between actors and audience to the first mutually unintelligible coastal encounter between Europeans and "savages." The intelligibility of this trope depended on the privileged place that coasts and islands occupied in the European imaginary—where they were the privileged sites of knowledge (Carter 1996, 261). The coast—the figure of differentiation and linear progress—and the island—the consummation of the line as differentiated datum—these were as much conceptual desiderata as physical facts. Alfred Russel Wallace, a confirmed islomaniac whose evolutionary theory outlined in *The Malay Archipelago* managed entirely to ignore the natural history of the seas in between (and this for theoretical, not practical, reasons), could yet define true Malays as "never building a house on dry land if they can find water to set it in, and never going anywhere on foot if they can reach the place in a boat" (Wallace 1986, 133). What would they have made of Cook's shoreline antics?

Geography's myth, then, was and often continues to be that its knowledge is not mythic, that it offers an impartial description of the earth's surface. As already hinted, it is no advantage to geography to deny this; for by denying it geographers deny the specificity of their discourse, the origins of it which might still permit a critical geography to emerge, one capable of engaging with that metaphysical geography which currently has us all in thrall. This point is easier to accept if it is also emphasized that a description of geographical knowledge as rooted in mythical concepts is not intended as fancily pejorative. Both these points lay, of course, at the heart of Giambattista Vico's method employed in his treatise *La Scienza Nuova*, published in 1725 (see also Mills 1982). A brief examination of his thesis reinforces the point that a recognition of the mythic basis of our geography is not only historically justified—that, in advance of the geographical Enlightenment (conveniently—mythically?—centering on Cook's great voyages), Vico had already grasped the mythopoeic character of knowledge and its consanguinity with the interests of empire. It is also critically urgent: geography continues to provide the maps of empire, the paradigms of power. Can it, could it, describe things differently?

10.3. Vico's Science: A Sketch-Map

Verum ipsum factum, Vico's methodological assumption that he amplified as "the truth beyond all question: that the world of civil society has certainly been made by men, and that it is in our ability to retrieve its principles from within the modifications of our human mind" (Mali 1992, 60–61; Vico 1984, 331) was hardly original. Most Enlightenment philosophers would have subscribed to it, while "social and political theorists since antiquity have commonly predicated their notions of *homo sapiens* on a particular image of man as *homo faber*" (Mali 1992, 61). Nor was Vico's fascination with myth eccentric. Men like Bayle, Hume, Montesquieu, and Fontenelle, responding to the rapidly enlarged geographic and ethnographic knowledge flowing into Europe from her American, Indian, and Far Eastern commercial and colonial interests, were confronted with an array of mythic systems that inevitably recalled their own Greco-Roman heritage; and the challenge was clear: to differentiate the logical workings of their own minds from those of the myth-bound primitive. For Enlightenment thinkers, as Peter Gay has remarked in a telling image, "myth could be sympathetically understood only after it had been fully conquered, but in the course of its conquest it had to be faced as the enemy" (Mali 1992, 149). This was Vico's oddity, his primitiveness: to appear to side with the enemy. Rejecting the Enlightenment myth of a unilinear progress from *mythos* to *logos*, Vico maintained—and this was the original inflection he lent his *verum-factum* principle—"the essentially mythopoeic constitution of humanity." All cultural creations, including Descartes's much vaunted geometrical method of reasoning, including even *La*

Scienza Nuova, "are recreations of myths" (Mali 1992, 5, 9; Vico 1984, 127).

Without going into the details of Vico's philological method of recovering the "poetic wisdom" of the ancients, the salient point to note is that the process of recovery employed three interrelated mental faculties: *memoria, fantasia,* and *ingegno.* Verene, paraphrasing Vico's method of reading, explains, "memory has 'three different aspects'; memory *(memoria)* when it 'remembers things' (this is parallel to grasping the composition as a whole, i.e., holding the whole work in mind); imagination *(fantasia)* when it 'alters or imitates them' (the reader closely follows, but alters into his own mind the connections and sequence of things in the text); and ingenuity *(ingegno)* when it 'gives them a new turn into proper arrangement and relationship'" (Verene 1991, 198). On the *verum-factum* principle, Vico's imaginative reconstruction of the ancient sources is undertaken in the same spirit that those sources and the events they describe were originally created. Thus, for example, "poetic geography," the mythopoeic creation of an inhabitable land or territory, employed the same three faculties Vico invoked in his own reconstruction. Indeed, an exact, philologically attested parallel exists: the rhetorical *topoi* have their counterpart in the physical *topoi* or places which the poetic wisdom of the ancients creates in the wilderness, thus slowly transforming it into a place for the habitation of men.

And more: the same *ingegno* also supplies the mythopoeic mechanism of imperial expansion and colonization. According to Vico, the Greeks named the parts of the world they colonized by analogy with the geography of Greece; they transplanted the distribution of *topoi* within their own homeland and, retaining their original spatial relationships with one another, lowered them over the new localities (Vico 1984, 285–291). As an aside, we may note that the same mythopoeic propensity was alive and well in the age of Enlightenment. A particularly elaborate illustration of Vico's threefold "memory" as an instrument of territorial expansion is offered by Matthew Flinders's naming of the islands and headlands in Spencer Gulf, South Australia. If memory recalled the spatial arrangement of the villages of his native Lincolnshire, then fantasy saw an imitation of that arrangement in the layout of islands and coasts. Finally, with the aid of *ingegno,* Flinders imposed his own geopoeic logic on the map, creating a new, pleasingly grouped and arranged *topos,* an object of knowledge fit for settlement (Carter 1987, 183–187).

Indeed a spatial metaphor is implicit throughout Vico's treatment of memory. If Descartes's geometrical or linear mode of reasoning "moves forward by a constant and gradual series of small steps," *ingegno*, Vico explains, "is the faculty that connects disparate and diverse things. The Latins called it acute or obtuse, both terms being derived from geometry. An acute wit penetrates more quickly and unites diverse things, just as two lines are conjoined at the point of an angle below 90 degrees. A wit is obtuse because it penetrates simple things more slowly and leaves diverse things far apart, just as two lines united at a point lie far apart at the base when their angle is greater than 90 degrees" (Vico 1988, 96–97). By contrast, Descartes's thought is singleminded and centered; it divorces knowledge from the faculties of *fantasia, ingegno* and *memoria:* "we are not really knowing while we are remembering, for knowledge is by definition an immediate intuition. To deduce something, and to know it 'means to arrive at it by passing from link to link of the whole chain with a continuous and unbroken movement of thought'" (Goetsch 1995, 91–92). Descartes's method "is apt to smother the student's specifically philosophic faculty, his capacity to perceive analogies existing between matters far apart and, apparently, most dissimilar," and Vico adds disparagingly, "That which is tenuous, delicately refined, may be represented by a single line, 'acute' by two. Metaphor, the greatest and brightest ornament of forceful, distinguished speech, undoubtedly plays the first role in acute, figurative expressions" (Vico 1990, 24).

Adumbrated here are two ways of traveling, and, if we allow that the report of traveling is a species of geography, also two kinds of geography. They are not diametrically opposed. Both involve exploration, an expansion of horizons, an enlargement of command. But they differ in the emphasis they place on the *process* of obtaining knowledge, and the role of the observer in this. If Vico attends to first impressions, recognizing the role of figurative expressions—those metaphors or analogies without which we could not bridge the little gaps in reason that novelties of nature necessarily present to the scientist—then Descartes, and the advocates of step-by-step progress, disregarding the roughness of the terrain, keep their mental gaze rigorously focused on the final goal: a knowledge where, taxonomically deprived of their glare, novelties are assimilated, and even far from home the philosophical traveler feels at home. Invoking again the *verum-factum* principle, we can posit a further connection—between the character of the observer and the character of the country he discovers. It may even be that his *peripateia* is an image of his way of knowing—a possibility already envisaged by Montaigne when, against the speculative geographies of empire, he advocated "topographers who would make detailed accounts of the places which they have actually been to" (Carter 1996, 180).

Descartes's geometrical method at once spatializes thought and detaches it from the ordinary hyperbolic space of everyday life. At the beginning of his enquiry into the grounds of indubitable truth, Descartes could compare his position with that of "travellers, who, finding themselves lost in a forest, know that they ought not to wander first to one side and then to the other, nor, still less, to stop in one place, but understand that they should continue to walk as straight as they can in one direction, not diverging for any slight reason, even though it was possibly chance alone that first determined their choice. By this means if they do not go exactly where they wish, they will at least arrive somewhere at the end, where probably they will be better off than in the middle of a forest" (Descartes 1972, I, 96). While this is a legitimate strategy when one is directionless and groundless, it no longer applies, Descartes maintains, once the philosopher has attained a critical perspective on his intellectual and historical surroundings. At this point it would be foolish to continue in a straight line without attending to the *direction*. The step-by-step logic of philosophers whose principles are ill-founded may be correct but it is wrongly directed: "just as in travelling while we turn our backs on the place which we desire to go, the longer and quicker we walk the further we recede from the place we are making for" (Descartes 1972, I, 208), so it is with wrongly founded thought.

The environment Vico's method subtends is, as his disparagement of Descartes suggests, rather different. As a science of origins Vico's enquiry depends on turning one's back on the place one desires to reach: in order to ascend, to acquire a superior understanding of the world we have made for ourselves, it is necessary to descend—to go back and down via the ladder of etymology to find the poetic figures that formed the basis of the ancient wisdom (see Goetsch 1995, 1). The double-movement implied here is not dialectical if that means an eventual synthesis, a secure passage from *mythos* to *logos;* it is continuous and inescapable. Knowledge is recursive just as nations do not grow progressively more civilized but suffer periods of social, political, and intellectual regression. However, Vico's idea of intellectual progress is not wholly opposed to that advanced by Descartes: like the Cartesians, he wants to draw a map; it is only that his method of survey is different. *Ingegno* takes the data of memory and *le contorna e pone in acconezza ed assettamento.* The verb *contornare* not only means to give a new turn, but also "to surround; to go round; to outline; to border." A *contorno* can be an "outline" or "or-namental border"; and, in the plural, "surroundings" or "environ-ment" (Goetsch 1995, 41). *Ingegno* connects disparate and diverse things, then, "by *framing* one image in its relation to another, *out-lining* in this way a master context of the whole. *Ingegno* 'goes round' the primary sensory topics, creating the human 'surroundings' or 'environment' by 'noting the commonplaces that must all be run over in order to know all there is in a thing that one desires to know well; that is, completely.'" *Ingegno,* Goetsch sums up, "environs the whole world" (Goetsch 1995, 42–43).

It also defines the relationship of the observer to his world, the sense in which inner and outer worlds construct and mutually in-form each other. Introducing Vico's scientific autobiography in the first issue of *Raccolta d'opusculi scientifici e filologici,* Calogerà ex-plained the plan to produce a series of essays in which leading Ital-ian intellectuals reflected on the origin and development of their ideas; it was Carlo Lodoli, he said, who had coined the term *periau-tografia* to characterize this new genre, a term which notably omit-ted the *bios* of modern autobiography. Lodoli's term admirably de-scribed Vico's conception of knowledge as self-knowledge: to write about oneself was, Vico understood, to write *round about,* to outline the environment of one's ideas. A knowing self existing anterior to the history of its sense perceptions, as Descartes imagined, was, lit-erally and figuratively, senseless: "*Peri* is a life-term, often used as a prefix to anatomical terms to characterise what surrounds a desig-nated organ," Verene notes (Verene 1991, 66). There may have been another analogy at work, between the microcosm of the body and the macrocosm of the physical space produced and occupied by the social body. Lodoli was an architectural theorist: in influencing Pi-ranesi to adopt the view that the Doric architecture of ancient Greek settlements like Poseidonia (south of Naples) were the products of an indigenous Italian tradition, he drew on Vico's theory of the lo-cal origins of ancient Italian wisdom (Polizzo and Valone 1982, 256ff). While that theory may have been fantastic, as one of his translators avers (Vico 1988, 154 n-8), it contained a deeper truth: by returning over and over again to great memory sites of antiquity, the Italians had made them their own. *Memoria,* which Vico thought nearly identical with *immaginazione* as a faculty of image-making and keeping, had combined with *fantasia* and *ingegno* to create a na-tional consciousness where, formerly, there had only been an ar-chitectural ruin. *Peripateia: periautografia*—the two were closely re-lated, and not only philologically.

10.4. Geography's *Ingegno:* James Hutton

Both Vico and Descartes—to stay with our paradigmatic representations of (respectively) inductive and deductive reasoning—were
concerned to map an intellectual terrain; evidently, though, their
methods were different, if complementary. If Descartes was a metaphysical Mercator wanting to establish a universally applicable projection, a longitude and latitude of reasoning independent of local
content, Vico was like the philosophical traveler—J. G. Forster, perhaps, or his son, Georg—who tempered his speculations with facts.
Such a traveler was a step ahead of the humble ship's captain. While
"instructive particulars thrive in the proper field of the navigator,"
Georg Forster explained in his redaction of his father's *Resolution*
Journals:

> I have studiously avoided nautical details both at sea and in har-
> bour, nor ventured to determine how often we reefed, or split a sail
> in a storm, how many times we tacked to weather a point, and how
> often our refractory bark disobeyed her Palinurus, and missed stays.
> The bearings and distances of projecting capes, of peaks, hills, and
> hummocks, of bays, harbours, ports, and coves, at different hours
> of the day, have likewise been omitted (Forster 1777, I, ix).

Philosophical peripatetics like the Forsters aspired to an arrangement
of facts that went beyond the facts and found out general laws. They
would have subscribed to the view later expressed by William
Whewell in his *History of the Inductive Sciences:* "[Only] when our
speculations are duly fed from the spring-heads of Observation, and
frequently drawn off into the region of Applied Science, we may have
a living stream of consistent and progressive knowledge" (Whewell
1984, 14).

The topographical analogy Whewell uses to characterize inductive reasoning is very suggestive, and we will return to it. Here,
though, the point to make is that inductive reasoning was notoriously unreliable: to extrapolate from the local to the general was to
court misunderstanding. And it was Hume's recognition of just this
issue, of course—that inductive reasoning is no logical basis to empirical procedure—that further highlights this difficulty for enlightenment thinkers and for geography (Oldroyd 1986, 100–141). It
might be true, as Whewell thought, that inductive reasoning flowed
well *downstream* as it consolidated isolated facts and massed them
into a living stream of knowledge, but attempts to use the same techniques to navigate upstream illustrated the tenuousness of its logic.
In the geographical field, Whewell's topographical analogy was not

merely an analogy: it could be a principle of reasoning. Thus Georg Forster, puzzled by the absence of major river mouths along the hitherto explored coasts of Australia, speculated that any great river must enter the sea towards the southwest of that continent. Thereby, he thought, access might be had to "a vast interior space of ground, equal to the continent of Europe . . . entirely unknown, and perhaps uninhabited." Forster lent his speculation an inductive veneer—the "variety of animal and vegetable productions" which the *Endeavour* had found along the east coast suggested to him that "the inner countries contain immense treasures of natural knowledge which must of course become of infinite use to the civilised nation, which shall first attempt to go in search of them"—but his urge to find a river where there was none betrays a *deductive* temper. Forster looked forward to the time when Australia could be reduced to general laws, and to speed this time it must be shown to consist of familiar *topoi* (Forster 1777, I, 197–198).

In what way, then, was a geographical knowledge capable of meeting the demands of Enlightenment *logos* to proceed? Where advance by analogy was fraught with error, it seemed logical to attempt to ground knowledge of the earth's surface deductively. But, as we have seen, a step-by-step accumulation of details in the field was nothing without an overall direction, a prior conceptual longitude and latitude. And yet was this not precisely geography's field, its subject matter, to fill in that conceptual grid—and how could this error-fraught business of coastal first impressions ever represent itself as a methodical science? The only way in which geography could achieve deductive certainty, in fact, was by making its focus historical: it was as geology, as a chronology of events underlying the present appearance and constitution of the earth's surface, that geography could elevate itself to deductive dignity. To be sure, its deductions had at first to be fed from the springheads of observation, but in due course it should be able to reduce the world's appearance to a set of universal principles. One professional consequence of this would be to establish the dependent nature of other knowledges. With a hint of complacency Humphry Davy remarks, "All the rocks and stones that have been as yet procured from New Holland, a country containing so many novelties in botany and zoology, have not any of the characters of new mineralogical species and precisely resemble those of the old continent" (Davy 1980, 60). By contrast, botany and zoology must, it seems, remain inductive, elaborating their principles inside the universal borders of time and space which geography has established. In this context the evolutionary embryologist

PAUL CARTER

Ernst Haeckel's proud boast sounds almost plaintive: "Among the chief phenomena that bear witness to the inductive law of evolution we have the geographical distribution of the various species of animals and plants over the surface of the earth, and their topographical distribution on the summits of mountains and in the depths of the ocean." (Haeckel 1874, I, 88).

Nowhere is the problematic nature of geography's *ingegno* more clearly illustrated than in the work of James Hutton. A lively debate about the relative importance of empirical observation and theoretical speculation in the emergence of Hutton's theory of the earth's physical history need not detain us (but see Dean 1992, 115ff). In any case, a Vichian view would suggest that the debate is academic. It is odd, though, that historians of science generally fail to see that Hutton's physics and his metaphysics are two aspects of one intellectual enterprise. Thus, in his recent book *James Hutton and the History of Geology*, Dennis Dean manages to pass over Hutton's longest book, *An Investigation of the Principles of Knowledge, and of the Progress of Reason, from Sense to Science and Philosophy* (1794), in less than half a page—even though this monumental work was compiled and written in the period between the first announcement of Hutton's "Theory of the Earth" in 1785 and the formal publication of that treatise "with Proofs and Illustrations" ten years later. In the *Investigation* Hutton explains that reasonable progress occurs step by step: "In scientific reasoning, knowledge is produced by steps; in each of these a principle is acquired on which to proceed in reasoning, and thus acquire a further step. It will then appear, that, in science, there is no certainty without seeing every step" (Hutton 1794, II, 285). This was the basis of Descartes's geometrical method, with its attempt to detach reason from the operations of memory, imagination, and *ingegno*. But what could it mean in the context of evolving a distinctively geographical mode of reasoning? If one object of Hutton's uniformitarian thesis was to overturn the abyssal model of catastrophists—with its origins in the deductionist vision of Burnet—then it was essential that, as Hutton put it, we find "no vestige of a beginning, no prospect of an end" (Dean 1992, 266).

Further, and here Hutton is every inch a soulmate of Vico's, these gradualist principles apply to the inner as well as outer world. As he acknowledges, "the growth of the mind does not, perhaps, proceed by any precise number of steps that we may be able to observe. There is a gradation, from a beginning which may be definite, although perhaps we do not define it precisely, to an end, which seems to be indefinite, at least we cannot define its utmost extent" (Hutton 1794,

II, 527). It seems, then, that even the intuition of step-by-step progress is arrived at by analogy, by extrapolation from the particular to the general. It cannot be derived from general principles: if there is a universal hopscotch figure within whose bounds we perform our mental and physical lives, then it is not one we can ever know or survey. In this situation, the philosopher is still defined by his *ingegno*, only now it expresses itself as a fantasy of continuity. Empirically this finds expression in an obsessive attention to "gradation," as if by an ever more rigorous examination of apparent gaps in nature he can reveal their ultimate connectedness. But as this is only an application of Zeno's method of proving the illusoriness of movement to the process of induction in the natural sciences, it is bound to fail ultimately: ultimately yet smaller gaps inside the gaps will continue to appear and gradation will turn out once again to be a series of discrete steps. Hence the force of Vico's appeal to analogy: only by metaphorical "leaps" can a true continuity be preserved.

Hence, too, the ambiguity of geographical *ingegno:* to borrow the kind of topographical metaphor commonplace in the figurative expression of Enlightenment philosophers—was the geographer in search of general laws to proceed step by step, as if the earth were a featureless, edgeless table—a seemingly self-contradictory premise; or was he obliged to posit an arrangement of individually distinct islands or *topoi* between which it was his task to navigate, thus closing up the gaps in nature? Was he "to make known the roads and to repair them: and finally to travel slowly, but with a firm unwavering tread, by the side of that pure and living stream of clear and simple knowledge, which has its source among us, which can serve as a comfort on our painful march, and as a thread in the labyrinth of these vast overgrown territories," as the author of the dictum *In natura non datur saltus,* Leibniz, maintained? (Leibniz 1961, 237–238). Or, admitting the little leap that links one unwavering footstep to the next, was one frankly to acknowledge the constructed (even mythopoeic) motive of one's knowledge, admitting as Jeremy Bentham did that "The footing on which . . . [Utility] rests every dispute, is that of . . . future fact—the probability of certain future contingencies" (Mack 1962, 275). Writing half a century after this, the author of *The Philosophy of the Unconscious,* Von Hartmann, attempted to rationalize this dilemma. While, according to him, inductive logic, entails what he calls "rational intuition . . . the Pegasus flight of the Unconscious, which carries in a moment from earth to heaven," the "deductive method" is "only the lame walking on stilts of conscious logic." As he explains, "Each putting of the foot to the ground forms a point of rest, a station . . . which produce[s] a conscious idea . . . The leaping or stepping itself, on the other hand, is . . . something momentary, timeless, because empirically falling into the Unconscious" (Hartmann 1884, I, 316ff).

One begins to see why, without the convenient category of the unconscious to hand, Enlightenment theorists of reason were so drawn to rivers. Presenting an unsegmented advance, a pure and chanceless transition from one place to the next, they seemed to take the chance out of induction; at the same time they allied deduction to the lie of the land, the world of the senses. Rivers encouraged retrospective as well as prospective reverie, and in between to the sailor they suggested the transport of reason, suspension of doubt. In this way they appealed physically as well as figuratively to the field geographer, to whom they showed the way in which knowledge was subtly compounded of *memoria, fantasia,* and *ingegno:* "rivers were desired, not merely to assuage physical need, but as geographical objects satisfying intellectual thirst. . . . The water offered not only the sensation of weightlessness but also, for [Australian explorer] Sturt at least, a sense of direction, a sensation of mental and geographical expectations fusing" (Carter 1987, 54).

10.5. Playfair's "Abyss"

Dean acutely notes that in his *Illustrations of the Huttonian System* (1802) Playfair did not simply paraphrase and exemplify Hutton's theory of the earth's history: Playfair, he says, "regards Hutton's theory as a human construct, not as a system of nature" (Dean 1992, 115). This can be taken further: in subjecting Hutton's ideas to a *verum-factum* principle, Playfair famously gave free rein to his *ingegno,* reorganizing Hutton's sometimes hard-to-follow argument into a smoothly flowing, logically comprehensible thesis. Hutton's theory was, Playfair explained self-deprecatingly, "proposed too briefly, and with too little detail of facts, for a system which involved so much that was new, and opposite to the opinions generally received. The descriptions which it contains of the phenomena of geology, suppose in the reader too great a knowledge of the things described. The reasoning is sometimes embarrassed by the care taken to render it strictly logical; and the transitions, from the author's peculiar notions of arrangement, are often unexpected and abrupt" (Playfair 1973, 165). Playfair undertook to fill in the gaps, to ensure that as far as possible Hutton's theory met his own stated criterion of scientific progress: "there is no certainty without seeing every step."

To this end, Playfair recovered Hutton's meaning somewhat in the way Vico went about deciphering the wisdom of the ancients. He made his case persuasive by asserting an identity between the philosopher and his philosophy. This assertion constituted Playfair's own contribution to knowledge, the application of his own *ingegno*. If the criterion of intellectual respectability was an ability to reason logically, combined with a willingness to let speculation be fed from "the spring-heads of Observation," then it was in this image that Hutton must be reminted. This is the context in which to read Playfair's claim that, "with an accurate eye for perceiving the characters of natural objects, he [Hutton] had in equal perfection the power of interpreting their signification, and of deciphering those ancient hieroglyphics which record the revolutions of the globe." Few mineralogists, Playfair declares, "[have] equalled him in reading the characters, which tell not only what a fossil *is,* but what it *has been,* and declare the series of changes through which it has passed" (Playfair 1973, 193). And this ability to differentiate ever more finely with a view to elaborating ever more completely a record of continuity in change suggests a spatial metaphor we are by now familiar with: "None was more skilful in marking the gradations of nature, as she passes from one extreme to another; more diligent in observing the *continuity* [italics in original] of her proceedings, or more sagacious in tracing her footsteps, even where they were most lightly impressed" (Playfair 1973, 193). Passages like these not only assert something about the subject: they perform what they assert rhetorically, eloquently establishing the *environment* of a life and its work.

The same principle of eloquence explains the "surprizing amount of space [Playfair devoted in his "Biographical Account"] to Hutton's philosophizing on physics and metaphysics" (Dean 1992, 123). To grasp the constructedness of Hutton's knowledge was to grasp the intellectual gradation that existed between Hutton's view of the physical world and his philosophy of mind; as Vico maintained, to understand the nature of a science it was necessary to know its origins; from a knowledge of its wellsprings everything followed. But Playfair's biographical approach had a more urgent motive: the demonstrated symmetry between Hutton's physical and metaphysical theories was important rhetorically, if not logically, in substantiating Hutton's reputation as a scientific theorist. Hutton's peculiar version of idealism had the effect of dissolving, or at least suspending, the induction-deduction dilemma. Although Hutton held that "there [was] no resemblance between the world without us, and the notions that we form of it," he also allowed that "The world . . . as

PAUL CARTER

conceived by us, is the creation of the mind itself, [it is] of the mind acted on from without, and receiving information from some external power." Thus, "our perceptions being consistent, and regulated by uniform and constant laws, are as much realities to us, as if they were the exact copies of things really existing" (Playfair 1973, 186). Our knowledge of the world, Hutton seems to say, stands in relation to the unknowable world as a parallel universe might stand to our own: perfectly fitting at every point but created according to entirely different principles. We know via a form of mental mimicry in which the mind copies to the best of its ability (according to its own constitution) what it receives via our external perceptions: we cannot know whether our knowledge of the world is a true one (we cannot escape our mimic representations); all we can judge is their internal, or rational, coherence, their appearance of creating a reasonable picture. In this way, induction and deduction must coincide.

But what was the result of this massive engineering of intellectual bridgeheads? It was to allow the scientist astride his Pegasus to see ever deeper into the abysses that opened up underneath his feet. It enabled him to give the gaps in reason a positive value; to ascribe to the voids a logical role. Presupposing the high road of reason, he could make the evidence of cracks and gaping discontinuities further support for his system. The gaps in reason came to have the same theoretical convenience as the chemist's Ether, as (later) the Structuralist linguist's phoneme—as Hutton's own "principle of levity," phlogiston: they were the featureless medium of change, unchangeable in themselves and unobservable except through their environmental effects. They resembled in a way the unconscious, those steps beyond number, that self-consciousness had to presuppose if it was to have any reason to reason. One senses a significant biographical motif contained in this unresolved tension between a desire to demonstrate seamless continuity and the necessity to acknowledge and negotiate abrupt transitions. It seems likely that Hutton's phlogiston theory is related to these issues. As a "principle of levity," phlogiston was analogous to the unexplained forces that pushed up the earth's mantle. More important, it was a "secret principle, or fugitive element" which accounted for sudden changes of state without leaving any trace of its own presence, and without suffering any change. Hutton's views can be usefully compared with the contemporary doctrine of elective affinities, a theory formalized by Humphry Davy, who showed that chemical affinities were electrical in nature. Goethe, of course, realized the psychological possibilities of this idea (Partington and McKie 1981, 367–369).

It is in this context that the fuller meaning of Playfair's famous account of the expedition to Siccar Point emerges; for here the continuity, the reasonableness of Hutton's system was grasped through observational proof of geological discontinuity; and Playfair's rhetorically ingenious description of his reaction—which might be compared with the headlong plunge Vico suffered as a child, as a result of which descent into unconsciousness he awoke to self-consciousness—is similarly mythopoeic, staging, in fact, that primary theatrical trick of shifting the spatial across into the temporal.

As Playfair inspects the physical ground, it progressively metamorphoses into the sands of time. Seeing the "immediate contact" of micaceous schistus and sandstone—"palpable evidence [giving] reality and substance to those theoretical speculations":

> We felt ourselves necessarily carried back to the time when the schistus on which we stood was yet at the bottom of the sea, and when the sandstone before us was only beginning to be deposited, in the shape of sand or mud, from the waters of a superincumbent ocean. An epocha [sic] still more remote presented itself, when even the most ancient of these rocks, instead of standing upright in vertical beds, lay in horizontal planes at the bottom of the sea, and was not yet disturbed by that immeasurable force which has burst asunder the solid pavement of the globe. Revolutions still more remote appeared in the distance of this extraordinary perspective. The mind seemed to grow giddy by looking so far into the abyss of time (Playfair 1973, 176–177).

The cause of Playfair's giddiness is not what is suddenly present to his external eye: it is what passes before his mental eye—*and* the vast discontinuity between the two visions. The source of his wonder is the *lack* of resemblance between the "palpable evidence" and the "theoretical speculations" they have occasioned; it is only the ingenuity of Hutton that can persuade him of their ultimate connection. It is the visible void which, carried over into the fluviatile economy of time, proves the validity of Hutton's inductive logic.

This necessary detour into the invisible has its psychological counterpart: in order to achieve the firm unwavering tread associated with traveling alongside the stream of clear knowledge, it is necessary also to depart from oneself, to experience the sensation of falling, of descent into the void. He who advances without reason is immediately lost, plunging, Leibniz says, "into morasses and shifting sands of doubts without end, wherein is nothing solid nor firm" (Leibniz 1961, 285). But this descent into the erosive realms of conspicuous change the enlightened geographer could hardly avoid.

PAUL CARTER

How were Hutton's geological interests first aroused if not by "studying the surface of the earth . . . looking with anxious curiosity into every pit, or ditch, or bed of a river that fell in his way"? "If he did not always avoid the fate of Thales [the Greek sage who fell into a well while walking about looking at the stars], his misfortune was certainly not owing to the same cause" (Playfair 1973, 148). Hutton might have distinguished himself from Thales on another count: although in no sense a vulgar Plutonian, Hutton was certainly not, like Thales, a Neptunian. Hutton kept his feet firmly on the ground *and for this reason* fell into the river. There remained after all in Leibniz's formulation of reasonable progress a mythic residue: to travel slowly *alongside* the river was only to make progress by analogy. A little gap still divided the philosopher from his goal, and really there was no way to overcome it except by stepping aside and plunging in.

10.6. By Analogy

Having asserted as axiomatic that geographical logic supplies our preeminent contemporary *mythos,* my remarks have spiraled rather rapidly downwards into particulars. A fuller exposition of the mythopoeic origins of Enlightenment geography and its characteristic objects of knowledge would need to attempt a graded return, an inductively managed ascent towards at least a provisional thesis based on more facts than those provided by one highly compressed account of a scientist and his biographer. But the flavor of such an investigation may have been suggested. It becomes clear that an understanding of the mythopoeic nature of geographical discourse involves an attention to the rhetoric of reason, its commonplaces of expression, unusual in most histories of science. As Vico clearly understood, myth's "vivid representations, images, similes, comparisons, metaphors, circumlocutions, phrases explaining things by their natural properties" were its meaning. Such figures of speech, he thought, "were the actual linguistic means by which humankind established itself and its world against the natural world" (Mali 1992, 165). The habitual spatialization of thought processes in Enlightenment writing about logic is not, then, trivial: it expresses the mythic underpinning of Enlightenment thought. When Whewell observed "The Table of the progress of any science would thus resemble the Map of a River, in which the waters from separate sources unite and make rivulets, which again meet with rivulets from other fountains, and thus go on forming by their junction trunks of a higher and higher order," he uttered a figurative commonplace (Whewell 1984, 11). In the realm of mental geography reasonable men lived like slave traders, where the largest rivers entered the sea.

In constituting the field of geographical enquiry, a fuller study would show, the prevalence of geographically derived figures of speech assumed a critical importance; for, as already emphasized, the philosophical traveler did not merely dream rivers, he hypothesized them as objects capable of observation; and to this end he looked for them, using the instruments of analogy to hand. This attempt to have the world conform to his mental geography was motivated: indeed, it had as its intention the concealment of its own motivation—it was Vico's insight to insist on the continuity between the nature of a science and the manner of its birth. In geographical logic, as elsewhere, inductive reasoning industriously colonized new shores of knowledge. But the imperium on whose behalf it labored was deduction. "I have formed a plan of universal conquest," explained Jeremy Bentham, "I intend to govern all the nations in the habitable globe after my death—With what weapons? With rhetoric? With fine speeches? With prohibitive and irritant clauses?—No: but with reasons, with a chain of . . . articulate and connected reasons, all depending upon one principle" (Mack 1962, 273–274). But in reality, as Mary Mack showed in her fine study, Bentham's imperium could be achieved only with the assistance of analogy (Mack 1962, 275ff). Unlike Vico, though, who saw the use of analogy archaeologically, as a way of relocating connections lost, Bentham saw it as the tool of invention. Hence Utilitarianism: "If analogy was a key means to the end of creative invention, creative invention was the key to the ultimate end, the greatest happiness of the greatest number" (Mack 1962, 277).

Such a mythopoeic investigation would also have to make a little leap of its own: out of the discourse of imperial geography into an examination of its environmental effects—which *pace* certain cultural theorists—are decidedly physical, bloody, and wordless. Again it was Vico's prescient genius to insist on the embeddedness of the production of meaning in the subjectivity of historical experience. But here a certain caution is needed lest we step in out of our depth: the very power that geographical discourse enjoys as our dominant mythos prevents us from feeling at once its constructedness; we lazily identify its *topoi* with real places. Thus even seasoned practitioners of cultural theory are apt to suspect an ironic joke at their expense when informed that the coasts, the rivers, islands, archipelagoes, ranges, and roads of the world should be placed inside quotation marks to indicate their culture-bound character. Much as the meaning of a text may be deconstructed, the obstinate places of the world cannot be so easily deferred. As the tacit theatre of all our cultural operations they, perhaps of necessity, have to be taken as given. But, as I say, the literalism of this approach only attests to the mythic role geography plays in our lives.

A mythopoeic examination of the Enlightenment origins of modern geographical discourse is, following Vico, the essential preliminary to the development of a genuinely critical geographical discourse, one able to remake itself, and in reconsequence, remake the world. To consolidate this development a variety of approaches might be needed, the convergence of insights from widely different directions. An ethnography of meeting places, for example, would help show how the coastlines and islands of the European imaginary, together with their ancillary tropes of shipwreck and disaster, were projections of mythic reasoning. The effects were real, fatal for those who received the foreigners, but they were nonetheless the products of mythopoeic logic: the coasts with their lure of a decisive step, islands with their temptation of self-evident factuality, were passages over the gaps of reason, rafts against plunging into the abyss of the unconscious. The progress of Western logos was not, then, inevitable—that has been reason's self-serving mythos—like the man in the desert who armed himself with two sheets of galvanized iron to drive through the sands, geographical thinking, no less than other forms of inductive science, has planted its own high road ahead of itself.

With Vico's mythopoeically inflected *verum-factum* principle in place, some subtler feedback effects also become noticeable and significant. It is not that a poetic metaphysics is imposed on a world out there: the physical world of the senses is, as Hutton implied, assimilated to reason in so far as it shows the stirrings of logical consciousness. Thus, in the natural sciences at least, there is a feedback relationship between the theorization of knowledge and the maplike organization and rationalization of the observational data on which it is based. As suggested earlier, Wallace's working-up of his field journals to produce *The Malay Archipelago* is an exemplary case in point. In place of a journal of first impressions, provisional hypotheses, mental and physical backtrackings, erasures, and revisions, Wallace presents to the public a fully rounded argument. No harm in

that, one might think, except that the organization of his field experience into a connected narrative composed of connected *topoi*
exactly parallels his geographical reorganization of the Malay "archipelago" in the interests of his evolutionary theory. Wallace's scientific argument is rhetorically reinforced, its persuasiveness derives
at least in part from Wallace's own creative *ingegno*. In this case, mental and physical geography exist symbiotically—or, to be more critical (and borrowing one of Wallace's own explanatory mechanisms
of survival), indulge in protective mimicry.

Can a mythopoeic enquiry of the kind outlined here be reduced
to rules? Vico clearly thought so. But the method he advocated was
by definition ruleless: to advance by way of analogy or metaphor
was to embrace the discontinuities embedded in continuity, but in
the hands of the ruthless inventor, or self-styled genius like Bentham, it was as liable as plodding reason to flatten out a world of differences. Poetic wisdom produced its Frankensteins and panopti-
cons as readily as it mapped new constitutions (and coastlines). In
stressing the mythopoeic character of knowledge, Vico emphasized
the ethical character of poetic wisdom: metaphors were actions.
Similarly, Arrowsmith's maps, faithfully reproducing a growing maze
of reef-encrusted coastlines, subtended new environments. These
were not natural but historical, mythopoeic, and the site of future
contingencies. We live in their shadow and wake. The first step, as
Hutton might say, is to realize this, even if we cannot define precisely its beginning.

References

Albritton, Claude C. 1989. *Catastrophic Episodes in Earth History.* London:
    Chapman & Hall.
Barnes, T., and Derek Gregory. 1997. *Reading Human Geography: The Poetics
    and Politics of Inquiry.* London: Arnold.
Barnes, Trevor R., and James S. Duncan, eds. 1992. *Writings Worlds: Dis-
    course, Text and Metaphor in the Representation of Landscape.* London and
    New York: Routledge.
Belyea, Barbara. 1992. "Amerindian Maps: the explorer as translator." *Jour-
    nal of Historical Geography* 18 (3), 267–277.
Buttimer, Anne. 1993. *Geography and the Human Spirit.* Baltimore, MD: The
    Johns Hopkins University Press.
Carter, Paul. 1996. *The Lie of the Land.* London: Faber & Faber.
———. 1987. *The Road to Botany Bay.* London: Faber & Faber.
Corbin, Alain. 1994. *The Lure of the Sea: The Discovery of the Seaside in the
    Western World, 1750–1840.* Cambridge: Polity Press.
Cosgrove, D. E. 1984. *Social Formation and Symbolic Landscape.* London:
    Croom Helm.

Cosgrove, D. E., and S. J. Daniels, eds. 1988. *The Iconography of Landscape.* Cambridge: Cambridge University Press.

Daniels, Stephen. 1993. *Fields of Vision: Landscape Imagery and National Identity in England and the United States.* Cambridge: Polity Press.

Davy, Humphry. 1980. *Humphry Davy on Geology: The 1805 Lectures for the General Audience,* edited by R. Siegfried and R. H. Dott. Madison: University of Wisconsin Press.

Dean, Dennis R. 1992. *James Hutton and the History of Geology.* Ithaca, NY: Cornell University Press.

Descartes, René. 1972. "Discourse on the Method of Rightly Conducting the Reason and Seeking for Truth in the Sciences" in *The Philosophical Works of Descartes,* translated by E. S. Haldane and G. R. T. Ross. Cambridge: Cambridge University Press, 2 vols.

Entrikin, Nicholas. 1991. *The Betweenness of Place: Towards a Geography of Modernity.* London and New York: Macmillan.

Forster, J. G. A. 1777. *A Voyage Round the World in His Britannic Majesty's Sloop* Resolution, *Commanded by Capt James Cook, during the years 1772, 3, 4 and 5.* London; B. White, J. Robson, P. Elmsby, and G. Robinson. 2 vols.

Goetsch, James Robert. 1995. *Vico's Axioms, The Geometry of the Human World.* New Haven: Yale University Press.

Gregory, Derek. 1994. *Geographical Imaginations.* Oxford: Blackwell.

Haeckel, Ernst. 1874. *The Evolution of Man,* translated by J. McCabe. New York: G. P. Putnam's. 2 vols.

Harley, J. B., and David Woodward, eds. 1994. *The History of Cartography Volume Two Book Two: Cartography in the Traditional East and South-East Asian Societies.* Chicago: University of Chicago Press.

Hartmann, Eduard von. 1884. *Philosophy of the Unconscious,* translated by W. C. Coupland. 3 vols. London: Trubner & Co.

Hutton, James. 1794. *An Investigation of the Principles of Knowledge and of the Progress of Reason.* 3 vols. Edinburgh: Strahan and Cadell.

Kafka, Franz. 1954. "Third Octavo Notebook," in *Wedding Preparations in the Country and Other Posthumous Writings,* translated by E. Kaiser and E. Wilkins. London: Secker & Warburg.

Leibniz, G. W. 1961. *Philosophical Writings,* translated by M. Morris. London: Dent.

Lewis, Martin, and Karen Wigen. 1997. *The Myth of Continents: A Critique of Metageography.* Berkeley, CA: University of California Press.

Mack, Mary P. 1962. *Jeremy Bentham, an Odyssey of Ideas, 1748–1792.* London: Heinemann.

Mali, Joseph. 1992. *The Rehabilitation of Myth: Vico's New Science.* Cambridge: Cambridge University Press.

Mayhew, Robert. 1998. "Was William Shakespeare an Eighteenth-Century Geographer? Constructing Histories of Geographical Knowledge." *Transactions of the Institute of British Geographers* 23 (1), 21–38.

Mills, William J. 1982. "Positivism Reversed: The Relevance of Giambattista Vico." *Transactions of the Institute of British Geographers* 7 (1), 1–14.

Obeyesekere, Gananath. c. 1992. *The Apotheosis of Captain Cook: European Mythmaking in the Pacific.* Princeton, NJ: Princeton University Press.

Oldroyd, David. 1986. *The Arch of Knowledge: An Introductory Study of the*

History of the Philosophy and Methodology of Science. London and New York: Methuen.

Partington, J. R., and D. McKie. 1981. *Historical Studies on the Phlogiston Theory*. New York: Arno Press.

Paulston, Rolland G., ed. 1996. *Social Cartography: Mapping Ways of Seeing Social and Educational Change*. New York: Garland Publishing.

Playfair, John. 1973. "Life of Dr. Hutton" in *James Hutton*, vol. 5, *Contributions to the History of Geology*, edited by G. W. White. New York: Hafner Press.

Polizzo, J. A. and J. Valone. 1982. "L'Influsso delle Idee Vichiane nell'Ispirazione di Piranesi" in *Vico e Venezia*, edited by Cesare de Michelis and G. Pizzamiglio. Florence: Leo S. Olschki.

Ryan, Judith. c. 1990. *Spirit in Land: Bark Paintings from Arnhem Land*. Melbourne: National Gallery of Victoria.

Sack, Robert. 1997. *Homo Geographicus*. Baltimore, MD: The Johns Hopkins University Press.

Sahlins, Marshall. 1995. *How "Natives" Think: About Captain Cook, for Example*. Chicago: University of Chicago Press.

Semple, Ellen C. 1911. *Influences of Geographic Environment*. New York: Henry Holt & Co.

Tuan, Yi-Fu. 1974. *Topophilia: A Study of Environmental Perception, Attitudes and Values*. Englewood Cliffs, NJ: Prentice-Hall.

Turnbull, David. 1989. *Maps Are Territories: Science Is an Atlas*. Chicago: University of Chicago Press.

Verene, Donald Phillip. 1991. *The New Art of Autobiography: An Essay on the Life of Giambattista Vico Written by Himself*. Oxford: Clarendon Press.

Vico, Giambattista. 1984. *The New Science of Giambattista Vico*, translated by M. H. Wallace and T. G. Bergin. Ithaca, NY: Cornell University Press.

———. 1988. *On the Most Ancient Wisdom of the Italians*, translated by L. M. Palmer. Ithaca, NY: Cornell University Press.

———. 1990a. *On the Study Methods of Our Time*, translated by Elio Gianturco. Ithaca, NY: Cornell University Press.

Wallace, Alfred Russel. 1986. *The Malay Archipelago, the Land of the Orang-Utan, and the Bird of Paradise*. Kuala Lumpur: Oxford University Press.

Whewell, William. 1984. "History of the inductive sciences," in *Selected Writings on the History of Science*, edited by Y. Elkana. Chicago: University of Chicago Press.

Chapter Eleven

A Geography of Enlightenment: The Critical Reception of Alexander von Humboldt's Mexico Work

NICOLAAS RUPKE

11.1. Humboldt's Place: The Enlightenment Tradition, Romanticism, or Classicism?

In writing a chapter on Alexander von Humboldt for a volume carrying the title *Geography and Enlightenment*, no apology is needed for connecting Humboldt with the first of the two title nouns. Anne Buttimer, among others, regards Humboldt as one of two founding fathers of geography (the other being Carl Ritter) (Buttimer 1993, 59), and although not everyone would go as far as this in appraising Humboldt, his credentials as a geographer are beyond dispute. By contrast, connecting Humboldt with the second of the two title nouns is highly controversial. After all, to many historians Humboldt is the archetypal representative of the very reaction against Enlightenment culture—of Romanticism. Was Humboldt not the Romantic naturalist *par excellence?*

The interpretation of Humboldt and his work as an integral part of the European, Romantic movement is to be found in a broad spectrum of secondary literature, extending from Alexander Gode–von Aesch's early classic, *Natural Science in German Romanticism* (1941), to Theodore Ziolkowski's more recent *German Romanticism and Its Institutions* (1990). The claim that Humboldt was a Romantic has been based on both his "Jugendarbeiten" such as the vitalistic "Die Lebenskraft oder der Rhodische Genius" (Gode–von Aesch 1941, 193–197) and his "Spaetwerke," in particular on the holistic *Kosmos* (1845–62) (Ziolkowski 1990, 253). To some authors, Humboldt, as the explorer of the Americas, is the very embodiment of Romanticism, and the cover illustration of *Romanticism and the Sciences* (1990), edited by Andrew Cunningham and Nicholas Jardine, shows Humboldt in his jungle hut on the Orinoco. Malcolm Nicolson, in his contribution to the Cunningham-Jardine volume, focuses on the holism and aesthetic concerns of Humboldt's plant geography and concludes that "Alexander von Humboldt may be seen as both a product of German Romanticism and an important exponent of a Romantic style within natural inquiry" (Nicolson 1990, 183). Mary Louise Pratt goes further and in her engaging study of European travel literature simply equates Humboldt's American "Reisewerk" with Romanticism (Pratt 1992, 137–138).

Equally broad, however, is the range of secondary literature in which Humboldt is seen as a representative of Enlightenment ideals. More than four decades ago, Erwin Ackerknecht categorically denied that Humboldt ever was a Romantic (Ackerknecht 1955, 92). Among German Humboldt scholars, Hans-Joachim Waschkies contends that Humboldt promoted a 'Weltbild' or 'Weltanschauung' that was characterized by the heterodox persuasion and the empiricism of the Enlightenment (Waschkies 1990). More recently, David Livingstone, in his authoritative survey of the geographical tradition, posits Humboldt at the apex of "Geography in the Enlightenment" (Livingstone 1992, 134–138). Thus, although the larger and productive part of Humboldt's life followed the close of the Enlightenment period, he is interpreted as having substantially contributed to an Enlightenment project, adding precision of measurement, conceptual sophistication, and instrumentation to the eighteenth-century geographical traditions of Europe.

One could take yet a third stance by arguing that Humboldt was representative of neither an Enlightenment tradition nor of Romanticism but should be interpreted, along with his brother Wilhelm von Humboldt, in the cultural context of German Classicism and in particular in the setting of Classicist humanism. Such an Humboldt interpretation has been favored by many Marxist-Leninist geographers and historians of what used to be East Germany, among whom are Gerhard Harig (1995) and Heinz Sanke (1969). In their perception, Humboldt, by pitching *Kosmos* at a level that made the book accessible to nonexperts, contributed to the emancipation of the working classes; moreover, his emphasis on the aesthetic aspects of nature was a pedagogical means—they maintained—to help the uneducated masses acquire scientific literacy. To these socialist scholars, Humboldt was a hero of the liberation struggles of the proletariat.

The purpose of this chapter is not to attempt to settle the controversy over whether Humboldt was an heir to the Enlightenment, a Romantic, or a representative of German Classicism. Given the conceptual fluidity of these categories on the one hand and the richness of Humboldt's work on the other, it is unlikely that the issue can ever be definitively and satisfactorily settled. In fact, a good case can be made for each of the three interpretations, depending on which of Humboldt's many activities, letters, friendships, articles, books, or proclivities one selects or emphasizes; and it is likely that Humboldt, who famously maneuvered to keep a middle road of nonpartisan moderation, would have been gratified by such exegetical pluralism of his life and work.

The strongest case for an Enlightenment interpretation of Humboldt can be made from his writings on the political economy of the Americas, in particular his 'political essay' on Mexico. Lutz Raphael argues that Humboldt's analysis of the political economy of Mexico and other Latin American regions carried the imprint of especially the Scottish Enlightenment, in that Humboldt abandoned the government- and state-heeding orientation of German cameralism, turning instead towards Enlightenment notions of economic and individual freedom which he borrowed from Adam Smith (1723–1790). For example, in the Mexico book Humboldt corroborated the belief that the relaxation of trade monopolies had produced an upturn in economic growth (Raphael 1995; for an earlier discussion see Brand 1959). In this chapter I do not elaborate on Raphael's thesis, taking for granted that Humboldt's work on Mexico reflected, in important ways, Enlightenment thought about the freedom and wealth of nations.

My purpose here is first, to document the contemporary reception of Humboldt's *oeuvre* and in particular to gauge the importance of the Mexico book relative to that of Humboldt's many other publications; and second, to sketch a European geography of this book's critical reception, asking the question: "How was the Humboldtian knowledge of the isthmus of the New World mapped onto the Old?" The interest of such a "geography of knowledge" (Livingstone 1995) lies in the fact that it may show the extent to which 'local' circumstances were involved in the production of the diverse meanings that the *Essai politique* acquired.

11.2. The Contemporary Reception of Humboldt

To assess the relative significance of Humboldt's *Essai politique,* I have carried out a systematic analysis of reviews of Humboldt's many publications. Reviews of Humboldt's publications that appeared in the British periodical press have previously been studied by William H. Brock (1993) and by Calvin P. Jones (1972–1973). In this study, a total of fifteen periodicals that published substantial reviews of

Table 11.1. Reviews of Humboldt's Publications in European Periodicals, 1790–1865

British	Years Covered
British Quarterly Review	1845–1865
Edinburgh Review	1802–1865
Monthly Review	1790–1845
North British Review	1844–1865
Quarterly Review	1809–1865
(London and) Westminster Review	1824–1865

French	
(Nouvelles) Annales des voyages	1808–1814
	1819–1965
Le Moniteur universelle	1790–1865
Revue des deux mondes	1829–1865

German	
(Neue) Allgemeine geographische Ephemeriden	1798–1831
Allgemeine Literatur-Zeitung	1790–1849
Göttingische gelehrte Anzeigen	1790–1865
Heidelbergische Jahrbuecher der Literatur	1808–1865
(Neue) Jenaische Allgemeine Literatur-Zeitung	1804–1848
Monatliche Correspondenz zur Befoerderung der Erd- und Himmelskunde	1800–1813

scientific works—six British, three French, and six German—were selected (table 11.1). An effort was made to choose periodicals that were in existence during as large a stretch as possible of the period of Humboldt's activity as an author of books, i.e., from 1790, the year in which Humboldt's first book appeared, the *Mineralogische Beobachtungen ueber einige Basalte am Rhein,* until 1862, when the fifth, posthumous volume of *Kosmos* was published. To ensure that late reviews would not slip the net, the year 1865 was chosen as actual cutoff date.

Only one French and one German periodical of our sample covered the entire period 1790–1865 (table 11.1). To compensate for periodicals that ceased to exist before 1865, others were chosen that started after the year 1790. These periodicals were systematically checked for reviews of Humboldt's works. An exact comparability in source magazines, from year to year, for the entire period 1790–1865, was not achieved, and during the early part of the period, the increase in the number of reviews was in part due to the increase in the number of new periodicals. The effect of this is that for the first circa fifteen years of Humboldt's authorship of books, reviews are potentially and comparatively underrepresented.

Humboldt published many and varied books and articles, which can be grouped into four categories (an early bibliography of Humboldt's writings was by Julius Loewenberg [1872]; a recent one, discussing the difficulties and pitfalls of the Humboldt bibliography is by Ulrike Leitner [1992]). First, there were his early writings, most famously a work on organic electricity, *Versuche über die gereizte Muskel- und Nervenfaser* (2 vols., 1797 [1798]). During the period of these 'Jugendarbeiten,' Humboldt studied at several German institutions of higher education in succession; his time at Göttingen is customarily highlighted, but more significant for understanding Humboldt's views on political economy may have been his period of study at the Hamburg School of Commerce (Beck 1959, vol. 1, 31–35). Subsequently, he was employed in the Prussian mining service, and, in 1795, conducted an extensive tour of the Swiss, French, and Italian Alps.

Second, there were the many and in part very voluminous publications that resulted from Humboldt's renowned journey of exploration of equatorial America (1799–1804), on which journey he was accompanied by the French botanist Aimé Bonpland (1773–1858). Most of these publications were written during Humboldt's Parisian period (1808–26), when he lived as an independently wealthy, private scholar. The production of the American *oeuvre* (Humboldt's "amerikanisches Reisewerk") was a major undertaking that exhausted his personal fortune. The thirty-volume work carried the collective title: *Voyage aux Regions équinoxiales du Nouveau Continent fait en 1799, 1800, 1801, 1802, 1803, et 1804, par Al. de Humboldt et A. Bonpland;* it dealt with botany, plant geography, zoology, physical geography, political economy, and included such classics as the *Essai politique sur le royaume de la Nouvelle-Espagne* (2 vols., 1808–11), the *Vues des Cordillères et monumens des peuples indigènes de l'Amerique* (1810–13) and the *Relation historique du voyage aux regions equinoxiales du Nouveau Continent* (3 vols., 1814–31).

Third, there were the publications that resulted from Humboldt's relatively short Russian journey of exploration (1829), e.g., *Asie centrale. Recherches sur les chaines de montagnes et la climatologie comparée* (3 vols., 1843). These belonged to his Berlin period (1827–59), during which he was employed as a royal chamberlain at the Prussian Court.

The fourth and last category included the publications of Humboldt's old age, most famously *Kosmos* (4 vols., 1845–62), a work that constituted both the summary of many—but by no means all—of Humboldt's lifelong interests and an holistic digest of the scientific study of celestial and terrestrial phenomena (for a recent discussion of *Kosmos*, see Michael Dettelbach [1997] and Nicolaas Rupke [1997]). On the coat-tails of its success, a new, third German edition of *Ansichten der Natur* was produced (1849; first and second edition, 1808, 1826, respectively), followed by translations into English (1849, 1850), and a new French translation (1851; first French version, 1808; second, 1828).

One could argue that reviews are by no means the only standard by which reception and relative success of books can be measured. Other indicators include the number of new editions or reprints of a book, the size of the print runs, or the number of translations. These indicators could indeed be applied to good effect in the case of Humboldt's *Ansichten* and *Kosmos*, but they would fail to register the importance of Humboldt's monumental American *Reisewerk*. Its twenty imperial folio volumes and ten quarto ones were notoriously expensive, and not many passed into circulation (see for example Robert Avé-Lallemant [1872, 23–24]); a contemporary author, writing in the *North American Review* (Anon. 1823, 1), complained: "Various circumstances have prevented a large number of copies of the different publications of M. de Humboldt from passing into circulation, either in Europe or America." Their bibliographical statistics would therefore fail to register the importance they actually had. Yet they were widely reviewed and excerpted; as an indicator of how Humboldt's many publications were received, 'reviews' would appear as effective a means as any.

Nearly all of Humboldt's books were reviewed in the fifteen periodicals of table 11.1. A total of 157 essays were found and used for the purpose of a bibliometric analysis. Only reviews concerned with a single book were counted, and a few composite reviews that were retrospective on Humboldt's writings up to a certain date were excluded because no accurately quantifiable part of these essays could be apportioned to any one publication (examples of 'composite reviews' occur in the *Annales des voyages* [Anon., 1809], the *Moniteur universel* [Anon. 1815] or also the *North American Review* [Anon. 1823]). The amount of reviewing attention paid to each book ('reviewing intensity') can be measured by counting the number of reviews, adding up the total number of pages of these, or by an accurate word count of the review essays. In this study, word counts were used in view of the extreme differences in words per page between periodicals. Word counts were then plotted against year of publication of the reviews (fig. 11.1).

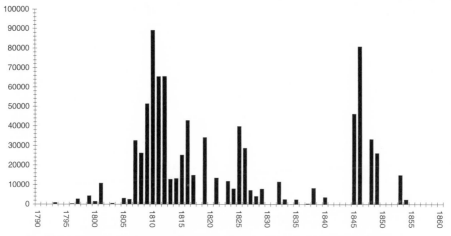

Fig. 11.1. Frequency diagram of reviews of Alexander von Humboldt's books in the British, French, and German periodicals, listed in table 11.1. The total number of words of the reviews are plotted against the years in which these were published. Note three crests: one around 1810, a diffuse one from 1815 to 1826, and a third around 1846.

The resulting graph shows that across learned Europe an enormous number of words were spent on Humboldt's writings, one year's word count rising as high as nearly 100,000, i.e., that of a modest book. Because the graph is based on a limited number of periodicals, in only three different languages—albeit those of the leading cultural powers of Europe—the full word count was much higher. After all, there were more than the fifteen magazines of this study that published reviews of Humboldt's books—even though mostly short ones. Moreover, essays on Humboldt's writings were carried by Dutch, Italian, and increasingly also American periodicals, as well as by others, such as (yet further afield) Indian publications.

The earliest review appeared in 1793 and was a discussion of Humboldt's *Florae Fribergensis Specimen* (Anon. 1793). The last reviews were of volume 3 of Humboldt's *Kosmos* ([Holland] 1853; for a discussion of the *Kosmos* reviews, see Rupke [1997, xv–xx]) and the first volume of *Kleinere Schriften von Alexander von Humboldt* (Anon. 1854). The fourth and fifth volumes of *Kosmos* were ignored by the periodicals listed in table 11.1, although volume 4 was discussed in other magazines such as the *Athenaeum* (Anon. 1858) and the *Wiener Zeitung* ([Grailich] 1858).

Three crests of reviewing intensity can be recognized (fig. 11.1): two fairly sharp ones, centered on 1810 and 1846, respectively, and a more diffuse one, made up of smaller peaks from 1815 until 1826.

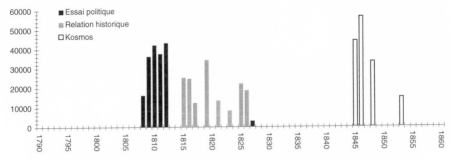

Fig. 11.2. Reviewing intensities of Alexander von Humboldt's three most extensively discussed books, the *Essai politique sur la Nouvelle-Espagne*, the *Relation historique*, and *Kosmos*.

The question arises: which of Humboldt's books produced these pinnacles? In figure 11.2, the word count of the main contributing book for each crest has been plotted. Unsurprisingly, the third was almost entirely generated by essays on *Kosmos* (table 11.2). Although the attention paid to *Kosmos* was considerable, it would be a crass instance of looking through the wrong end of the telescope if one were to identify Humboldt's contemporary renown with his last major work. After all, the 1846 *Kosmos* crest was preceded by some four decades during which Humboldt's writings received a great deal of international attention.

The main component of the diffuse, middle bulge was the *Relation historique* (fig. 11.2; table 11.2). The spread of its individual peaks was the result of the spread of the dates of publication of the different volumes of the work and of their translations (French original: 3 vols., 1814–17, 1819–21, 1825–31; German translation: 1815–32; English translation: 1814–29). Humboldt became notorious for both delaying the publication of promised writings and dragging out their completion. The publication of the *Relation historique*, in abbreviated form, had been advertised in the *Allgemeine geographische Ephemeriden* for as early as 1805 (Anon. 1805, 122). The planned fourth volume never saw the light. As in the case of *Kosmos*, reviewer's fatigue may have set in, and in our sample of periodicals no reviews appeared of volume three of the *Relation historique*.

The greatest reviewing intensity took place relatively early, around the year 1810, when Humboldt turned 41. The first and largest peak of reviewing success was produced by Humboldt's *Essai politique sur le royaume de la Nouvelle-Espagne* (fig. 11.2; table 11.2; for contrast see fig. 11.3). This is the book that gave content to Humboldt's international fame—the treatise that made his name a household word in the educated circles of the Western world. The original French

NICOLAAS RUPKE

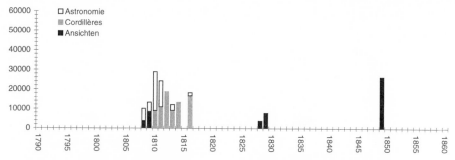

Fig. 11.3. Relatively low reviewing intensities of the three editions of Alexander von Humboldt's *Ansichten der Natur,* of the *Receuil d'observations astronomiques,* and the *Vues des Cordillères.*

text appeared in two quarto volumes during the period 1808–11 and was accompanied by a folio *Atlas géographique et physique du royaume de la Nouvelle-Espagne* (1808–12). In 1811, a French octavo edition in five volumes was also published (there was another, four-volume edition in 1828). The German translation, *Versuch über den politischen Zustand des Königreichs Neu-Spanien,* came out during 1809–14, and the English translation, *Political Essay on the Kingdom of New-Spain,* in the years 1810–11 (for a detailed bibliographical commentary on Humboldt's Mexican writings, see Hanno Beck [1991, 527–578]).

11.3. A Geography of Reception of the Mexico Book

The amount of attention paid to Humboldt's books differed from one region of Europe to another. In comparing the French, German, and British reviews, clear differences can be spotted, in length and format as well as in content. This becomes immediately evident when one separately plots the British, French, and German contributions to the cumulative totals for individual books (figs. 11.4, 11.5). In the case of the *Vues des Cordillères,* for example, the reviews for 1810 were exclusively French, for 1812 and 1813 German, and for 1816 English. The main reason for this difference probably was the date of publication of the several language versions: Humboldt wrote the book in French, and parts of it were available for review before the work was completed in 1810; the German translation, *Piktoreske Ansichten der Cordilleren,* was next off the press (1810); and the English translation, *Researches Concerning the Institutions and Monuments of the Ancient Inhabitants of America,* was the last to appear (1814).

In the case of the *Ansichten der Natur* (and its French and English

Table 11.2 Review Essays of Humboldt's Three most Reviewed Books (from the periodicals in table 11.1)*

Essai politique

Allgemeine geographische Ephemeriden, 1809, 28: 195–212, 317–331, 438–457; 1810, 31: 154–177, 333–343; 1811, 34: 75–77. *Allgemeine Literatur-Zeitung,* 5 Jan. 1810: 42–48; 6 Jan. 1810: 50–55; 8 Jan. 1810: 58–64; 9 Jan. 1810: 66–70; 9 March 1812: 466–472; 10 March 1812: 474–480; 11 March 1812: 482–488; 12 March 1812: 490–496; 13 March 1812: 498–504. *Annales des Voyages,* 1808, 5: 245–264. *Edinburgh Review,* 1810, 16: 62–102 [John Allen]; 1811, 19: 164–198 [John Allen]. *Göttingische gelehrte Anzeigen,* 14 March 1811: 401–407 [Johann Georg Heinrich Feder]. *Heidelbergische Jahrbücher fuer Literatur,* 1810, 3: 9–23. *Monatliche Correspondenz,* 1808, 18: 201–226, 312–330 [F. von Zach]; 1809, 19: 61–75, 141–155 [Zach]; 1809, 20: 461–485, 523–540 [Zach]; 1812, 25: 63–84, 159–182, 245–264, 273–282 [Zach]. *Moniteur universelle,* 30 June 1808: 714–715 (Jean-Baptiste Biot); 16 Febr. 1809: 183–184; 17 Febr. 1809: 186–188; 18 Febr. 1809: 191–192; 27 July 1809: 821–822 (Jean-Baptiste Biot); 27 April 1811: 451–453 (Jacques Peuchet); 1 Dec. 1811: 1281–1282 (Jacques Peuchet); 14 Jan. 1812: 54–56 (Jacques Peuchet); 28 June 1827: 1008–1009. *Monthly Review,* 1811, 66: 353–65; 1812, 67: 35–43; 1812, 69: 34–54.

Relation historique

Allgemeine geographische Ephemeriden, 1815, 46: 425–455. *Allgemeine Literatur-Zeitung,* March 1825: 497–504, 505–512 ('Ergaenzungsblaetter'); Jan. 1826: 50–56, 58–63, 66–70, 74–78. *Edinburgh Review,* 1815, 25: 86–11 [John Leslie]. *Moniteur universelle,* 22 Jan. 1815: 86–87. *Monthly Review,* 1816, 79: 1–17; 1819, 88: 234–46; 1819, 90: 14–24; 1823, 100: 264–281. *Nouvelles annales des voyages,* 1819, 3: 202–222 (Conrad Malte-Brun); 1825, 27: 369–390 (Conrad Malte-Brun); 1825, 28: 81–100 (Conrad Malte-Brun); 1826, 32: 81–94, 223–234. *Quarterly Review,* 1816, 14: 368–402 [John Barrow]; 1817, 18: 134–159 [John Barrow]; 1819, 21: 320–352 [John Barrow]; 1821, 25: 365–392 [John Barrow].

Kosmos

British Quarterly Review, 1846, 3: 320–357 [T. Hussey]. *Edinburgh Review,* 1848, 87: 170–229 [John Herschel]. *Heidelbergische Jahrbücher fuer Literatur,* Dec. 1845: 810–817 (Georg Wilhelm Munke). *North British Review,* 1845, 4: 202–54 [David Brewster]. *Quarterly Review,* 1845, 77: 154–191 [J. D. Forbes]; 1853, 94: 49–97 [Henry Holland]. *Revue des deux mondes,* 1846, 14: 742–782 (Armand de Quatrefages). *Westminster Review,* 1845, 44: 152–203 [John Crosse].

*The names of authors of signed reviews are indicated in parentheses, and those of authors of anonymously published reviews in square brackets.

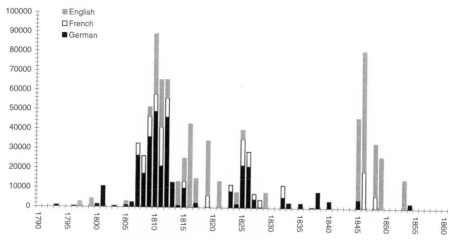

Fig. 11.4. Reviewing intensities of fig. 11.1, broken down according to the geography of language: French, German, and English.

translations), however, the difference may well reflect deeper causes, related to national-cultural dissimilarities: in our sample of periodicals, the first edition of *Ansichten* (translated into French but not into English) received German, French, and English reviews; the second (again, in German and French) French and English; and the third (with a French and two English translations) exclusively English (fig. 11.5). *Kosmos,* too, received more English than French or German attention (fig. 11.5). One might conclude that the appeal of these two books was greater among British readers than among Humboldt's fellow continentals; and it is true that the British reviews gave *Kosmos* a twist in support of natural theology—a feature that sharply distinguished the British from the continental reception (Rupke 1997, xix). A geographically circumscribed difference in appreciation may also lie behind the fact that the reviewing bulge for the *Relation historique* consisted predominantly of English essays (fig. 11.5). Obviously, a primary reason for any relative lack of French attention is the circumstance that in our sample of periodicals there are twice as many English and German titles as there are French. Yet the comparative scarcity of French titles itself reflects a characteristic difference in reviewing culture (see below).

Admittedly, the national differences in reception came on top of a certain unanimity about why Humboldt's Mexico book was worth so much attention. The reviewers were in agreement that the *Essai politique* was exciting, because it offered in scientific, geographical, and statistical terms a well-informed account of the geopolitical and commercial significance of Central America. Humboldt's detailed and

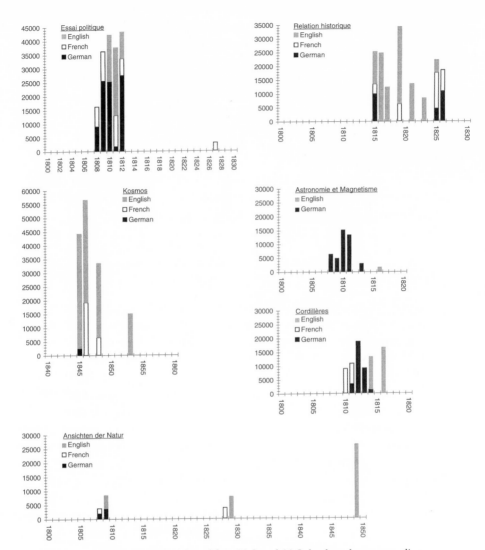

Fig. 11.5. Reviewing intensities of figs. 11.2 and 11.3, broken down according to the geography of language: French, German, and English.

comprehensive description—for all that Godlewska shows Humboldt's descriptive practices to be more complex than at first hand appears—made this previously opaque Spanish possession transparent, instilling the belief that Mexico was pivotal to the control of world trade.

Reception differences were greatest between the British on the one hand and the French and Germans on the other. English-language reviews tended to be longer and have a more critical, analytical

NICOLAAS RUPKE

character; the book's contents were discussed, not merely summa-rized, as continental reviewers tended to do. Such discussions pro-vided an opportunity for promoting particular, 'local' causes such as natural theology in the reviews of *Kosmos*. In the case of the *Essai politique*, the *Edinburgh Review* turned its discussion of the book into an occasion to deplore the 'lamentable' uprising against the Span-ish Crown that had taken place in Mexico, and into a sustained ar-gument in favor of keeping European, imperial control of overseas colonies ([John Allen] 1811, 164–182).

Typical, too, of British reviewers was that they expressed criticism of Humboldt, something very rarely done by their continental counterparts of our sample (Rupke 1997, xviii–xx). Such criticism primarily focused on what were judged to be the poor arrangement, the verbosity, and the excess of trivial detail in Humboldt's account. The *Monthly Review*, for example, having found "solidity of thought and novelty of information" in the *Political Essay*, then continued:

> If, however, he [the reader] expects perspicuity of arrangement, or condensation of language, his chance of gratification will be greatly impaired; the author not having given, either in the present or in his former works, any evidence of his title in these respects to lit-erary renown. He furnishes a striking example of the difference of the habits which are required for the observations of active travel-ling, and the patient labour of in-door composition. Haste, and its usual concomitants, diffuseness and repetition, are the great blem-ishes of the book before us, and prevent us from finding amuse-ment in that which we are satisfied beforehand will not fail to con-vey instruction. (Anon. 1811, 353)

The *Essai politique* was given considerable attention across Europe; yet different emphases were laid by British, French, and German re-viewers. For instance, in French and German periodicals it was ad-miringly related that Humboldt had solidified the map of Mexico with new, improved determinations of latitude, longitude, and alti-tude; the *Allgemeine geographische Ephemeriden*, for instance, in a 23.5-page review, printed a lengthy table of no fewer than 8.5 pages of "Astronomisch bestimmte geographische Puncte in Neu-Spanien" ("Astronomically determined geographical points in New Spain") (Anon. 1810). By contrast, in British reviews the commercially strate-gic aspect of this new, more reliable map was stressed: it served the transformation of Mexico into the world's most significant transit-area for global, commercial traffic, nearly equidistant between Eu-rope and the Far East. Harbors on the east and west coasts, relatively short land connections between these, possibly by means of camels (to be introduced), would improve commercial traffic with Asia and with the west coast of North America, including the fur trade with Vancouver. This particular British–Continental contrast is further evident from the reviews of Humboldt's publications that deal with his measurements to determine precise geographic locations, such as the *Recueil d'observations astronomiques, d'operations trigonométriques et de mesures barométriques* (1808–1811), written with the assistance of the astronomer Jabbo Oltmanns (1783–1833); this book, together with related publications, received serious attention (fig. 11.3), but nearly all of it was German (fig. 11.5).

Humboldt's detailed suggestions for digging a canal to create a navigable route through the Americas, and thus improve Europe's trade with Asia, proved characteristically captivating to the British. Much discussion was generated by the eight different possible waterways, suggested and illustrated by Humboldt in the *Essai politique,* ranging from two across North America, two across South America, and four through the Central American isthmus. The possible connection of the Atlantic and the Pacific that Humboldt preferred was not through the narrow Panama part of Central America, but through the part of Mexico where the Rio Huasacualco, which empties into the Gulf of Mexico, and the Rio Chicapa, which flows into the Pacific Ocean, could be connected by a relatively short canal (Humboldt 1811a, 11–28). Much was written on the subject—even entire books, e.g., Robert Birks Pitman's substantial treatise "on the practicability of joining the Atlantic and Pacific Oceans, by means of a Ship-Canal across the isthmus of America" (Pitman 1825; see further Biermann 1990, 69–72; Frankel 1959).

One of the illustrations that accompanied Humboldt's book was a cross-sectional profile of Mexico, traced along the road from Acapulco to Veracruz, via Mexico City. It demonstrated the elevated nature of much of the Mexican mainland, showing sharp drops in altitude towards the coasts (Humboldt 1811b, plates 12, 13). Humboldt's account of the Mexican plateau, too, was much discussed. The perceived significance of this tableland was less its physiographical characteristics, however, than the implications of these for European, colonial settlement. A substantial part of New Spain was located in the tropics, of which the climate was considered dangerously unhealthy to Caucasians. The relatively high altitude of the Mexican plateau, however, gave it climatic conditions comparable to those of the temperate regions of southern Europe, making it possible for European settlers to live there. In the low-lying coastal

NICOLAAS RUPKE

regions with their harsh tropical conditions, Indians and other non-European races could perform necessary labor. In other words, the physiographical profile through Mexico, praised by later geographers for its scientific originality (Beck 1985, 23), was interpreted by British reviewers first and foremost as underscoring a point of commercial place, not of archetypal physiography.

A further reason for the popularity of Humboldt's account of Mexico was his statistical documentation of its mineral and agricultural wealth. Several of the review essays included tables showing the enormous quantities of gold and silver that were extracted from Mexican mines. These data were widely copied for mercantile purposes, especially by the British. For example, extracts from the *Essai politique* were used as publicity material to encourage British investment in Mexican mining. In 1824, John Taylor, Treasurer to the Geological Society of London, published Selections from Humboldt's *Essai politique* in support of an investment company that Taylor directed, "The Adventurers in the Mines of Real del Monte." He cited the success of two earlier such companies, the "Anglo-Mexican Association" (capital: GBP 1.000.000—in GBP 100—shares), and the United Mexican Association (capital: GBP 240.000, in GBP 40). Taylor argued that the introduction of English steam engines (for crushing rocks) would greatly enhance the productivity of Mexico's rich mines (Burt 1977, 39–47; Taylor 1824).

These geographical contrasts in the European reception of Humboldt's work on Mexico not merely reflected national differences in the appreciation of his books, but also disparities of reviewing cultures. They also highlight, albeit in a particular way, the varying geographies of reception of natural science (see also Gould's chapter here). National reviewing intensities as shown in figures 11.4 and 11.5 not only express the relative popularity of Humboldt and his writings, but also a diversity of reviewing practices. Scholarly attention has been paid to the periodical literature of eighteenth-century Germany by Ulrike Enke (1995) and by Ute Schneider (1995); also—but to a lesser degree—to those of the first half of the nineteenth century, in particular by Sibylle Obenaus (1974; see also Kirchner [1958–1962] and Kronick [1991]). A number of British reviewing periodicals have been studied—a by now classic source of information with extensive references to secondary literature on individual periodicals is the *Wellesley Index to Victorian Periodicals* (Houghton [1966–1987])—but virtually no comparisons have been made to date, of British with French or German reviewing practices.

There were major differences that significantly influenced the individual national contributions to the graphs of figures 11.1 through 11.5. The leading British reviewing periodicals such as the *Edinburgh Review*, the *Quarterly Review*—commercial ventures, publishing exclusively reviews, in many instances large essays, on a broad range of literature—had no equivalents in either France or Germany. In France, a majority of reviews of Humboldt's books was published in specialized, 'professional' magazines, in particular in the *Annales des voyages*, edited by the geographer Conrad Malte Brun (1775–1826), or in the daily *Moniteur universel* by, among others, the physicist Jean-Baptiste Biot (1774–1862). As mentioned above, there simply did not exist as many French outlets for reviews of Humboldt's oeuvre as there were British, or, for that matter, German.

In the German-speaking lands the leading reviews had a different place again from both the British and the French. The *Göttingische gelehrte Anzeigen*, for example, was a noncommercial periodical that served in part as an acquisitions register for the library of Goettingen University. It published many reviews on a broad range of books, but individual contributions tended to be shorter than those in the British periodicals. Both German and British reviewers wrote anonymously, but authorship of reviews mattered less in the case of the *Anzeigen*, and it was not uncommon that books were 'announced' by their own authors (on who the reviewers were, see Fambach [1976]; one of several surviving manuscript lists of the authors of the reviews is available in Göttingen's *Niedersächsische Staats- und Universitätsbibliothek*).

Many more differences in national reviewing culture can be cited, especially when periodicals from other European language areas are included in the comparison. Italian reviews, such as those in the *Giornale Enciclopedico di Napoli*, not uncommonly consisted of translated extracts from the books in question. In the Netherlands there were several major review magazines, such as the *Algemeene Vaderlandsche Letter-Oefeningen*, but less attention was paid in these to scientific literature than in the British reviews. A study of national reviewing cultures is needed before data as represented in figures 11.4 and 11.5 can be adequately interpreted.

11.4. Conclusion

This chapter highlights the importance of Humboldt's Mexico work for an understanding of his contemporary impact. Many authors make the mistake of a selective and anachronistic emphasis on a particular work by Humboldt. Even an account as first-rate as Margareta Bowen's *Empiricism and Geographical Thought* (1981) discusses Humboldt primarily in terms of *Kosmos* (Bowen 1981, 240–59). Mary Louise Pratt points to three other, different books, asserting that the most widely read, discussed, reviewed, and excerpted instances of Humboldt's many publications were the *Ansichten der Natur,* the *Vues des Cordillères,* and the *Relation historique* (Pratt 1992, 119). Recent Darwin scholarship, too, has pointed to the importance of the *Relation historique* (see, for example, Browne [1995, 133] and Desmond and Moore [1992, 91, 108, 468]).

Humboldt's rise to international fame should not be ascribed to *Kosmos.* It would be wrong, too, to attribute his initial renown to the personal narrative of his American journey of exploration, the *Relation historique.* Judging by 157 reviews of Humboldt's publications that appeared in a representative sample of six British, three French, and six German reviewing periodicals, Humboldt's early, international acclaim was more the result of his account of the political economy of Mexico, the *Essai politique sur le royaume de la Nouvelle-Espagne.* The nature of the international response to this major work indicates that Humboldt rose to prominence in his role as a colonial surveyor of Spanish Central America, appealing to Eurocentric political and economic interests. The Mexico book was dedicated to the Spanish king: "A sa majesté catholique Charles IV, roi d'Espagne et des Indes." The sanitized image of Humboldt as a cosmopolitan universalist is of a later date and does not fit the original Humboldt.

From our bibliometric analysis, it would appear that it is correct to cite the *Relation historique,* too, as a widely discussed book; but the contributions of the *Ansichten* and the *Vues des Cordillères* to the cumulative graph are modest (fig. 11.3). The low reviewing intensity of the *Ansichten* is notable, the more so if each of the three editions is considered in isolation. The only fairly substantial reviews were of the third edition and thus came late, in the wake of *Kosmos,* and they were British (fig. 11.5). The *Vues des Cordillères* attracted more comment than the *Ansichten,* yet also this did not add up to the reviewing intensities of the three books of fig. 11.2. These bibliometric facts would appear to narrow the basis on which Pratt puts forward her thesis that during the early part of the nineteenth century, Spanish America was reinvented from Humboldt's writings in terms of the Romantic images of luxuriant tropical forests, snow-capped volcanoes, and vast interior plains (Pratt 1992, 125–126).

The Humboldt phenomenon was more complex than commonly indicated. The Mexico book paints a different Humboldt from the customary portrait. A diagnostic feature of what has been called "Humboldtian science" (Cannon 1978, 73–110; Dettelbach 1996; Home 1995; Nicolson 1987, 1996; see also Rupke 1996), is the use of isolines on world distribution maps, on which point, see also Godlewska's chapter here. Humboldt introduced isolines in a classic 1817 paper on the spread of heat across the globe (it was multiply published, in French, German, and British magazines; one of the French versions included Humboldt's original illustration of isolines [Humboldt 1817]). Long before Humboldt drew global isolines, however, he produced in the atlas to his book on Mexico a world map with trade lines, showing the dispersal of mineral wealth across the globe (Humboldt 1811b, plate 19, no. I). This fact neatly captures the Humboldt of the *Essai politique:* his famous holism was commercially focused, before it acquired the Romanticism of *Ansichten* and long before it took on the cleansed, scientific character of *Kosmos.*

The geographic approach, moreover, comparing and contrasting the varying receptions of Humboldt's writings from across Europe, brings into clear view this complexity of the Humboldt phenomenon, demonstrating the constitutive significance of place in the production of the various meanings that became attached to even a single work such as the *Essai politique.* This promises fruitful results to a further study of Humboldt, his many works, and the very large body of secondary literature on him, looking at these from the very geographic perspective that he himself so successfully adopted (Rupke, in preparation).

References

Ackerknecht, Erwin H. 1955. "Georg Forster, Alexander von Humboldt, and ethnology." *Isis* 46: 83–95.

[Allen, John]. 1810. "Humboldt—Essai Politique sur la Nouvelle Espagne." *Edinburgh Review* 16: 62–102.

———. 1811. *Edinburgh Review* 19: 164–98.

Anon. 1793. *Allgemeine Literatur-Zeitung* no. 250: 477–79.

———. 1805. "Anzeige von der Herausgabe der Schriften der Herren Alexander von Humboldt und Aimé Bonpland." *Allgemeine geographische Ephemeriden* 17: 122–124.

———. 1809. *Annales des voyages* 4: 114–35.

———. 1810. *Allgemeine geographische Ephemeriden* 31: 154–177.

———. 1811. "Humboldt's Political Essay on New Spain." *Monthly Review* 66: 353–365.

———. 1815. *Moniteur universelle* 22 Jan.: 86–87.

———. 1923. "Humboldt's Works." *North American Review* 16: 1–30.

———. 1854. *Heidelberger Jahrbuecher fuer Literatur* 47: 823–829.

———. 1858. *Athenaeum* 6 Nov. 1858: 589–90.

Avé-Lallement, Robert. 1872. "Alexander von Humboldt. Sein Aufenthalt in Paris (1808–1826)," in *Alexander von Humboldt. Eine wissenschaftliche Biographie*, vol. 2, ed. Karl Bruhns. Leipzig: Brockhaus, pp. 1–92.

Beck, Hanno. 1959. *Alexander von Humboldt*. 2 vols. Wiesbaden: Steiner Verlag.

———. 1985. "Die Geographie Alexander von Humboldts," in *Alexander von Humboldt: Leben und Werk*, ed. Wolfgang-Hagen Hein. Frankfurt am Main: Weisbecker Verlag, pp. 221–238.

———. 1991. "Zu dieser Ausgabe des Mexico-Werkes," in *Alexander von Humboldt. Mexico-Werk*, ed. Hanno Beck. Darmstadt: Wissenshaftliche Buchgesellschaft, pp. 527–78.

Biermann, Kurt-R. 1990. *Miscellanea Humboldtiana*. Berlin: Akademie-Verlag.

Bowen, Margaret, 1981. *Empiricism and Geographical Thought. From Francis Bacon to Alexander von Humboldt*. Cambridge: Cambridge University Press.

Brand, Donald D. 1959. "Humboldt's Essai Politique Sur Le Royaume De La Nouvelle Espagne," in *Alexander von Humboldt. Studien zu seiner universalen Geisteshaltung*, ed. Joachim H. Schultze. Berlin: de Gruyter, pp. 123–141.

Brock, William H. 1993. "Humboldt and the British: A Note on the Character of British Science." *Annals of Science* 50: 365–72.

Browne, Janet. 1995. *Charles Darwin Voyaging*. Princeton: Princeton University Press.

Burt, Roger. 1977. *John Taylor: Mining Entrepreneur and Engineer, 1799–1863*. Buxton, Derbys: Moorland Publishing Company.

Buttimer, Anne. 1993. *Geography and the Human Spirit*. Baltimore and London: Johns Hopkins University Press.

Cannon, Susan F. 1978. *Science in Culture: The Early Victorian Period*. New York: Dawson and Science History Publications.

Desmond, Adrian, and James Moore. 1992. *Darwin*. London: Penguin.

Dettelbach, Michael. 1996. "Humboldtian Science," in *Cultures of Natural History*, eds. Nicholas Jardine, James A. Secord, and Emma C. Spary. Cambridge: Cambridge University Press, pp. 287–304.

———. 1997. "Introduction," in Alexander von Humboldt, *Cosmos: A Sketch of the Physical Description of the Universe*. Baltimore: Johns Hopkins University Press, vol. 2, pp. vii–xxxix.

Enke, Ulrike. 1995. *Samuel Thomas Sömmering. Rezensionen fuer die Göttingische gelehrten Anzeigen. Erster Teil: Rezensionen 1780–1801*. Stuttgart: Fischer.

Fambach, Oscar. 1976. *Die Mitarbeiter der Göttingischen gelehrten Anzeigen*. 1769–1836. Tuebingen: Universitaetsbibliothek.

Frankel, Walter K. 1959. "Alexander von Humboldt und der Panamakanal," in *Alexander von Humboldt. Studien zu seiner uinversalen Geisteshaltung*, ed. Joachim H. Schultze. Berlin: de Gruyter, pp. 235–242.

Gode–von Aesch, Alexander. 1941. *Natural Science in German Romanticism*. New York: Columbia University Press.

[Grailich, Joseph]. 1858. "Kosmos von Alexander v. Humboldt." *Wiener Zeitung* 1: 311–12, 325–26.

Harig, Gerhard. 1959. "Alexander von Humboldt, der Naturforscher des deutschen Humanismus," in *Alexander von Humboldt. Eine Auswahl*, ed. Gerhard Harig. Leipzig and Jena: Urania, pp. 9–38.

[Holland, Henry]. 1853. "Humboldt's Cosmos—sidereal astronomy." *Quarterly Review* 94: 49–79.

Home, R. W. 1995. "Humboldtian Science Revisited: an Australian Case Study." *History of Science* 33: 1–22.

Houghton, Walther E. ed. 1966–1987. *The Wellesley Index to Victorian Periodicals, 1824–1900.* 4 vols. Toronto: University of Toronto Press.

Humboldt, Alexander von. [1811a] 1971. *Essai politique sur la Nouvelle-Espagne.* Quarto edition, 2 vols. Paris: Schoell. Facsimile edition, Hildesheim and New York: Olms.

———. [1811b] 1971. *Atlas geographique et physique du royaume de la Nouvelle-Espagne.* Paris: Schoell. Facsimile edition, Hildesheim and New York: Olms.

———. 1817. "Sur des lignes isothermes." *Annales de chimie et de physique* 5: 102–112.

Jones, Calvin P. 1972–1973. "The Spanish-American works of Alexander von Humboldt as viewed by leading British Periodicals, 1800–1830." *The Americas* 29: 442–48.

Kirchner, Joachim, 1958–1962. *Das deutsche Zeitschriftenwesen, seine Geschichte und seine Probleme.* 2 vols. Wiesbaden: Harrassowitz.

Kronick, David A. 1991. *Scientific and Technical Periodicals of the Seventeenth and Eighteenth Centuries: A Guide.* Metuchen, NJ: The Scarecrow Press.

Leitner, Ulrike. 1992. *Alexander von Humboldts Werk. Probleme damaliger Publikation und heutiger Bibliographie.* Berlin: Alexander-von-Humboldt-Forschungsstelle.

Livingstone, David N. 1992. *The Geographical Tradition. Episodes in the History of a Contested Enterprise.* Oxford: Blackwell.

———. 1995. "The Spaces of Knowledge: Contributions Towards a Historical Geography of Science." *Environment and Planning D: Society and Space* 13: 5–34.

Loewenberg, Julius. 1872. "Alexander von Humboldt. Bibliographische uebersicht seiner Werke, Schriften und zerstreuten Abhandlungen," in *Alexander von Humboldt. Eine wissenschaftliche Biographie*, vol. 2, ed. Karl Bruhns. Leipzig: Brockhaus, pp. 485–552.

Nicolson, Malcolm. 1987. "Alexander von Humboldt, Humboldtian Science and the Origins of the Study of Vegetation." *History of Science* 25: pp. 167–94.

———. 1990. "Alexander von Humboldt and the Geography of Vegetation," in *Romanticism and the Sciences*, eds. Andrew Cunningham and Nicholas Jardine. Cambridge: Cambridge University Press, pp. 169–185.

———. 1996. "Humboldtian Plant Geography after Humboldt: The Link to Ecology." *British Journal for the History of Science* 29: 289–310.

Obenaus, Sibylle. 1974. "Die deutschen allgemeinen kritischen Zeitschriften in der ersten Haelfte des 19. Jahrhunderts. Entwurf einer Gesamtdarstellung." *Archiv für Geschichte des Buchwesens* 14: 1–122.

Pitman, Robert B. 1825. *A Succinct View and Analysis of Authentic Information Extant, in Original Works, on the Practicability of Joining the Atlantic and*

Pacific Oceans, by Means of a Ship-Canal Across the Isthmus of America. London: Richardson.

Pratt, Mary Louise. 1992. *Imperial Eyes. Travel Writing and Transculturation.* London and New York: Routledge.

Raphael, Lutz. 1995. "Freiheit und Wohlstand der Nationen. Alexander von Humboldts Analysen der politischen Zustaende Amerikas und das politische Denken seiner Zeit." *Historische Zeitschrift* 260: 749–776.

Rupke, Nicolaas A. 1996. "Humboldtian Medicine." *Medical History* 40: 293–310.

———. 1997. "Introduction," in Alexander von Humboldt, *Cosmos: A Sketch of the Physical Description of the Universe.* Vol. 1. Baltimore: Johns Hopkins University Press, pp. vii–xxxv.

———. In preparation. *The Many Lives of Alexander von Humboldt.*

Sanke, Heinz. 1969. "Alexander von Humboldt." *Petermanns Geographische Mitteilungen* 1969: 81–88.

Schneider, Ute. 1995. *Friedrich Nicolais Allgemeine Deutsche Bibliothek als Integrationsmedium der Gelehrtenrepublik.* Wiesbaden: Harrassowitz.

Taylor, John. 1824. *Selections from the Works of the Baron de Humboldt, Relating to the Climate, Inhabitants, Productions and Mines of Mexico.* London: Longman.

Waschkies, Hans-Joachim, 1990. "Alexander von Humboldt's aufklaererisches Weltbild," in *Alexander von Humboldt. Weltbild und Wirkung auf die Wissenschaften.* ed. Uta Lindgren. Koeln: Boehlau, pp. 169–186.

Ziolkowski, Theodore. 1990. *German Romanticism and Its Institutions.* Princeton: Princeton University Press.

Placings

As noted at the outset, one of the central concerns of this collection as a whole is to suggest the *situated nature* of geographical knowledge and of Enlightenment as a process. In these terms, we would want to consider ideas about rationality, about moral philosophy, and about the different nature of the human condition not as abstract historical terms "floating free" but, rather, as Livingstone's chapter discusses, as ideas grounded in and even shaped by certain places and specific settings. And this is to claim that there are new insights to be had in considering what might collectively be termed the (historical) geographies of Enlightenment ideas.

The chapter by Stephen Daniels, Susanne Seymour, and Charles Watkins helps turn our gaze as historians and geographers of *the* Enlightenment away from the urban spaces of metropolitan knowledge towards the perhaps more 'earthy' sites and spaces of rural Enlightenment. Here, in a variety of ways, concerns about natural order and the condition of nature were all closely associated with questions of improvement, moral and economic, local and national. Enlightenment ideas in England were, of course, sited in the salons and scientific bodies of the urban public sphere, and they also traveled between such sites in the form of those books, maps, and globes fueling the widespread civic and sociable interests in geography as a means to national and global knowledge. But we must not forget that the Enlightenment was discussed and produced in the houses and gardens of rural landowners. For men such as Humphry Repton at Woburn, and for Thomas Andrew Knight, putting nature to order in Georgian England demanded those same skills of observation, classification, systematic measurement, and representation that Joseph Banks (himself an estate owner in Lincolnshire and in Derbyshire) and others brought to the geographies of the Pacific. Scale is important here.

Geographers and natural historians in what Banks and others called "this enlightened age" moved across and sought to understand the globe: Knight's Enlightenment came from meticulous observation of the everyday, from local field work, from his own hothouse. This is not to separate and categorize as distinct the sites of enlightenment knowledge. For all that Knight conducted and promoted local work, he was at the same time part of the networks of those international scientific circles centered, for Englishmen anyway, on London.

There is, then, a geography to Enlightenment knowledge not just in the places of its making but in its public circulation between private rooms, public spaces, and the world 'out there' that was to be understood. It is just such "alertness to space" as Michel Foucault put it, that informs Chris Philo's engagement with the sites of reason and of unreason in Enlightenment Edinburgh. Philo's concern is with the historical geography of an idea, Reason. The term is clearly central to any historiographical treatment of the Enlightenment. Yet, it must not only be understood in the context of its time: it has to be seen as dependent upon the parallel 'invention' of Unreason. And both ideas had related geographies. The Age of Reason, as contemporaries understood and termed it, was thus an age for the construction of sites of Unreason, notably madhouses, and for the intellectual placing of ideas of madness (and of illness and sexuality) in the context of their time. This is not, Philo argues, to see matters of space and place as, simply, the geographical reflection of enlightenment ideas. Rather, it is to treat such sites and places as constitutive expressions of such ideas and, like Daniels et al., again to stress the importance of the local and the particular. Siting (and sighting) places of Unreason allows us to go some way toward a geographical account of the Enlightenment's shadows, the 'dark side' of those notions of urban design which sought to promote, through the 'geometric impulse' typical of the age, a new rational order both in society and in the built form of cities and in their individual institutions.

Peter Gould's paper on the 1755 destruction of Lisbon by earthquake and tidal wave sketches the outlines of what we might understand as the geography of reception of Enlightenment news. Of course, how the news of Lisbon's circumstances on All Saint's Day 1755 traveled around Europe to be discussed in print and debated from pulpits may not be typical of the geography of Enlightenment information flows. But, for one spectacular example, it illustrates how we can conceive of and place different geographies of Enlightenment knowledge: time geographies that have to do with the reception of news; social geographies of information circulation that have to do with who had access to what newspaper and other written sources in the public sphere or with written correspondence and word-of-mouth contact between individuals; local geographies of Enlightenment that have to do with why, at particular times in particular ways in particular places, the production and reception of knowledge took the forms it did.

Chapter Twelve

Enlightenment, Improvement, and the Geographies of Horticulture in Later Georgian England

Stephen Daniels, Susanne Seymour, and Charles Watkins

12.1. Introduction

The term "enlightenment" was not naturalized into the English language until the late nineteenth century, and then pejoratively to caricature rationalist currents of continental European culture. Ever since, scholars from many places, of all persuasions, have found it difficult to find an accommodation between England and the Enlightenment. As Roy Porter has pointed out, the irony is that throughout the eighteenth century, enlightened *philosophes* from throughout Europe were Anglophile, admiring England's freethinking, empiricism, social openness and prosperity, and its intellectual heroes, Newton, Bacon, and Locke (Porter 1981). "Enlightened men" as they called themselves were conscious of living in enlightened times. Joseph Banks, the impresario of enlightened England, recalled his achievements on the South Seas: "I may flatter myself that being the first man of scientific education who undertook a voyage of discovery and that voyage of discovery being the first which turned out satisfactorily in this enlightened age" (Gascoigne 1994, 32).

Recent scholars have recovered the power of the English Enlightenment in its peculiarity, less oppositional than its continental counterpart, more comprehensive, practical rather than speculative, not so much the preserve of an intellectual coterie as the possession of a range of people from printers to parsons in a highly commercial culture. The English Enlightenment was less a shining path through a dark forest of reaction than an integral part of the cultural landscape, materially so in the many sites of knowledge from the British Museum and Royal Exchange to various clubs and coffeehouses and in the transformations of town and countryside, such as street lighting and enclosure, which followed the application of enlightened knowledge (Porter 1981).

In his recent biography of Banks, John Gascoigne (1994) has suggested that the Whiggish term "improvement" is a more accurate concept to explain the progressive culture of the period:

> Britain's elite embraced the ideal of improvement as a means of reconciling the possibility of economic progress with the maintenance of a well regulated society. . . . The very word 'improvement' with its connotations of building on what was already there was reassuring, while the practice of improvement with its attention to the reassuringly mundane—turnips, manure and the better breeds of stock—drained it of any unsettling associations. (Gascoigne 1994, 235)

An emphasis on the idea of improvement usefully repositions enlightened culture, away from a prime focus on the metropolitan, libertarian world of London to the more regulated rural world of the landed estate, from the freemarket of ideas to the world of patronage and pragmatism.

The landed estate was still the basis of social, political, and economic authority in Britain. The larger landowners, those who could afford to undertake more innovative regimes, owned lands in a variety of regions, industrial, urban, and agricultural, as well as plantations overseas (Seymour, Daniels, and Watkins 1998). Aristocrats owned large tracts of London and controlled development there as carefully as they did around their country houses. A second-generation landowner, Joseph Banks promoted the patrician version of improvement. His estates at Revesby, Lincolnshire, and Overton, Derbyshire; his suburban gardens at Spring Grove, Isleworth; his house at 32 Soho Square (on the Portland estate); his management of the Royal Gardens at Kew; his direction from Kew of improvements in plantation agriculture overseas; these form the framework for his adventurous voyage with the *Endeavour* (Banks et al. 1994). The agriculturalist Arthur Young found in Banks's estate office at Revesby a model alliance of power and knowledge:

> There is a catalogue of names and subjects in every drawer so that whether the enquiry concerned a man or drainage, or an enclosure, or a farm, or a wood, the request was scarce named before a mass of information was before me. Such an apartment and such apparatus must be of incomparable use in the management of every great estate or indeed in any circumstances. (quoted in Miller and Reill 1996, 33)

As Young's description implies, Banks's estate office was not typical of others he came across in his tour of Lincolnshire, or even

S. DANIELS, S. SEYMOUR, C. WATKINS

elsewhere. What mattered at Revesby, and on the other such estates, is that they were models to be publicized, to be emulated.

The problem of focusing on improvement is reproducing its consensus ideology and giving the impression that enlightened culture radiated from model estates through the networks of polite society. It is not just that the flows were always more complicated but that the discourse and practice of improvement were complicated, even contentious. The very power and range of the term "improvement," from reading to statecraft, in an increasingly complex society dispersed and destabilized its meaning. There were strong, self-consciously regional strains to the culture of even agricultural improvement, for example in Norfolk and Herefordshire. Within that progressive, imperial notion of Britain, there were strong, self-consciously national strains to the culture of improvement, prominently Scottish, but also Welsh (Daniels, Seymour, and Watkins 1997). Those radical, speculative, cosmopolitan strains of progressive culture, which we might be tempted to refer to as an enlightenment, did not politely accede to patrician hegemony; Erasmus Darwin's botany, not less than Joseph Priestley's pneumatics, put pressure on conservative notions of improvement, even the very term itself. During the Napoleonic Wars there arose a powerful anti-improvement literature initiated by defenders of the landed estate as a stable and stabilizing domain. Enclosure was censured for greed and grinding the poor; model farms and gardens dismissed as useless showpieces. If improvement implied a harmonious estate, it became an ideological battleground, a contested terrain (Daniels 1993, 41–111; Daniels and Seymour 1990; Wilmot 1990).

In this essay we examine the landed estate as a site for horticulture, a pursuit that blossomed throughout Britain in the early nineteenth century and became a paradigm of enlightened knowledge. We focus on two figures: Humphry Repton, the leading landscape gardener of the time, and Thomas Andrew Knight, President of the London Horticultural Society. We examine Repton's commission for the sixth Duke of Bedford at Woburn, Bedfordshire, and Knight's experimentation on his own estates at Elton and Downton in Herefordshire.

12.2. Horticulture

Horticulture in England was not a new enthusiasm. The collection, study, and cultivation of various plants, wild and luxuriant, native and exotic, useful and ornamental, was well established, sufficiently so to seed an interest which spread through all ranks of society, to most places and into many sites and spaces, some adapted to the invasion of plants, some purpose built: the artisan's pub, the lady's breakfast room, the schoolroom, the clipper ship, the cottager's allotment, the parson's pleasure ground, the gentleman's estate (Cooke 1992; Miller and Reill 1996; Schteir 1988; Secord 1994). The very term 'garden' extended its meaning to cover a variety of enclosures, and traditional types of garden were subject to shifts in status and meaning (Ritvo 1992). Writing from Jamaica, Thomas Dancer declared "a botanical garden is not now, as formerly, considered merely as an appendage to a college or university but is become an object of concern with enlightened men of every description, even the mercantile class, in maritime and manufacturing towns" (quoted in Gascoigne 1994, 33).

These spaces didn't fit together neatly or harmoniously. Horticulture was framed by a variety of pursuits, by husbandry, herbalism, botany, vegetable physiology, and aesthetics. And within each of these pursuits there were a variety of contending interests.

A commitment to agriculture and forestry, overlapped with interests in horticulture. There was a practical dimension to this, but no less prominently a commercial and aesthetic dimension too. Well-managed orchards and vegetable gardens, no less than well-managed fields and plantations, produced heavy yields of pleasure, profit, and patriotism. Moreover at this time, the close and careful attention to smaller plots was esteemed as a model for managing the country as a whole (Daniels 1992). From the 1790s a new optic emerges, or re-emerges, focused on foregrounds, and the detail to be discerned there. Roots, tussocks, mosses, lichens, ferns, rocks, soil strata, trailing plants, pot herbs, birds nests, and beehives come into focus not only as individual specimens but as ingredients in local ecologies (Allen 1976, 26–51; Murdoch 1986). The cult of native plants and local attachments was paralleled, perhaps in some measure provoked by, the cult of the exotic, the import and circulation of new varieties of plants, some from the furthest reaches of the empire. In 1782 a nurseryman in Dorset was advertising a range of exotics, including *Rosa Pennsylvanica,* Hottentot Cherry, Mexican Lily, and Madagaskar Periwinkle. And hybridization produced an astonishing range of domestic varieties, including 320 different kinds of cherry, 575 kinds of hyacinths. Between 1789 and 1813 the number of species at Kew, the nation's botanical entrepot, doubled from 5,500 to 11,000. Certain plants became wildly fashionable. The number of rose varieties in 1800 totaled less than 100; by 1826 there were 1400. Planting

S. DANIELS, S. SEYMOUR, C. WATKINS

styles multiplied. Some reflected taxonomic schema, including spatial or ecological systems, others projected scenic images of far away places (Morley 1993, 48–58).

12.3. Woburn

Woburn, the family seat of the Dukes of Bedford, was the centerpiece of one of the largest and most valuable estates in Britain. At its center was one of the largest parks in the kingdom. Like many aristocratic parks, it was a complex landscape with a variety of zones, for livestock, forestry, watercourses, and game. Like some, such as Holkham, Petworth, Welbeck, and Wimpole, it was also a showpiece of progressive, indeed avant garde agriculture, in the words of John Lawrence's *Modern Land Steward* (1801) a "theatre . . . for the display of *all* the notable varieties of experimental husbandry" (quoted in Williamson 1995, 122). The fifth Duke instituted week-long agricultural shows known as "sheep shearings," held at the model farm, with exhibitions of state-of-the-art machinery and livestock. Bedford employed a succession of men at Woburn, like the geologist John Farey, as steward, and the planter William Pontey, noted for their experimental methods. A former assistant to the architect Henry Holland, Robert Salmon was employed at Woburn as clerk of the works, and as well as designing a model farm within the park, he turned his hand to a series of award-winning mechanical inventions, including machines for ploughing, sowing, reaping, pruning, haymaking, canal equipment, surgical instruments, and a "humane man trap" (Smith 1983, 112–191).

In the 1790s, the fifth Duke's patronage of both agricultural experiment and political radicalism alarmed conservative commentators. Burke warned him that French revolutionaries wanted "new lands for new trials" and while their "geographers and geometricians . . . have an eye on his grace's lands, their chemists are no less taken with his buildings." Gillray's cartoon *The Generae of Patriotism* depicts the Duke sowing gold coins which yield a crop of Jacobin wheat, complete with liberty caps. The Duke's forestry consultant William Pontey was denounced in the *Anti-Jacobin Review* for his "theoretical speculations" on the growing of fir trees (Daniels 1988, 51; Wilmot 1990, 51–53).

Inheriting the estate in 1803, upon the death of his elder brother, the sixth Duke extended the experimental regime at Woburn and instituted a new focus of innovation in horticultural science. Among the sixth Duke's new appointments at Woburn were George Sinclair, Edmund Cartwright, and James Forbes, each charged with investigating a variety of plants, from kale to heather, conducting trials in glasshouses and experimental plots, writing articles in learned journals and in lavish, limited edition volumes. Among the hired consultants were chemist Humphry Davy, botanist James Edward Smith, seedsman Thomas Gibbs, glasshouse technologist William Atkinson, and architect Jeffry Wyatt. Under the sixth Duke, Woburn was developed as a leading research institution, arguably after the decline of the Royal Gardens at Kew from the 1820s, the preeminent one.

While the sixth Duke continued the family campaign for political reform and was keen to diffuse the benefits of horticultural innovation widely, starting with the tenants and cottage gardens on his own estate, his regime at Woburn served to ratify the aristocratic, paternalist model of enlightenment promoted by Banks and instituted in such metropolitan venues as the Royal Institution. Indeed, there is a parallel between the design and management of spaces. The combining of laboratory and lecture theater, the expensive range of instruments, and the select audience, at Humphry Davy's demonstrations of agricultural chemistry at the Royal Institution were ingredients of the Sheep Shearings at Woburn (Golinski 1992, 188–235). Horticulture, more than agriculture, was congenial to the sixth Duke because he was a virtuoso, a collector of artworks, antiquities, books, and sculpture as well as plants and shrubs. Horticulture was a form of connoisseurship. Arthur Young recognized this early on. Visiting one Sunday, for a dinner with the Duke's associates in 1804, Young noted:

> Several apartments newly furnished, and many very expensive articles, clocks, &c from Paris to the amount of 2,000l. Much done to the greenhouse, and everywhere a profusion of expense. . . . An extravagant duchess, Paris toys, a great farm, little economy and great debts, will prove a canker in all the rosebuds of his garden of life . . . What has a Christian to do with such scenes? (Young 1967, 396)

12.4. Humphry Repton

From 1788, Humphry Repton began his career as a landscape gardener. To prepare for this he read among other texts on aesthetics, Erasmus Darwin's *Botanic Garden* (1789), and one of his early correspondents, Anna Seward commended Darwin's Linnaean "picturesque garden of botanic science" at Lichfield. But for long Repton's designs had little botanical or horticultural content, displaying little information on exactly what, or indeed how, to plant. He took a largely aesthetic view of his vocation. His principle influence was William Mason's poem *The Garden*, which unlike Erasmus Darwin's saw no aesthetic implications in Linnaeus, merely a marshalling of "vegetation's verdant brood . . . in order'd file" (quoted in Laird 1996).

After the turn of the century, Repton shifted his view, focusing more on gardens (rather than parks), bringing the flower closer to the house, indeed assimilating the garden to domestic architecture. Repton also expressed a greater concern with botanical knowledge, or rather his lack of it. In November 1800, while he was traveling to a commission at the Duke of Portland's at Bulstrode, he took with him a bundle of old Reviews to read, one of which, the *Monthly Review* for August 1794, had an account of botanical discoveries in the Alps by James Edward Smith, founder of the Linnaean Society. Smith and Repton had been school friends in Norwich but had lost touch when Smith went to Edinburgh to study medicine, and Repton was keen to renew their acquaintance. The need to do so much have struck Repton when he arrived at Bulstrode. The Portlands were patrons of horticulture and botany—Bulstrode possessed a leading botanic collection—and one of the guests when he arrived was Samuel Goodenough, founder member of the Linnaean Society and a close correspondent of Smith's. Repton wrote to Smith recalling a childhood walk when, sheltering from a thunderstorm in a hollow tree near a bog, they found "an unexpected comfort in the most perfect specimen of Helvellus Acaulsis." Repton thinks that is the name but "I betray my ignorance in Botany"—"Dr Goodenough has kindly corrected my Helvellus to Helvella." "You have reached the summit of the Science you delight in," Repton told Smith, "whilst I more humble have been content to climb to the top of my Art." It is clear that Repton's clients were demanding more botanically informed designs. He wrote again to Smith about one who wanted to arrange flowers to open in a certain order at certain times of the day as Linnaeus had done. "I am not sufficient Botanist to give him any instruction," Repton told Smith, "I often regret I have dedicated my life to the canvas on which nature stretches her wonders—rather than to the individual wonders separately" (Repton 1800, 1811, 1815a).

Repton was commissioned at Woburn in 1804. His Red Book of designs was his longest and most lavish, one to rival some of the other luxury volumes Bedford had collected and commissioned. Repton tried to make it worthy of its setting, Henry Holland's "magnificent library in which this volume aspires to hold a place" (Repton 1805).

The focus of the Woburn Red Book is the pleasure ground. The scheme is highly architectural:

> The gardens or pleasure-grounds near a house may be considered as so many different apartments belonging to its state, its comfort, and its pleasure. The magnificence of the house depends on the number, as well as the size of its rooms; and the similitude between the house and the garden may be justly extended to the mode of decoration . . . If in its unfurnished state there chance to be a looking-glass without a frame, it can only reflect the bare walls; and in like manner a pool of water, without surrounding plantations or other features, reflects only the nakedness of the scene. This similitude might be extended to all articles of furniture for use of ornament required in an apartment, comparing them with the seats and buildings and sculpture appropriate to a garden. Thus the Pleasure Ground at Woburn requires to be enriched and furnished like its Palace, where good taste is everywhere conspicuous.

Repton listed the gardens, a rosery, American garden, Chinese garden, botanic garden for the classification of plants, an "animated garden" or menagerie, and lastly an "English garden," a shady walk "enriched with flowers and creeping plants" "connecting the whole." For the garden around the Duchess's Chinese dairy, Repton suggested planting some recent arrivals which had been "naturalized in England": *Hydrangea macropphylla* (introduced by Banks in 1789, identified by Smith in 1799), *Aucuba japonica,* and the repeat flowering China Rose.[1] The "Nucleus, that combines the several parts into one magnificent whole" was the Forcing Garden. Repton shows this illuminated on a frosty night, with gardeners working and family or guests promenading past.

The appointment of George Sinclair (one of the best gardeners in the south of Scotland who worked for the Gordons) as "botanist gardener" at Woburn, a position he retained until 1824, had a decisive influence on the shaping and reshaping of Repton's design for the pleasure grounds. In a report on the progress of his design in 1816, Repton indicates that while the "disposition" of the gardens had been completed, a number, including the Forcing Garden and American Garden, had yet to be constructed and a new one, a botanic garden for grasses, had been included. Sinclair undertook a program of investigating which grasses were profitable for different kinds of pasture, conducting field trials, and writing up his findings in *Hortus Gramineus Woburniensis* (1816), illustrated by actual specimens, dried and colored, with examples of seed (Smith 1983, 142–150). By the time Woburn's gardener James Forbes described the pleasure grounds in *Hortus Woburniensis* (1833), Repton's design had been further

S. DANIELS, S. SEYMOUR, C. WATKINS

adapted to the Duke's botanical demands, with a heathery, pinery, holly garden, willow garden, and camellia house. The space which had been the "nucleus" of Repton's design, the walled enclave of the Forcing Garden, was now turfed and planted like a piece of parkland. But the overall shape and many of the elements of Repton's plan of thirty years before had been retained, and Forbes was keen to give Repton credit for the authorship of the design (Forbes 1833).

In 1810–11 a further plan added into the Red Book reflected the Duke of Bedford's antiquarian interests. Repton reported Bedford

> One day observed, that out of the numerous cottages called Gothic, which everywhere present themselves near the high roads, he had never seen one which did not betray its modern character and recent date. At the same time, his Grace expresses a desire to have a cottage of the style and date of buildings prior to the reign of Henry VIII, of which only some imperfect fragments now remain. (Repton 1840, 421–422)

The cottage Repton designed was a keeper's house on the estate at Apsley Wood, about three miles from Woburn Abbey. John Adey Repton was largely responsible for the design, which was a capriccio of "some curious specimens of timber buildings," printed in *Archaeologia,* the journal of the Society of Antiquaries. Among the sources were fragments of fifteen buildings from throughout the eastern counties. The ornaments painted on posts and rails were taken from a portrait of Henry VIII and his family purchased by the Society of Antiquaries. The Reptons attempted to "assimilate a garden to the same character," taking hints from various Tudor paintings and engravings, especially by Holbein. The selection of flowers "has been taken from those represented in the nosegays of old portraits of the same period, preserved in the Picture Gallery at Woburn," including "rosemary, columbine, double-crowfoot, clove-pinks, marigold, double-daisy, monks wood, southernwood, pansies, white rose, yellow lillies, turk's cap etc" (Repton 1840, 424).

Repton used his visits to Woburn to advance his botanical reputation. From information gathered in the winter of 1808–9 he presented a paper in April 1810 to the Linnaean Society, dedicated to Smith, supporting Linnaeus against the opinion expressed by most other authorities (from Evelyn to Curtis to Joseph Banks), that ivy was a destroyer of trees. Like a roving agricultural reporter, a Young or Marshall, he used his traveling to test his observations, observing the relation between trees and ivy from Warwickshire to Middlesex, Norfolk to Herefordshire, Cambridgeshire to Yorkshire, by high roads

and country lanes as well as within the parks of his patrons. The manuscript was written at Woburn which was the key site for Repton's
observations:

> At Woburn Abbey the timber has so generally been denuded of Ivy,
> that I despaired of finding any example . . . I afterwards discovered
> in the park a remarkable specimen, which is the outermost tree of
> a grove, and the most exposed to the southwest. The tree nearest
> to it has some dead branches, and seems evidently to have yielded
> to its neighbour's superior vigour. As this is an example obvious to
> all the agriculturalists who attend the Woburn sheep shearing, I
> have, with the Duke's permission, marked a drive very near this
> specimen, which may serve to call the attention of the curious to
> this subject. I should further add the result of some experiments
> made by Mr Salmon, who is well known for his mechanic ingenu-
> ity, and who has the superintendence of His Grace's woods at
> Woburn. He tried to the comparative substance and strength of
> several kinds of timber with the same kinds *Ivy-bound* as he calls it;
> but he could not find any difference, and is of the opinion "that in
> old trees it does not harm . . . but he is still convinced that he has
> seen young trees killed by the Ivy . . . But experience has discov-
> ered that the destruction of turnips and other plants while young,
> and the thinning of green fruit from trees, is a part of the economy
> of nature." (Repton 1815b, 32–33)

Repton demanded "a less rigorous persecution of the plant" (he
told Smith Banks called it a "vegetable beast of prey" [Repton 1809]),
for the leaves and berries of ivy provided food in winter for livestock
and game, and it did much to "improve the beauty of our winter
scenery." The Picturesque, so long supported and criticized on artistic, painterly grounds, is here given a scientific warrant. Repton used
his new botanical interest, both in the pleasure grounds and the
plantations, to affirm a Burkean view of landscaping and estate management, mansions, and parks as places where the gentry should re-
side and exercise a benign authority.

Much to his anxiety, Repton's paper took years to appear in the
Society's *Transactions,* and then without the illustrations he offered.
When it was eventually published in 1815, Repton's career had collapsed, a consequence of the economic and social conditions of
wartime and Repton's own crippled state after a carriage accident. In
a long letter to Smith, shot through with social and spiritual anxieties, Repton asked if he should ever be at a loss for a name,

> And could affix it to some plant of the Ivy tribe, or of any climbing
> genus which, like myself, want to be supported, I should rejoice to
> have my name recorded by your power of conferring immortality.
> My great predecessor *Adam* would never have been able to find

names for a hundreth part of your vocabulary; but he lived in a garden with one friend and one enemy, who, like Buonaparte in our days, was the enemy of peace . . . Is it possible for you, who know so much more of the created evidences of the Deity, to feel satiety here, and a wish to enlarge your scene of observation? I have no doubt this will happen to all whose active minds lead them to wish for such enlargement; and then you and I shall meet, and compare our ideas on that and a great many other subjects. (Repton 1815a)

Repton asked Smith if he could reprint his paper on ivy in his last book, and moral testament, *Fragments on . . . Landscape Gardening* (1816) but permission arrived after the book had gone to press. Repton also failed in his appeal to attach his name to ivy.

12.5. Thomas Andrew Knight and Horticultural Science

Thomas Andrew Knight (1759–1838) is best known as the President of the Horticultural Society of London (later to become the Royal Horticultural Society) and as the younger brother of Richard Payne Knight, author of the picturesque tract, *The Landscape: A Didactic Poem* (1794). Knight became President of the Horticultural Society in 1811, a position he held until his death in 1838. A protégé of Joseph Banks, who secured him a position as a founding member of the Society when it was established in 1804, and sometimes dismissed as Banks's front man (Elliott 1994), Knight was famous in his own right as one of Europe's leading protagonists in the field of vegetable physiology (Kendrick 1981, 250; Spary 1996, 189). He was one of the first to practice the artificial hybridization of plants, and his ideas, for example, on the circulation of sap, were adopted by respected scientists across Europe, including Henri Dutrochet, the discoverer of osmosis, and Humphry Davy, who adopted Knight's ideas in his *Elements of Agricultural Chemistry* (1813) and his earlier lectures commissioned by the Board of Agriculture (1802–1812) (Knight 1992; Stackhouse-Acton 1841, v–vi; Wilmot 1990, 23).

Knight held a strong belief in the notion of human progress and strove for this through the derivation of understanding from first-hand experience of the world and the application of this to practical ends, while at the same time placing stringent restraints on the role of theoretical speculation. With such views he reflects the key roles of experience and utility characteristic of both Lockean and Baconian thought. In his prospectus for the Horticultural Society in 1812, Knight set out his agenda for progress in his chosen field. Referring to fruit, he argued that "almost every ameliorated variety . . .

appears to have been the offspring of accident, or of culture applied to other purposes." His aim was to bring a greater rigor to the subject:

> An ample and unexplored field for future discovery and improvement lies before us, in which nature does not appear to have formed any limits to the success of our labours, if properly applied. (quoted in Elliott 1994, 127)

Knight's field of exploration and discovery did not take him across the globe as it had his mentor and friend Banks, or into a metropolitan laboratory. Instead, he centered his scientific endeavors on the meticulous observation of his everyday surroundings, most notably 'in the field' on his family's landed estates in Herefordshire. In this section of the chapter we will consider the generation of Knight's horticultural knowledge in terms of the different sites in which he operated and the different roles he played: as father and estate owner in the family garden, hothouse, home farm, and orchards of his estates in Herefordshire; as President of the London Horticultural Society and a renowned scientist in the capital, with its international community of specialists; and as correspondent with leading scientists and representatives of horticultural and agricultural societies in Europe, the Americas, and Australia.

12.6. At Home in Herefordshire: Knight as 'Improving' Estate Owner

It was in his home county of Herefordshire that Knight's scientific curiosity was first aroused and his reputation forged, rather than through a formal education in biology or botany. Knight rarely left the county, going abroad only once during his lifetime—a short and disturbing visit to Paris in 1790 with Richard, his more radical brother. He paid only a brief visit to London each year to fulfill his duties to the Horticultural Society by attending its anniversary meeting.

Recognized as the home of the English Georgic, Herefordshire was a county of varied landscape, production, and landholdings (Barrell 1980, 117, 173–4; Daniels and Watkins 1994). Its estates were generally of only modest extent, and estate owners traditionally celebrated the combination of pleasure, profit, and patriotism on their properties in archetypical georgic fashion. Herefordshire also had an explicitly experimental tradition. Benjamin Stillingfleet, the author of the first English translation of Linnaeus, undertook experiments

S. DANIELS, S. SEYMOUR, C. WATKINS

with grasses in the 1750s at Foxley, the home of Knight's neighbor Uvedale Price (Daniels and Watkins 1991).

Knight's own scientific inquiries were stimulated through his intimate observation of the Herefordshire countryside and its produce. His daughter reports in her *Life* of her father (Stackhouse-Acton 1841, 9):

> Mr. Knight's mind, from the earliest dawn of his understanding, seemed peculiarly formed for the enjoyment of a country life; and the part of England on which his lot had fallen was eminently calculated to draw forth and exercise the latent faculties of his mind. Its hills, its valleys, its rivers, its vegetable productions, its geological structure, and its meteorological changes, were to him objects of philosophical investigation.

Although he experimented with plants from the potato to the pineapple, with little interest in flowers, Knight's major focus was on orchard fruits, produce for which Herefordshire was famous. Knight himself recounted this debt to his home country:

> Being born in the midst of orchards I was early led to ask whence the varieties of fruit I saw came, and how they were produced; I could obtain no satisfactory answer and was thence led to commence experiments. (Letter from Knight to *The Cottage Gardener* in 1828, quoted in Bagenal 1933, 457)

Knight's horticultural interests and experimentations formed just one theme in an array of concerns connected with his role as a landed gentleman. These ranged from silviculture, agriculture, and shooting, to the welfare of tenants, the operation of the poor law, the commutation of tithes and general issues of political economy. Unlike his brother Richard, he appears to have been little interested in aesthetics but was fond of poetry, sharing with his favorite poet, Crabbe, "no great 'love for painting, or music, or architecture, and little for what a painter's eye considers the beauty of landscape'" (Stackhouse-Acton 1841, 73–4).

During his life, Knight developed a formidable reputation as an agricultural improver. His initial introduction to Banks, made by his brother, Richard, in the early 1790s, was in fact due to Knight's reputation as a successful breeder of Hereford cattle, Banks being in search of a Herefordshire correspondent to the Board of Agriculture (established 1793) (Stackhouse-Acton 1841, quoted in Anon. 1841, 351). In the second of the Board of Agriculture reports for Herefordshire, produced by Duncumb in 1805, Knight is the most quoted authority on the state of agriculture in the county. The site for most of his agricultural experiments was the modest farm he kept at Elton, near Downton, his residence after his marriage in 1791. There, he experimented with and bred 'improved' livestock, developed new hybrid crops and undertook trials with agricultural implements. His experimentation with animals figures prominently in his correspondence with Banks and forms the subject of a number of his published papers (Knight 1797–1819, 1841). Prizes for his improved cattle were received in the national forum of the Smithfield Show as well as in the county context of the Herefordshire Agricultural Society (of which Knight was a founder member in 1797). (Garrold 1898, 7–9; Knight 1800; Stackhouse-Acton 1841, 19)

Knight's early horticultural experiments were made at Wormsley Grange, his childhood home near Foxley, a modest house, later let to a gentleman farmer. The walled garden, nursery, and orchards where Knight conducted his early experiments lay immediately adjacent to the house which Knight shared with his mother. Within this strongly domestic and estate setting, Knight in the 1780s began the experiments on orchard trees which informed his 1797 *Treatise on the Culture of the Apple and the Pear*. The original trees of two of his early new varieties, the Grange apple and the Downton pippin, were reported by Knight still to be growing at Wormsley in 1811 (Knight 1811, viii).

But it was in the north of the county, first at Elton Hall, where he lived from his marriage in 1791, and subsequently at Downton Castle, where he moved in 1808, that most of Knight's experiments were undertaken. The main site for his experiments at Elton was the walled garden which he made himself, about a quarter of a mile from the small Queen Anne house (Bagenal 1938, 319; Inglis-Jones 1968, 366). Here he installed a hothouse, built to his own design, undertook "some of his most important early work on the breeding of new varieties of fruit trees," and established an orchard of these new varieties (Bagenal 1933, 458; 1938, 319). His daughter reported how "the acquisition of a hothouse and a farm now enabled him to prosecute his experiments in horticulture and agriculture with more advantage than heretofore," despite his modest income as a younger brother. Knight also constructed "all the machinery he required for conducting the most elaborate experiments," assisted only by "a common carpenter or blacksmith" (Stackhouse-Acton 1841, quoted in Anon. 1841, 351). Another of his hothouse designs was built at Downton in 1818 (Bagenal 1938, 324; Knight 1841, 242–245). This was located in the kitchen garden which a visitor of 1837 remarked was "the principle seat of Mr. Knight's horticultural experiments."

S. DANIELS, S. SEYMOUR, C. WATKINS

Despite this, the garden was only a modest affair, located a few hundred yards north of Downton house (Anon. 1838, 209–212).

One of Knight's notebooks, in which he recorded his experiments, survives, dating from 1797. Although the observations are not particularly full, his methods were rigorous. For example, he not only developed new hybrids but recorded the rates and timings of germination from seed. A rough sketch in the notebook indicates that the young plants produced were systematically planted out in trial plots, with their location and aspect carefully recorded (Kendrick 1981, 251). Another striking aspect is how agricultural, horticultural, and orchard crops were mixed in the garden, reflecting the range of Knight's interests. Peas for experimentation were sown in front of the vinery in 1798 and pears "in the Vinery in the Garden" on Christmas Day 1810 (Knight 1798–1821).

A visitor to Downton in the late 1830s, when Knight was in his 78th year, was Sir George Stewart Mackenzie, himself a notable horticulturalist (Elliott 1994, 125). In the garden he found "no display—nothing for show—all was perfectly business-like, and full of experiment. Various modes of culture were in progress with everything; and reasons were given for commencing every experiment." Mackenzie also found Knight's keen horticultural insight mirrored in his political judgement, describing him as "one who can discern rottenness in church and state, as well as canker in a fruit-tree, and can fathom both" (account in *Edinburgh Chronicle* 1838, quoted in Stackhouse-Acton 1841, 57–58).

The gardens at both Elton and Downton also provided a site where Knight's roles as scientific experimenter, estate owner, and father overlapped as he soon had a young family (three daughters and one son). His eldest daughter, Frances (born in 1793), herself noted how her father was "anxious to cultivate" in his children "a taste for horticulture, natural history and other rational pursuits," and looked back to "the hours spent with him in his study, or in his garden, as among the happiest recollections" (Stackhouse-Acton 1841, 73). Frances continued to be actively involved in experiments at Downton. Her obituary in *The Gardeners' Chronicle* (Anon. 1881, 182–3) records that she was Knight's "constant associate in the many experiments carried out at Downton Castle, which gave such an impetus to scientific horticulture in the first quarter of the present century," and pays a great tribute to her own experimental skills and her "wide knowledge of geological, botanical, horticultural and antiquarian lore." It is probable that Frances not only produced the account of her father's life, but also edited the *Selection of Knight's Papers* with which the *Life* was published in 1841, although this task is usually attributed to George Bentham, Secretary of the Horticultural Society, and John Lindley, Professor of Botany at London University College and the Royal Institution. As an accomplished artist she also illustrated a number of Knight's new fruit varieties in his 1811 publication, *Pomona Herefordiensis*. Knight's encouragement of Frances may go some way towards explaining why the Horticultural Society was among the first of London's scientific societies to admit women. The first woman Fellow, the Countess of Radnor, was elected in 1830, "it having been agreed that there was nothing in the charter or byelaws disqualifying women" (Hadfield 1969, 270). Whether this move relates to a more enlightened attitude towards women, encouraged by Knight's personal views on education and by the well-known botanical interests of a number of aristocratic women, most notably the second Duchess of Portland in the first half of the eighteenth century (Gascoigne 1994, 80–81), to a traditional association of women with horticulture (and particularly flowers), or to the Society's campaign to encourage more noble patronage is uncertain.

Knight was not just an experimenter but attempted to put the knowledge gained from his experiments into practice on his estates. By 1827 his Wormsley estate covered 1,128 acres on which there were 18 orchards, most no more than a few acres in size, clustered around the farmsteads. Although he did not take these orchards in hand and manage them himself, Knight actively improved them by paying for new trees to be planted. An account of 1821 indicates that 331 apple and pear trees had been freshly planted in four orchards on the Wormsley estate at Knight's request (Hereford County Record Office [H.C.R.O.] T74/728). In addition, Duncumb (1805, 91) related how Knight made judicious use of some less valuable estate lands, worth no more than 12s an acre as pasture, by planting orchards and raising the land value to £3 an acre. Knight also encouraged the further development of an experimental culture in Herefordshire by distributing specimens of his new varieties to friends and neighbors as gifts, a practice he also adopted with correspondents across the world (Biddulph 1824; Birt 1812).

The local context but national concerns behind Knight's experiments are reflected in the naming of his new fruit varieties. Several were named after local figures or their estates, for example, the Foxley Apple (named after the estate of his neighbor and friend Uvedale Price). But a more imperial note was struck when in 1815 Knight called one of his new cherry varieties "Waterloo," it having "ripened fruit for the first time a few weeks after the battle to which it owes its name" (Bagenal 1933, 458). This is one of Knight's few enduring varieties. Knight, it should be acknowledged, was working on the mistaken premise that varieties had limited lifespans, thereby making it necessary to develop new ones. It was nonetheless his own interest and activity in breeding new varieties rather than his motives or the particular new fruits and vegetables which he himself produced which were most important in stimulating the production of further varieties (Elliott 1994, 128).

12.7. Knight in London and International Circles

Another key arena for the development and exercise of Knight's horticultural knowledge was London. Most important was Banks's house, No. 32 Soho Square, with its collection of botanical specimens and unrivaled natural history library. These, together with the figure of Banks himself, attracted "many of the most distinguished men in science and literature" from across Europe. It was here, his daughter reported, Knight "had occasionally opportunities of comparing his own observations and theories with those of many of the most celebrated naturalists of all countries." But she, like others, felt it would have been better for him if the occasions had been more frequent, "for it would have saved him trouble in working out facts which cost all the labour and time of original discoveries, and which labour would have been more profitably employed in building on the sub-structure already laid by other hands." Knight was from an early period suspicious of scientific literature, and it appears that he "for some years purposely avoided to read the works of his precursors in the field of vegetable physiology, from an idea that, by the study of nature, unbiased by the opinions of others, he should be most likely to arrive at truth" (Stackhouse-Acton 1841, quoted in Anon. 1841, 351). Such scepticism might be related to embarrassment at his lack of formal schooling. Knight was "unable to write, and scarcely able to read" by the age of nine (Anon. 1838, 99) and he may have also been defensive about his lack of education or interest in the classics, particularly in light of his brother Richard's reputation as a classical scholar (Clarke and Penny 1982). Knight turned this lack of formal education into a virtue and, as late as 1828, claimed "if I have made any discoveries, I am indebted to my ignorance, and to the advantage of having no lights, instead of such as would have misled me" (Letter from Knight to *The Cottage Gardener* in 1828, quoted in Bagenal 1933, 457). His obituary writer explicitly related his subsequent skills as an experimenter to this lack of early education:

> It was in the idle days of his childhood, when he could derive no assistance from books, that his active mind was first directed to the contemplation of the phenomena of vegetable life; and he then acquired that fixed habit of thinking and judging for himself, which laid the foundation of his reputation as an original observer and experimentalist. (Anon. 1838, 99)

Banks, however, encouraged Knight to read the scientific canon with, it appears, some success as Knight did become more concerned with theories and published knowledge. In his experiments on sap, begun by 1799, it is apparent that Knight was aware of the work by Hales on capillary attraction and Hunter's theory that the movement of sap was caused by the pressure of air (Kendrick 1981, 251). Likewise during Dutrochet's two-week visit in 1827 (see below), Knight reported they had "made some experiments together, and investigated the hypotheses of different writers" (Letter from Knight to Mr. Williams, quoted in Stackhouse-Acton 1841, 40). Banks also assisted Knight to publish the results of his experiments, as well as securing his position in the Horticultural Society, thereby helping to integrate him into London scientific culture. By his death Knight had published almost 150 articles in the Horticultural Society's *Transactions* alone, an indication that he came to regard the scientific literature as an important means of disseminating the discoveries of empirical enquiry (Elliott 1994, 123).

London was also the venue for a number of attacks upon Knight and his empirical knowledge. Knight's scepticism over scientific theories was fueled in particular by the extravagant claims made during his lifetime by gardeners and nurserymen in advertisements for plant cures. In the early years of the nineteenth century, the normally temperate Knight became involved in a very public and acrimonious confrontation which threatened his emerging reputation in London horticultural circles. The disagreement arose when as reputable a figure as Mr. William Forsyth, former Curator of the Chelsea Physic Garden and by 1801 royal gardener to George III at Kensington, advertised his "plaister" to treat tree wounds (Elliott 1994, 123). Knight challenged the exaggerated claims, made more by Forsyth's friends than by Forsyth himself, that whole boughs were restored by the "plaister." One of Forsyth's friends, a Dr. Anderson, claimed that the treatment would restore holes in trees and severely rotted trees. Knight publicly dismissed this claim and challenged either Anderson or Forsyth to repeat what he described as this "miracle" "in any garden, or piece of ground, which shall be under the care and inspection of the President of the Royal Society, or of such Persons as he shall appoint" (Knight 1803). Anderson in turn accused Knight of ungentlemanly behavior, but more poignantly attacked him for lacking the skill he most admired (and for which he was most admired): meticulous observation. Somewhat cuttingly he proposed that when Knight was next in London, he come to Kensington Gardens, "accompanied by some Gentlemen of sound understanding, capable of observing facts, and of drawing just conclusions from them" to settle the dispute (Anderson 1802).

Knight's professed mistrust of the published claims of many professional gardeners and his scepticism towards theoretical knowledge probably contributed to the uneasy relations which developed between him and J. C. Loudon in the 1820s (Simo 1988, 147–164). Loudon, an advocate of well-educated and well-paid professionals, recorded that in a paper of 1820 Knight had boasted that a man who "neither knew a letter nor a figure" had grown pineapples in a "far superior way" to the generality of professional gardeners. When Loudon subsequently discovered Knight employing a gardener, Mr. Richard Williams, who had recently published a paper (*An Account of an Easy Method of Destroying Caterpillars*, 1828) he announced, "We are very happy to learn that Mr. Knight has thought it worthwhile to keep a gardener who can not only read but write" (Loudon 1829, 87–88).

Loudon did not stop there but went on to make a direct attack on Knight's reputation as a utilitarian. Knight's obituary in the *Gentleman's Magazine* (Anon. 1838, 100) claimed that the driving force behind his experimentation was utility, "it was only when facts had some great practical bearing that he applied himself seriously to investigate the phenomena connected with them." This led Knight, according to the obituary writer, to focus on experiments "to improve the races of domesticated plants, to establish important points of cultivation upon sound physiological reasoning, to increase the amount of food which may be procured from a given space of land, all of them subjects closely connected with the welfare of his country." Although Loudon acknowledged "a very great personal regard" for Knight and thought his reputation in "physiological experiments" deserved, he questioned his credibility as a cultivator and the influence of his practical papers in the *Transactions of the Horticultural Society*. Offended at Knight's earlier criticism of professional, educated gardeners, the type he saw as crucial in the production of horticultural knowledge, Loudon announced:

> Mr. Knight, by showing the utility of general knowledge to gardeners, and advocating the cause of garden libraries as the means for acquiring this knowledge, might do more for the advancement of horticulture, than by all the practical papers that he has ever written, or ever will write. Among practical gardeners these papers go for nothing, and deservedly so; for what is the result of all that Mr. Knight has stated in regard to the culture of the pine-apple, the strawberry, or the mango? (Loudon 1829, 87).

While Loudon may have been justified in his attack on Knight's papers on exotic fruits, there are grounds for questioning his general criticism. On one hand, Knight's strawberries were among the ancestors of the modern large-fruited varieties (Elliott 1994, 128); on another, Loudon fails to refer to the orchard fruits or field crops for which Knight was more famous.

Loudon's statement caused great affront to the members of the Horticultural Society Council who refused the offending issue of the *Gardener's Magazine,* sent as a present to the Society library and agreed to send an official communication to Loudon in which "they do unhesitatingly declare their opinion that the assertion of the inutility of Mr. Knight's papers is decidedly unfounded and untrue" (quoted by Loudon 1829, 87). Knight's utilitarian views on horticultural science had led to work on landscape gardening being excluded from the Society's *Transactions,* and this may have annoyed Loudon, whose early career had been based around landscape husbandry (Jacques 1983, 190). But his attack on Knight is likely to have been part of a more general dispute which he had with the Horticultural Society. Loudon, a liberal, middle-class professional, objected to the Society's attempts in the 1820s to 'ennoble' what had originally been a socially mixed membership in order to secure greater financial security and social standing (Stackhouse-Acton 1841).

Despite the development of a considerable reputation in London and internationally, both as a botanist and agriculturalist, Knight remained unusual in that he spent most of his time in Herefordshire and even when President of the Horticultural Society was not involved in its routine Council meetings. However, Knight compensated for his absence from London and his lack of overseas travel by inviting people to visit him in Herefordshire. Humphry Davy, whom Knight met through Banks, became a particularly good friend, and although Knight occasionally had visits in Herefordshire "from the

S. DANIELS, S. SEYMOUR, C. WATKINS

scientific men he met in London," Davy was unusual in that he became a regular, twice-yearly visitor from 1806 until his marriage in 1812 (Stackhouse-Acton 1841, quoted in Bagenal 1933, 458). Another eminent visitor to Downton was the French physiologist, Dutrochet, a long-term correspondent of Knight who eventually visited him in 1817. Dutrochet was carrying out work on membranes and flows relevant to Knight's researches on the circulation of sap and spent a fortnight at Downton with Knight (letter from Knight to Mr. Williams, quoted in Stackhouse-Acton 1841, 40).

But more important in compensating for his reluctance to leave Herefordshire was Knight's development of an active correspondence. His daughter reported that he carried on a "very extensive correspondence, not only with many of the men most distinguished for their attainments in science in Great Britain, but with most of the writers on vegetable physiology and horticulture on the continents of Europe and America" (Stackhouse-Acton 1841, 64). Knight's correspondence with Banks is an extensive and well-documented example of this. As President of the Horticultural Society, Knight also fostered an international circle through the many similar organizations which were being established in other countries, especially in North America, but also in Europe and Australia. These frequently wished to associate with the London Society and Knight received many offers of honorary membership and awards. In his turn he sent out specimens of the fruits, trees, and other plants he had bred, which often led to plant exchanges being set up. For example, in 1828, George Foy, a diplomat in Stockholm, reported the safe arrival "in the best condition" of a bag of plants Knight had sent which he conveyed to the Stockholm Agricultural Society. This organization in turn appointed Knight as a member and, with reference to the "many rare Plants in the Northern Provinces of Sweden," requested an exchange with the Horticultural Society (Foy 1828). Such exchanges affirm Knight's connections with the imperial dimensions of English scientific horticulture, established by Banks and centered on Kew.

12.8. Conclusions

This chapter has considered the geographies of horticultural knowledge associated with two prominent figures of later Georgian Britain: Humphry Repton and Thomas Andrew Knight. A number of points are raised by these studies.

Our account serves to highlight the different types of horticulture and, indeed, of enlightenment, which were pursued. The work of Knight at Elton and Downton, with its concentration on orchard fruits and vegetables, shows a strong utilitarian bent. His approach was physiological with little concern for aesthetics. The few remaining illustrations of his horticultural interests are in the form of working sketches of experiments, estate plans, or carefully depicted plates of his new varieties. By contrast, the activities of the Duke of Bedford and Repton at Woburn placed a greater emphasis on flowers and display, as befits a prominent aristocrat. The approach at Woburn was characterized by connoisseurship as much as scientific experiment, and the gardens were designed as much to display scientific horticultural collections, which complemented collections of art and sculpture, as to facilitate scientific experimentation. Woburn, in contrast to Elton or Downton, is lavishly depicted in Repton's detailed watercolors and plans. Woburn, of course, was an aristocratic venue on a huge scale, whereas Elton, and even Downton, were much more modest establishments where the practices of horticultural research were geared to the concerns and resources of a county gentry family.

The chapter also highlights the importance of field studies in the development of horticultural science and, more generally, of enlightened knowledge, even if the sites of such work were less exotic than those documented by, for example, either Outram or Bravo. Knight was neither the traveling explorer-scientist, passing through an area, nor the sedentary laboratory scientist (Outram 1996) assembling into experimental collections specimens from around the world. Knight kept generally to the fields of his native Herefordshire. His was a well-traveled route, not one passed through only once, and his observations and experiments were strongly influenced by (and in the latter case sometimes carried out in) the field context of his everyday surroundings. Repton was a constant traveler (Daniels 1996). His field was defined by a network of commissions, and, in the case of his observations on ivy, by various pieces of woodland he passed on the way. Not every commission exercised his horticultural knowledge. At Woburn, he entered a complex and specialized arena of horticultural research as a professional consultant.

While gentlemanly patronage of horticulture was important, and some estate owners were themselves active experimentalists, there were other important patrons and practitioners. On landed estates themselves there were resident gardeners and experts drawn from the newly recognized sciences, while laborers and female family

S. DANIELS, S. SEYMOUR, C. WATKINS

members were often hidden figures in the production of horticultural knowledge. Beyond the landed estate, botanical and nursery gardens were important sources of knowledge, and as the nineteenth century proceeded, the sites of horticultural science dispersed from the landed estate to municipal and middle-class gardens and increasingly professional research establishments. The rise of the professional gardener, whose role focused as much on municipal and suburban gardens as on landed estates, was significant here. The most famous of these was J. C. Loudon. His was a much more democratic vision of horticultural knowledge. While Loudon acknowledged the role of the gentleman as patron, such patronage alone, he felt, would not lead to horticultural improvement. This required the insight and practices of well-educated professional gardeners who should be encouraged and fairly rewarded (Simo 1988).

The landed estate had a strong influence on the types of horticultural enquiry undertaken and the means through which they were pursued in later Georgian Britain. Kew, the preeminent site of British horticulture under Banks, was itself carved from an established royal estate, while eminent figures of British scientific pursuit, for example, Banks, Sinclair, and Knight, used their estates as sites for investigation. Serious scientific concerns merged with the general fashion for 'improvement' in polite society, prompting a wide range of horticultural interests. So while landed interests helped establish knowledges of horticultural science in later Georgian Britain, they also threatened to dissipate its underlying focus.

Notes

We would like to thank the Leverhulme Trust for its support in funding the work on Thomas Andrew Knight as part of a project on Picturesque Landscaping and Estate Management: The Regional Context of a Landscape Theory, Leverhulme Award F114AK, and that on Humphry Repton under the project on Humphry Repton, Landscape Improvement and the Geography of Georgian England, Leverhulme Award F/111/AV.

Our thanks also go to those who made useful comments on the paper at the Geography and Enlightenment conference (University of Edinburgh July 1996) and at an Eighteenth Century Studies Group seminar at the University of Warwick (November 1996) and in particular to the editors for comments on an earlier draft of this paper.

1. In his *Memoir* (n.d., c. 1814) Repton confessed he had "of late viewed with a jealous eye the irruption recently made by the new China Rose, which however valuable in winter from its dark glossy foliage and hardy flower is but like a rouged beauty—and must not attempt to vie with the genuine English scented Rose" (25).

References

Allen, David Alliston. 1976. *The Naturalist in Britain*. London: Allen Lane.

Anderson, James. 1802. Open Letter (Printed) from Anderson to Knight, 20 December. H.C.R.O. T74/545.

Anon. 1838. "Obituary—T. A. Knight." *The Gentleman's Magazine* 10(2): 99–100.

———. 1838. "Some Account of the Gardens of Herefordshire By J. B. W." *The Gardener's Magazine* 14: 209–212.

———. 1841. "A Selection from the Physiological and Horticultural Papers of the Late Thomas Andrew Knight, Esq. To which is Prefixed a Sketch of His Life." *The Gardeners' Chronicle* 1 (29 May): 351–352.

———. 1881. "Obituary of Frances Stackhouse-Acton." *The Gardeners' Chronicle* 15 (1): 182–183.

Bagenal, N. B. 1933. "Thomas Andrew Knight." *The Gardeners' Chronicle* 94 (16 December): 457–458.

———. 1938. "Thomas Andrew Knight, 1759–1838." *Journal of the Royal Horticultural Society* 63 (7): 319–324.

Banks, R. E. R., B. Elliott, J. G. Hawkes, G. King-Hele, and G. L. Lucas, eds. 1994. *Sir Joseph Banks: A Global Perspective*. Kew: Royal Botanic Gardens.

Barrell, John. 1980. *The Dark Side of the Landscape*. Cambridge: Cambridge University Press.

Biddulph, Benjamin. 1824. Letter to Thomas Andrew Knight, 8 March. H.C.R.O. T74/722.

Birt, Thomas. 1812. Letter to Thomas Andrew Knight, 2 November. H.C.R.O. T74/545.

Clarke, Michael, and Nicholas Penny. 1982. *The Arrogant Connoisseur: Richard Payne Knight 1751–1824*. Manchester: Manchester University Press.

Cooke, Ian K. S. 1992. "Whiteknights and the Marquis of Blandford." *Garden History* 20(1): 28–44.

Daniels, Stephen. 1988. "The Political Iconography of Woodland in Later Georgian England," in *The Iconography of Landscape: Essays on the Symbolic Representation, Design and Use of Past Environments*, eds. Denis Cosgrove and Stephen Daniels. Cambridge: Cambridge University Press, pp. 43–82.

———. 1992. "Love and Death Across an English Garden: Constable's Paintings of his Family's Flower and Kitchen Gardens." *Huntington Library Quarterly* 55(3): 433–458.

———. 1993. *Fields of Vision: Landscape Imagery and National Identity in England and the United States*. Cambridge: Polity Press.

———. 1996. "On the Road with Humphry Repton." *Journal of Garden History* 16(3): 170–191.

Daniels, Stephen, and Susanne Seymour. 1990. "Landscape Design and the Idea of Improvement, 1730–1900," in *An Historical Geography of England and Wales*, eds. R. A. Dodgson and R. A. Butlin. Second Edition. London: Academic Press, pp. 487–520.

Daniels, Stephen, Susanne Seymour, and Charles Watkins. 1997. "Border Country: the Politics of the Picturesque in the Middle Wye Valley," in *Prospects for the Nation: Recent Essays in British Landscape, 1750–1880*, eds.

Michael Rosenthal, Christiana Payne, and Scott Wilcox. New Haven and London: Yale University Press, pp. 157–181.

Daniels, Stephen, and Charles Watkins. 1991. "Picturesque Landscaping and Estate Management: Uvedale Price at Foxley, 1770–1829." *Rural History* 2(2): 141–169.

Daniels, Stephen, and Charles Watkins, eds. 1994. *The Picturesque Landscape: Visions of Georgian Herefordshire.* Nottingham: Department of Geography, University of Nottingham.

Davy, Humphry. 1813. *Elements of Agricultural Chemistry.* London: Longman.

Duncumb, John. 1805. *General View of the Agriculture of the County of Hereford.* London: Sherwood, Neeley and Jones.

Elliott, Brent. 1994. "The Promotion of Horticulture," in *Sir Joseph Banks: A Global Perspective,* eds. R. E. R. Banks et al. Kew: Royal Botanic Gardens, pp. 117–131.

Forbes, James. 1833. *Hortus Woburniensis.* London: James Ridgway.

Foy, George. 1828. Letter to Thomas Andrew Knight, 11 April. H.C.R.O. T74/474.

Garrold, T. W. 1898. *Transactions of the Herefordshire Agricultural Society from Its Institute in 1797 to 1809.* H.C.R.O. AF57/14/12.

Gascoigne, John. 1994. *Joseph Banks and the English Enlightenment: Useful Knowledge and Polite Culture.* Cambridge: Cambridge University Press.

Golinski, Jan. 1992. *Science as Public Culture: Chemistry and Enlightenment in Britain, 1760–1820.* Cambridge: Cambridge University Press.

Hadfield, Miles. 1969. *A History of British Gardening.* London: Spring Books.

Inglis-Jones, Elizabeth. 1968. "The Knights of Downton Castle." *The National Library of Wales Journal* 15: 237–288.

Jacques, David. 1983. *Georgian Gardens: The Reign of Nature.* Batsford: London.

Kendrick, F. M. 1981. "Some Notes on the Botanical Work of Thomas Andrew Knight of Downton." *Transactions of the Woolhope Naturalists' Field Club* 43(3): 249–253.

Knight, David. 1992. *Humphry Davy: Science and Power.* Cambridge: Cambridge University Press.

Knight, Richard Payne. 1794. *The Landscape, A Didactic Poem, in Three Books. Addressed to Uvedale Price Esq.* London: R. Faulder.

Knight, Thomas Andrew. 1797. *Treatise on the Culture of the Apple and the Pear and on the Manufacture of Cider and Perry.* Ludlow: H. Proctor.

———. 1797–1819. Letters to Joseph Banks. Natural History Museum General Library, London.

———. 1798–1821. Notebook in which he recorded his experiments. H.C.R.O. T74/437.

———. 1800. "Account of Herefordshire Breeds of Sheep, Cattle, Horses and Hogs," *Communications to the Board of Agriculture* 2: 172–191.

———. 1803. Printed reply to Dr Anderson, 7 January. Elton, Ludlow: H. Proctor. H.C.R.O. T74/545.

———. 1811. *Pomona Herefordiensis.* London: The Agricultural Society of Herefordshire.

———. 1841. "Upon the culture of the pine-apple, Read before the Horticultural Society, 1820," in *Selection from the Physiological and Horticultural Papers, Published in the Transactions of the Royal and Horticultural Societies by the Late Thomas Andrew Knight, to Which is Prefixed a Sketch of His Life.* London: Longman, pp. 242–245.

Laird, Mark. 1996. "Corbeille, Parterre and Treillage: The Case of Humphry Repton's Penchant for the French Style of Planting." *Journal of Garden History* 16(3): 153–169.

Loudon, J. C. 1829. *The Gardener's Magazine* 8: 87–88.

Miller, David Phillip, and Peter Hanns Reill, eds. 1996. *Visions of Empire: Voyages, Botany and Representations of Nature.* Cambridge: Cambridge University Press.

Morley, John. 1993. *Regency Design 1790–1840.* London: Zwemmer.

Murdoch, John. 1986. "Foregrounds and Focus: Changes in the Perception of Landscape," in *The Lake District: A Sort of National Property,* ed. Victoria and Albert Museum. London: Victoria and Albert Museum, pp. 43–59.

Outram, Dorinda. 1996. "New Spaces in Natural History," in *Cultures of Natural History,* eds. N. Jardine, J. A. Secord, and E. C. Spary. Cambridge: Cambridge University Press, pp. 249–265.

Porter, Roy. 1981. "The Enlightenment in England," in *The Enlightenment in National Context,* eds. Roy Porter and Mikulás Teich. Cambridge: Cambridge University Press, pp. 1–19.

Repton, Humphry. 1800. Letter to James Edward Smith, 5 November. *Smith Correspondence* 8:160. Linnean Society.

———. 1805. *Designs, Observations and Plans for the Improvement of the Grounds at Woburn Abbey.* Private Collection.

———. 1809. Letter to James Edward Smith, 19 June. *Smith Correspondence.* Linnaean Society.

———. 1811. Letter to James Edward Smith, 22 October. *Smith Correspondence* 8:166. Linnaean Society.

———. 1815a. Letter to James Edward Smith, 9 July. *Smith Correspondence* 8:169. Linnaean Society.

———. 1815b. "Observations on the Supposed Effects of Ivy Upon Trees." *Transactions Linnaean Society* 11: 10–34.

———. 1840. *The Landscape Gardening of the Late Humphry Repton Esq.,* ed. J. C. Loudon. London: Longman.

———. n.d. (c. 1814). *Memoir.* British Museum Add. Mss. 62112.

Ritvo, Harriet. 1992. "At the Edge of the Garden: Nature and Domestication in Eighteenth and Nineteenth Century Britain." *Huntington Library Quarterly* 55(3): 363–378.

Schteir, Ann B. 1988. "Botany in the Breakfast Room: Women and Early Nineteenth-Century British Plant Study," in *Uneasy Careers and Intimate Lives: Women in Science 1789–1979,* eds. Pnina G. Abir-Am and Dorinda Outram. New Brunswick and London: Rutgers University Press, pp. 31–43.

Secord, A. 1994. "Science in the Pub: Artisan Botanists in Early Nineteenth Century Lancashire." *History of Science* 32: 269–315.

Seymour, Susanne, Stephen Daniels, and Charles Watkins. 1998. "Estate and Empire: Sir George Cornewall's Management of Moccas, Herefordshire and La Taste, Grenada, 1771–1819." *Journal of Historical Geography* 24: 313–351.

Simo, Melanie Louise. 1988. *Loudon and the Landscape*. New Haven and London: Yale University Press.

Sinclair, George. 1816. *Hortus Gramineus Woburniensis*. London.

Smith, Paul. 1983. "The Landed Estate as Patron of Scientific Innovation: Horticulture at Woburn Abbey, 1802–39." Ph.D. thesis, Open University.

Spary, E. C. 1996. "Political, Natural and Bodily Economies," in *Cultures of Natural History*, eds. N. Jardine, J. A. Secord, and E. C. Spary. Cambridge: Cambridge University Press, pp. 178–196.

Stackhouse-Acton, Frances. 1841. "Life of Thomas Andrew Knight," in *A Selection from the Physiological and Horticultural Papers, Published in the Transactions of the Royal and Horticultural Societies by the Late Thomas Andrew Knight, to Which is Prefixed a Sketch of His Life*. London: Longman, pp. 1–77.

Williamson, Tom. 1995. *Polite Landscapes: Gardens and Society in Eighteenth-Century England*. Baltimore: John Hopkins University Press.

Wilmot, Sarah. 1990. *The Business of 'Improvement': Agriculture and Scientific Culture in Britain c. 1700–c. 1870*. Historical Geography Research Series Number 24. London: Historical Geography Research Group of Institute of British Geographers.

Young, Arthur. 1967. *The Autobiography of Arthur Young*, ed. M. Betham-Edwards. New York: Augustus M. Kelley.

Chapter Thirteen
Edinburgh, Enlightenment, and the Geographies of Unreason

Chris Philo

13.1. A Tale of Madness in Enlightenment Edinburgh

On 17 October 1774 the Scottish poet Robert Fergusson died, his final waking moments being spent confined to a cell in the Edinburgh City Bedlam, the only specialist facility then available in Edinburgh for the reception of mentally disordered individuals without means. Fergusson had been a well-known figure, a poet of ability, and also an individual bound up with the excitement of the so-called "Scottish Enlightenment" as played out in Edinburgh ("the Athens of the North," as it was on occasion described).[1] Indeed, during the period 1768 to 1774 when resident in Edinburgh working as a copying clerk in the Commissary Office to provide for his mother and younger sister, Fergusson had split his free time between writing poetry and associating himself with the ideas, personnel, and sites of Edinburgh's growing 'enlightened' community. As Allan Beveridge explains in a recent examination of Fergusson's case:

> Fergusson was now partaking in the social life of Edinburgh—that 'hotbed of genius,' as Smollett had dubbed the capital, where the leading figures of the Scottish Enlightenment enjoyed a 'familiar fraternity' in the clubs and taverns of the city. Fergusson himself was a member of the Cape Club which attracted a wide spectrum of Edinburgh's society—and his company was eagerly sought. (Beveridge 1990, 311)

But this happy state of affairs was to end all too soon, with the seemingly sudden onset of an acute mental disorder with physical correlates—itself spanning two brief phases, and possibly caused by venereal disease in conjunction with a blow on the head[2]—which led to "[t]he myth of 'the poor, white-faced drunken, vicious boy that raved himself to death in the Edinburgh madhouse'" (Beveridge 1990, 309–310).[3] And it was this descent into 'madness' that occasioned

Fergusson's brush with the despair, custodians, and sites of Edinburgh's neglected underworld.

The received wisdom of Fergusson's condition resulting from sinful dissipation is probably unjust, but for my purposes the details matter less than the outline tale of an individual who, in effect, joined together two parallel worlds between which there was only a limited amount of traffic. More specifically, Fergusson's 'madness' meant that his friends, while trying to do what they thought best, committed him to a species of institution—a general house of confinement, albeit one possessing separate accommodation for mentally disordered individuals—which, arguably, was both created and then clothed in secrecy by the Age of Enlightenment. It was a place apart, almost a polar extreme to the refined university buildings, salons and coffeehouses with which the Enlightenment is usually associated, and yet at the same time it was not entirely separate from a few people whose more 'natural' setting was in such sites. Thus, when in the Bedlam he came under the care of Dr. Alexander Wood, a respected if rather eccentric surgeon,[4] and was also visited by Dr. Andrew Duncan, then in his late twenties, who was so appalled by the lack of adequate provisions for the likes of Fergusson that he began a campaign which led eventually to the opening of first the Royal Asylum at Montrose (1779) and then the Royal Edinburgh Asylum (chartered in 1807).

The purpose of this chapter is to draw away from this small tale to consider further the separations and connections of the two very different worlds identified here: on the one hand, the sites of reason, the exalted and well-lit places of the Enlightenment from and through which its intellectual, artistic, and architectural brilliance was pieced together; and on the other hand, the sites of unreason, the despised and darkened places which in effect marked an underside to the Enlightenment largely if not entirely neglected by its representatives. My main intention is hence to investigate something of these various *spaces* involved in the constitution of the Enlightenment and its underside. It is to tease out—albeit in a preliminary fashion—the *geographies* of reason, unreason, and their encounter as key components of the Enlightenment, and in so doing to contribute to research in both the historical geography of ideas (with its concern for the spaces of knowledge production and consumption) and the historical geography of social institutions (with its concern for the spaces of control, correction, and care). It is to underline the extent to which the European Enlightenment was indeed situated, not just in named places such as Edinburgh but more particularly in certain grounded spaces such as lecture rooms and hospital wards. It is also to suggest that Enlightenment as a 'process' permitted of different geographies: some obvious, illuminated, and celebrated; others hidden, shadowy, and unannounced, but all in different ways integral to what this thing called Enlightenment entailed then and has since come to represent.

13.2. Reason and Unreason: Imaginative and Material Geographies

I wish to begin by revisiting a text which is pivotal to any account of the relations between reason and unreason in the Enlightenment: namely, Michel Foucault's *Madness and Civilization: A History of Insanity in the Age of Reason* (1967).[5] Far from being a narrow contribution to the critical history of psychiatry, although of course it does make such a contribution,[6] this text narrates what Foucault subsequently terms an 'archaeology'—a patient excavation—of the shifting relations between Reason and Unreason (or Madness) in Europe from ancient times into the twentieth century. In this work the phenomena of Reason and Unreason (or Madness), announced with capitalized first letters, are envisaged as elemental forces, positions, or possibilities striding down the ages, locked in deadly combat and pervading all facets of social life, religious thought, and political judgment. In this sense, the work is an epic, and also one with a clear temporal (even teleological) logic which scripts the long-term historical process whereby Unreason (with its own distinctive experiences, truths, and languages) is progressively trapped and effectively reduced to nothingness, to silence, by the machinations of Reason. The original preface, which Foucault was later to suppress, made this narrative logic explicit:

> We have yet to write the history of that other form of madness, by which [human beings], in an act of sovereign reason, confine their neighbours, and communicate and recognise each other through the merciless language of non-madness; . . . We must try to return, in history, to that zero point in the course of madness at which madness is an undifferentiated experience, a not yet divided experience of division itself. We must describe, from the start of its trajectory, that 'other form' which relegates Reason and Madness to one side or the other of its action as things henceforth external, deaf to all exchange, and as though dead to one another. (Foucault 1976, ix)

This passage is highly contentious, not least because of the phenomenological undertones present in imagining deep, unifying essences to both Reason and Unreason,[7] but perhaps it is feasible to strip the

CHRIS PHILO

claims here of their grander pretensions and pitfalls, and to follow instead the more grounded purpose energising most of the book's chapters. And this purpose involves searching for the historical eruption of that 'other form'—the will and associated technologies—through which the ideas and practices of reason were torn apart from the notions and behaviors of unreason,[8] with the latter increasingly being subjected to a 'confinement' which effectively removed it from the normal commerce of everyday life (rendering it 'external,' unheard and in effect dead to the world).

In the architecture of *Madness and Civilization* the medieval and early modern periods are depicted as ones when 'unreasonable' people of all kinds were very much present in regular public spaces alongside ordinary individuals and communities going about their daily business. Whatever warrant there may be for such a depiction, and I suggest that there is some (Philo 1997a), Foucault focuses attention on what he variously terms the "Classical Age" or "the Age of Reason"—the period of European Enlightenment from the later years of the seventeenth century into the majority of the eighteenth—as the key phase in his long-term history when the splitting apart of reason and unreason occurred with a vengeance. For Foucault, it was evidently the period when reason was in the throes of 'inventing' itself (deciding what it meant to possess and to exhibit the qualities of reason, to be reasonable) and was doing so in opposition to an unreason (its exact opposite) which it also 'invented' (identified, named, fixed) in parallel and duly consigned to spaces other than those to be filled with its own 'enlightenment.'

Foucault's attention is thereby drawn to what he terms "the great confinement," and in this connection he offers a deceptively simple picture of the Enlightenment instituting a whole geography of sequestered spaces for the housing of people apparently without reason, who were to be removed to "general houses of confinement" shut away from the spaces (the streets, squares, and many other public arenas) now to be occupied solely by those reckoned to possess reason. As Roy Porter puts it:

> [T]he French thinker, Michel Foucault, has contended that Enlightenment principles and absolutist policy focussed in the name of rational administration, to promote cruel social policies. For instance, various kinds of social misfits—the old, the sick, beggars, petty criminals, and the mad—were taken off the streets, lumped together as an 'unreasonable' social residue, and locked up in institutions. Here what purported to be 'enlightened' action was in reality repressive. (Porter 1990, 9)

Foucault delights in listing the ragbag of 'deviants' who ended up being confined, referencing "the debauched, spendthrift fathers, prodigal sons, blasphemers, men who 'seek to undo themselves,' libertines" (Foucault 1967, 65), and he might also have mentioned 'mad' poets such as Fergusson in the Edinburgh City Bedlam. Furthermore, he points out that, notwithstanding how chaotic this assemblage may look to today's eyes, the Enlightenment mentality could readily discern a unity, a singularity, a coherence which roped together all of these unreasonable subcitizens as the negativity of an emerging 'reason' (especially Foucault 1967, 45).

One dimension of this coherence was idleness, a growing belief that unreason and idleness went hand-in-glove, and—while I should say more about this elision, and about the strains of a Marxian argument which Foucault constructs about a link here to the first stirrings of European capitalism[9]—the point is that he recovers the origins of a perceived association (between unreason and idleness) which in various guises has featured in European social thought (and social policy) ever since. More saliently for this chapter, it should be noted that the 'mad,' those people who might today be termed mentally disordered, diseased, sick, or ill, comprised only *part* of the sequestered population of people without reason (an important qualification given standard critiques of Foucault as a historian of psychiatry).[10] This being said, it appears that the Enlightenment did increasingly equate unreason and madness, deciding that the state of being without reason entailed being without the normal faculties of mind, thereby establishing another association which has subsequently entered centrally into the European *Zeitgeist*.[11] In addition, there is also the suggestion that the disordered character of madness per se, the strange antics of the 'raving lunatic,' quickly became recognized as a source of severe disruption to the ordered regimes of the sequestered spaces: so much so that, by the mid- to late-eighteenth century, a new impulse was agitating for mad people to be decanted from these spaces into specialist spaces earmarked for them alone, the lunatic asylums.[12] This is part of quite another story, however, and one that is peripheral to the central concerns of the present chapter.

For my purposes, the most significant aspect of Foucault's window on reason's capturing of unreason is its alertness to space: an alertness reflecting what is for many a thoroughly 'spatial history' permeating the whole of *Madness and Civilization* (and most of his other major works as well).[13] His account of "the great confinement" is shot through with references to the fact of social exclusion being at one and the same time spatial exclusion, and there is a prevailing image of the reaches beyond the gates of cities being reserved for people regarded as marginal, unworthy, and threatening: from medieval lepers sent to the leprosarium, to the Enlightenment's legions of the unreasonable sent to the general house of confinement, to the modern era's mad people sent to the asylum. A key passage runs as follows:

> Between labour and idleness in the Classical world ran a line of demarcation that replaced the exclusion of leprosy. The asylum was substituted for the lazar house, in the geography of haunted places as in the landscape of the moral universe. The old rites of excommunication were revived, but in the world of production and commerce. It was in these places of doomed and despised idleness, in this space invented by a society which had derived an ethical transcendence from the law of work, that madness would appear and soon expand until it had annexed them. A day was to come when it could possess these sterile reaches by a sort of very old and very dim right of inheritance. The nineteenth century would consent, would even insist that to the mad and to them alone be transferred these lands on which, a hundred and fifty years before, [people] had sought to pen the poor, the vagabond, the unemployed. (Foucault 1967, 57)

There are certainly polemical-poetical layers to this passage, and I would also argue that it accents an *imaginative geography*[14]—an imagining in "the landscape of the moral universe" whereby particular human groups came to be envisaged as properly positioned outside of the normal round—as much as it does a *material geography* of actual banishment to the exurban wastes. And yet I would insist that Foucault understands these spaces in more than just a metaphorical fashion, and I would agree with his assessment that they really existed in considerable number spread across the lands of Enlightenment Europe: not just in the shape of the Parisian *Hôpital Général* (founded in 1656 via an administrative reform creating a web of institutions for receiving the nonproductive poor of the French capital), but as a host of houses of confinement (poorhouses, workhouses, correction houses, pseudohospitals) which not uncommonly did retreat towards and beyond the physical boundaries of populous settlements.[15]

What is more, it is vital to consider the manner in which, as Foucault explains, unreason became "sequestered and in the fortresses of confinement bound to [r]eason and to the rules of morality, and to their monotonous rights" (Foucault 1967, 64). The implication is that, ironically enough, these institutionalized spaces of unreason were in practice subjected to the tyrannies of reason in a curiously intense fashion, even if at the same time they remained shaded and shunned, left apart from the spaces of knowledge, commerce, and culture which reason was much more proudly and ostentatiously patrolling elsewhere:

> A sensibility was born which had drawn a line and laid a cornerstone, and which chose—only to banish. The concrete space of Classical society reserved a neutral region, a blank page where the real life of the city was suspended; here, order no longer freely confronted disorder, reason no longer tried to make its own way among all that might evade or seek to deny it. Here reason reigned in the pure state, in a triumph arranged for it in advance over a frenzied unreason. (Foucault 1967, 64)

Thus, from an open encounter and dialogue across a multitude of medieval and early modern spaces, unreason became walled away in the "blank regions," fortresses or cathedrals of moral order[16] where it was effectively reduced to an unheard emptiness and closely regulated by the moral-intellectual precepts of reason. The spaces concerned were not just offshoots of attitudes and policies towards awkward people, but were themselves supports integral to the emergence and refining of this whole parcel of meaning through which the Enlightenment was building itself and its own determinedly 'reasonable' self-identity.

13.3. The 'Reasoning' of Enlightenment Edinburgh

I now wish to offer further thoughts on the geographies of reason and unreason, partly to substantiate aspects of Foucault's arguments about reason, unreason, and their spatialized encounter, but also to engage with more substantive evidence pointing to ways in which this encounter was effected in, through and across a particular series of situated spaces: namely, the 'high' and the 'low' sites of Enlightenment Edinburgh.

13.3.1. Building "the City of Reason"

The Scottish Enlightenment was not solely an Edinburgh occurrence, since other cities such as Glasgow, Aberdeen, and St. Andrews figured heavily in the intellectual and practical advances (Campbell and Skinner 1982; Carter and Pittock 1987; Hook and Sher 1995; Wood 1988), and the agricultural improvements also regarded as part of this 'national' move to Enlightenment obviously diffused into the Scottish countryside (Withers 1988).[17] Even so, historians acknowledge that the Scottish Enlightenment was predominantly town-based, and that "the movement in its heyday was centered on Edinburgh" (Daiches 1986, 3):

> It was an urban movement and its intimacy was prompted, and its progress facilitated, by the forms of social and intellectual expression that towns and urban living encouraged. That the milieu of the Scottish Enlightenment lay in the universities and the legal and ecclesiastical professions not only emphasises its urban-ness, but also underlines the importance of Edinburgh over the other university towns. (Chitnis 1976, 4–5)

Moreover:

> Edinburgh was where the General Assembly of the Kirk was held and was the pivot of the Scottish legal system, where the courts were housed. The city also maintained the leading Scottish university and medical school of the day. Hence were congregated in the city particular professions, all of which had a university education in common. Shared intellectual interests and pursuits led to the formation of learned societies in eighteenth-century Edinburgh where divines, doctors, advocates and professors met for mutual cerebral stimulation. As they were similar kinds of people, they also enjoyed relaxing together and formed many a social club, alongside more serious corporate organisations. (Chitnis 1976, 36–37)

A. C. Chitnis and others expand on these claims by charting how the "aristocratic world" of Scotland's landowners linked into the orbits of the Kirk, law, education, and entertainment to secure "a small and coherent world, one which provided the encouragement and the security necessary for the writers of the Enlightenment" (Rendall 1978, 17–18). Reference is duly made to a diversity of societies and clubs for the intellectuals and *literati,* examples being the Philosophical Society, the Royal Society of Edinburgh, the Musical Society, the Select Society and the Cape Club, and occasionally notes are appended about the coincidence of these "social spaces" with physical meeting places in particular Edinburgh buildings.[18] Attention is also paid to the presence of libraries, museums, and even botanical gardens as conducive (and on occasion academically stimulating) sites crucial to Edinburgh's status as a foremost Enlightenment "city of learning" (Withers 1995a, 1995b). Moreover, the lens is sometimes turned on the city's medical establishment, and on the involvement of medical figures in the founding and running of hospitals, dispensaries, asylums, and other facilities, many of which— through being frequented by the sick poor—comprised unusual points of connection between 'high' Edinburgh society and its 'low' counterpart. Such a connection was manifested when Dr. Duncan visited Fergusson in the city Bedlam, and something similar arose whenever a well-known doctor such as William Cullen—whose approach to "medicine promoted ideas that the Enlightenment generated" (McGirr 1990, 11)—practiced in the hospitals as both a service to the ordinary citizens of the city and an opportunity for teaching its 'enlightened' citizens. A different point of connection is hinted at in Chitnis's statement that Edinburgh's "Old Town taverns became favourite haunts and meeting places" (Chitnis 1976, 37), and it is appropriate to consider how the networks of intellectuals, writers, and others—including Fergusson—spilled out into sites such as bawdy public houses in the chaos of the old city streets (see below), and in the process embraced and constituted an encounter between reason and unreason quite unlike that performed in the hospitals.[19]

Edinburgh's spaces of reason can begin to be mapped out, then, but there are further dimensions to the 'reasoning' of this eighteenth-century city which must also be considered. David Meakin repeats the claim that the *philosophes* of the Enlightenment "demolished the Heavenly City of St. Augustine only to build it with more up-to-date materials," suggesting that "[n]ew materials, however, mean a new city" (Meakin 1994, 5), but it is uncertain to what extent it can be claimed that the Enlightenment did produce a city—perhaps a more "secular city"—in its own image. Nonetheless, there is much in the well-known development of eighteenth-century Edinburgh which suggests the creation of a distinctive urban form, leading the historian A. J. Youngson (1967) expressly to term it "the city of reason" and the geographer J. Wreford Watson to declare that "Georgian Edinburgh was an outcome of the Age of Reason" (Watson 1989, 99: see also Reed 1982). Edinburgh in the first half of the century was actually a messy place, mixing up peoples, neighborhoods, occupations, and activities quite haphazardly, and the city's so-called "Old Town" could only have satisfied those with a preference for the "uneven" and the "irregular":

> This led to quite idiosyncratic arrangements of space: the castle mart was distinctly separate from, and immediately below, the castle; the cathedral market flowed all around it and, in the form of notorious 'luckenbooths,' actually impinged upon it; the palace market was its front yard. Each market had quite irregular entries and gates; each was made up of temporary booths or permanent buildings of quite irregular width, height, depth and appearance . . . . Ten-storey tenements were not uncommon, and the back gardens,

or burgage lots of all the town's burghers, became filled in with building until the alley-ways wound through one back courtyard after another, like a rabbit warren . . . . There was no space left for children to play in, except the narrow alleys and back yards: green and open space had been virtually abolished. The space left to women, even for hanging out clothes to dry, was next to nothing: clothes were hung out on poles extended from windows high above a street or a yard. Space for burning rubbish or disposing of garbage was not thought of. (Watson 1989, 98–99)

A commentary of 1752 spelled out at some length the difficulties of the Old Town, noting the "steepness," "narrowness," and "dirtiness" of an urban area hemmed in behind the ancient walls, and also how the presence of houses crowded together more so than in any other European city led to "a great want of free air, light, cleanliness, and every other comfortable accommodation" (in Youngson 1966, 6). The city was also compared unfavorably to London, with particular mention made of Edinburgh citizens throwing "foul water, filth, dirt and other nastiness" out of their windows, and of "urban refuse, butchers' offal and other town guano" flowing down the hill on which the city stood.

The transformation of the city was anticipated in a key pamphlet (the source of the above-mentioned criticisms) prepared by a body called the "Convention of Royal Burghs," made up of powerful commercial interests, and this document—the *Proposals for Carrying On Certain Public Works in the City of Edinburgh*—called for an enlarging of the wealthier residential areas, thus adding a "New Town" to the existing city layout, coupled to the "adorning" of the city with a fresh array of public buildings (such as an Exchange, law courts and a theater).[20] Chitnis (1976, especially 21–28) stresses the economic motivations propelling these proposals, the desire on the part of the powerful burgesses to create a more profitable commercial environment in which they could operate, and he thereby sees it as little more than a coincidence that the Scottish Enlightenment and Edinburgh's urban improvement were contemporaneous. David Daiches disagrees, however, arguing that the New Town "was planned to achieve in architecture and in the use of space the ideals of order, elegance, rationality, progress and proper social relationships represented by the Scottish Enlightenment" (Daiches 1986, 21). Watson extends this argument, referencing the relevance to the burgesses of Adam Smith's arguments about maximizing the "*economic* use of land" (Watson 1989, 99–100), and also positing the centrality of an approach to aesthetics through which Smith displayed his love for geometry and concern for the "systematical beauty" generated by "regularity, balance, conformity and uniformity" (Watson 1989, 100):[21]

> This strong prejudice among the leaders of the Enlightenment led to a very geometric layout of Edinburgh New Town in rectangles and squares with broad boulevards and ending in fine views. There was no obstructive building widening or narrowing the streets, no squares with unseen and unanticipated openings, no houses jutting out into the street or leaning over it from above, no sudden changes in colour or texture of building, no mystical symbols projecting like gargoyles from building facades, no windows of uneven height placed at uneven levels—nothing in fact that typified the Old Town: only the new rationality expressed in the new uniformity. (Watson 1989, 101)

This passage indicates something of the character of the New Town as it appeared on the landscape downslope from the Old Town, following the drawing up of James Craig's plans—a simple effort involving two squares joined by a straight street and flanked by two others—and then by the start of building in 1767 when builders began to buy up plots and to put up structures (in line with strict regulations specified by the municipal authorities).[22] The resulting geometric ordering of this New Town can perhaps be described as a paramount geography of reason, a starkly 'reasoned geography,' and the extent to which it entailed 'enlightenment' flowing into darkness is captured in Watson's further claims about it being "a town of light and air" where "the skies came right into, and were part of, the New Town" (Watson 1989, 103).[23]

13.3.2. Shutting up the City's Most 'Unreasonable' People

A revealing observation is made by two scholars on the limitations of the Enlightenment's experiments with urban planning:

> These improvements hardly touched the poorer areas of the old towns. Enlightenment writers were aware of the misery of urban life for the majority of the population . . . , but it was not until the next century [the nineteenth] that the whole civil order was threatened by riot and pestilence [and] that any genuine attempt was made at urban renewal. (Samuels and Samuels 1991, 535)

Reflecting specifically on the Scottish experience, Thomas Markus suggests that at the heart of Enlightenment planning and architecture there was a "discourse of silence" which took "the form of almost total inactivity in consideration of working-class urban housing"

CHRIS PHILO

(Markus 1982a, 8). The impression is of "silent spaces" or "spaces of silence" (to extend Markus's vocabulary) comprised by vast swathes of poor-quality housing scrunched together in the midst of narrow snaking streets, dimly lit courts, craft quarters, market booths, animals, hardship, and filth. These were specific spaces of unreason identified in outline by the texts of the Enlightenment, but then left to their own devices in tandem with an Enlightenment mentality which positioned their residents as nothing: as incidentals to the bigger picture of societal achievement except on occasions where their unreasonable behaviors became an impediment to the accumulation of both capital and virtue. When Enlightenment thinkers such as David Hume referred to the "common people," they normally did so in a negative and prejudicial fashion, describing the multitudes as being innately "disorderly," "superstitious, undisciplined, seditious, and ignorant" (Chisick 1989, 12), and also regarding them as incapable of participating in serious debate, decision making or political activity. Indeed, in some of Hume's writings "the people are presented as subrational, as children who have not yet reached the age of reason, and as madmen [sic] who are without, or who have lost, their reason" (Chisick 1989, 16). Such a stance suggests an equation of unreason with madness in precisely the emerging fashion of the Classical Age, as argued by Foucault, and in so doing it conceptually dismissed the spaces occupied by the common people as ones where there was no value in even trying to diffuse "the truth as it was perceived by the enlightened" (Chisick 1989, 17).

Even so, some attempts were made, in Edinburgh as elsewhere, to intervene in the excesses of the city's most unreasonable people. The objective was to introduce some systematicity into the points of connection between reason and unreason, creating key sites in which the most problematic of unreason's populations could be captured by the forces of reason, and there confronted by a whole series of demands on their bodies, minds, and souls. Markus argues that in Enlightenment architecture matters of function (as bound up with those of form and space) started to be specified in "[t]exts which use[d] . . . words as prescriptions, that is as instructions to the designer before a building exists, [and which] are called briefs today" (Markus 1988, 173), and the impression is hence of 'briefed spaces' (to extent Markus's vocabulary) which involved the construction or adaptation of built structures intended to perform definite functions in reason's struggle against unreason. Markus himself provides an exhaustive survey of such 'briefed spaces' as installed in the built landscape by the later (Scottish) Enlightenment, and his studies dovetail with Foucault's claims about the Classical Age and its confining geographies.

Enlightenment Edinburgh was certainly not left behind in the move to create institutional spaces in which 'the unreasonable' might be incarcerated, and it appears that during the eighteenth century the city experienced the growth and refining of its penal-welfare provisions through which the most problematic of its poor residents were shut up, retrained, or 'repaired.' From the previous century the city inherited a house of correction, and it is intriguing to learn that as early as 1675 the town council "ordain[ed] the Town Treasurer to build some little houses for keeping mad people on the south side of the correction house" (Hay 1979, 170), these being the structures later to be identified as the 'Old Bedlam'.[24] In 1698 the town council ordered the building of a charity workhouse just to the west of the house of correction (Smith 1978, vol. 1, 188), which meant the creation of a distinctive institutional cluster on the south-western fringe of the Old Town where all manner of unreasonable people were being confined (albeit with an early example of the mad per se being separated out from other species of unreason, even if still being retained on the same smallish plot of urban land). There is some evidence that by the mid-eighteenth century "the city was constantly plagued by small bands of professional beggars coming in from their dens to the south and east" (Gilhooley 1988, ix), and perhaps there was a link here to the building in 1743 of a sizable new charity workhouse adjacent to the house of correction, in that the authorities were seeking to set aside the latter for the undeserving poor (including convicted vagrants) by relieving it of the deserving poor who could be siphoned off into the former. In fact, this facility was put up on the precise spot of the Old Bedlam, and the new City Bedlam (where Fergusson ended his days) was then constructed in 1746 as a 21-cell facility just to the south of where its predecessor had stood (Beveridge 1990, 319–20). The result was a substantial amalgam of institutional land uses in proximity to one another in the vicinity of Bristow Port (see fig. 13.1), reinforcing the sense of an overall space which—its own internal differentiations notwithstanding—clearly illustrates Foucault's arguments about how the Classical Age united its unreasonable elements both imaginatively and materially.

The city's 'proper' criminals were imprisoned elsewhere, though, and by the second half of the eighteenth century disquiet was being expressed about the state of their accommodation in the Tollbooth off the High Street in the heart of the city. This building was de-

CHRIS PHILO

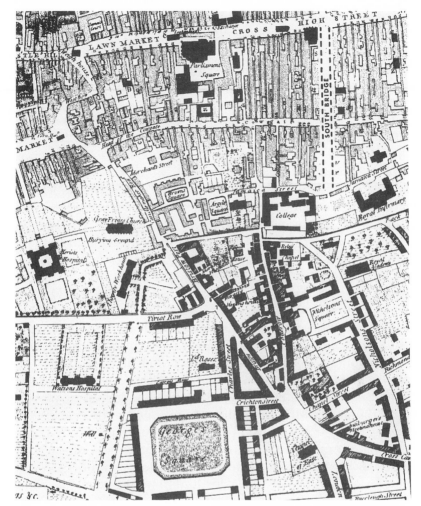

Fig. 13.1.   Map of central Edinburgh, circa 1780, showing part of the central Old Town just to the south of the New Town. Note the amalgam of institutional land uses present here, particularly adjacent to the Bristow Port thoroughfare, including the charity workhouse, two hospitals, and a burial ground. The new City Bedlam was probably sited behind the old city wall at the junction of Bristow Port and Teviot Row. Note too the location of the Royal Infirmary to the east of this institutional cluster.

nounced by a visitor in 1778, who found it to be squalid, filthy, insanitary, and "kept in a slovenly condition" (Daiches 1978, 170), while "[o]ne commentator at the time drew the comparison between prisoners and farm animals, and concluded that horses and pigs were better off" (Nimmo 1991, 59). In 1782 a plan was published by two Edinburgh worthies for a new composite prison and bridewell erected on the same site,[25] and in 1791 an Act of Parliament was passed for the building of such a facility, although it was not until 1795 that an unusual semicircular structure of some size appeared —not on the same site but on the brow of Calton Hill—to be the new City Bridewell[26] (and this structure was joined to the west after 1815 by the new jail, which was eventually referred to as the "West Division").

Enlightenment Edinburgh saw the emergence of several other institutionalized spaces of unreason in the sense meant here. One scholar records how "[t]he plight of orphans and the children of destitute parents . . . inspired some of the earliest charitable societies and institutions in and around Edinburgh," partly out of benevolence but also out of "a hope that habits of industry and the inculcating of Christian principles would save many of the young people from the pauper's frequent solution to the prospect of salvation—a life of petty crime" (Anon. n.d., 1). Towards the close of the century, concern about the state of the City Bedlam prompted agitation by the likes of Dr. Duncan for a proper asylum to house local pauper lunatics, and there is a story to tell about the founding in 1813 of the Royal Edinburgh or Morningside Lunatic Asylum, although I tend (with Foucault) to see this separation of madness per se from the wider universe of unreason as more a product of the modern era than of the Age of Reason. What Edinburgh also experienced in this period was the opening of medical facilities, and most notably the Royal Infirmary, which was founded in 1738 and completed in 1748 (Risse 1986, 25–29). Located in the southeastern part of the city (see fig. 13.1), surrounded by gardens and comprising a definite landmark "resembl[ing] large aristocratic townhouses or hotels" (Risse 1986, 30: see also fig. 13.2), this institution would seem to be better cast as a prominent space of reason (one frequented by leading medical figures associated with the Scottish Enlightenment) than as a space of unreason, but, in practice, it straddled both of these possibilities. It was a four-storey building, and on the ground floor there were "[t]welve separated cells . . . [which] were initially designed to shelter an equal number of 'lunaticks,' upper- or middle-class individuals judged to be 'mad'" (Risse 1986, 31).[27] On the top floor there was "the great central operating theatre" and spectating area for over two hundred students, all covered over by a dome or "cupola," and its designer—William Adam—intended this domed theatre to double as a "chapel" or even as an "astronomical observatory" (Markus 1982b, 34).[28] In other words, in the darkened recesses of the institution's ground floor were to be found the downcast representatives of unreason, while in the lighted rooms of the top floor was to

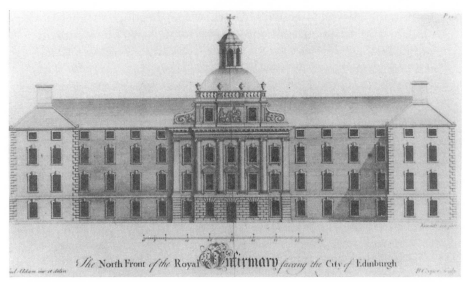

Fig. 13.2.   The front elevation of the Edinburgh Royal Infirmary.

be found a shrine to both the secular achievements of reason and the upwards gaze of those seeking knowledge: a dramatic expression perhaps of reason's capturing, setting apart and effective negation of unreason at this time.

13.4. "The Geometric Spirit" and the Triumph of Reason

The front cover of the Tavistock English version of Foucault's *Madness and Civilization* illustrates the theme of reason capturing unreason, in that here the depiction of a hand taken from one of Escher's etchings cradles an extract from Goya's *The Madhouse* painting showing a wildly staring naked madman. The latter conveys a sense of the passions unleashed, of the frenzies and terrors of a human being descending into irrationality, and the appearance is organic with blurry edges and splodges of color. The former, in contrast, conveys a sense of calm reflection, of the contemplative moment of a human being elevated into rationality, and the appearance is almost mechanistic with clear edges, sharp definition, and a subtle use of black-and-white shadings. The latter also suggests the disarray of a human world without reason, one in which disorder, indiscriminate mixings, and sociospatial chaos reign; whereas the former suggests the organization of a human world with reason, one in which order, strict classifications, and sociospatial purity reign. The latter hints at the crazy, fused geographies of unreason arguably typical of pre-Enlightenment European societies; whereas the former hints at the sane, dissected geographies of reason running through the thought-and-action of Enlightenment (and post-Enlightenment) European societies (see also Sack 1980). And the narrative of *Madness and Civilization* tells of how the latter replaces, takes over, starts to command, and reduces to nothingness the former; indeed, it tells of how the psychological, cultural, and sociospatial imperatives of reason triumph over unreason. Intriguingly, Markus also uses illustrations—Piranesi's *Carceri* series of etchings from the 1740s—to develop a similar line of argument about the role of Enlightenment architecture in allowing reason to capture unreason:

> Here there are vast subterranean spaces—ambiguous, paradoxical and dynamic—with staircases leading to nowhere, impossible perspectives and unfinished vaults. All the categories of classical form and space are dissolved. Above the ground, glimpses can be caught of a light, orderly, upper world, obeying all the rules of the Academies and Schools. It seems likely that Piranesi saw that world of reason, light and order as the real prison, which sits on the hidden, dark, disorganised, unclassified, creative forces of human nature and society. (Markus 1989, 106)

> The dark underworld of impossible perspective, unbounded space, undefinable functions, unruly forms and primitive technology, is weighed down from above by a world of light, reason, classical order, spatial division, clearly stated function and rational technology. The contrast could not be more extreme. That upper world, of the rule-governed architecture of power, is the true prison. The late-Enlightenment exploited this architecture to the full. (Markus 1988, 223)

> The erection of a superstructure of order, light and reason over a secret dark and chaotic netherworld never had such a good architectural exponent as Piranesi . . . . Piranesi saw that all formal architecture was, potentially at least, alienating in this way—its organised forms and static divisions of space denying, through the power of light, reason and order, the force of individual freedom, feeling, and germination in the darkness of a deeper order. Thus the *Carceri* are not *eccentric* architecture but the *essence* of architecture itself; a representation of a contradiction—it is the light of order and not the darkness of chaos which is truly imprisoning. (Markus 1982a, 106–107)

There are undoubtedly objections to be raised to these claims of both Foucault and Markus, but it is also evident that exploring their arguments through the sorts of visual images referenced here vividly brings out what might be termed the 'spatialities'[29] of reason, unreason and their encounter.

In her survey of what the Abbé de Condillac contributed to the French Enlightenment, Isabel Knight stresses what she terms "the geometric spirit" and observes that 'geometry' became "a talisman for the Age of Reason" (Knight 1968, 19). Condillac himself supposed that "one could reason in metaphysics and morals with as much precision as in geometry; that one could form ideas as exact as the geometers, and, like them, give precise and constant meanings of expression" (in Knight 1968, 15). Knight thereby summarizes as follows the inflection that Condillac gave to both the battle against the inexactitudes of seventeenth-century metaphysics and Enlightenment borrowings from the emerging quantitative-empirical sciences:

> The world, even as science from Galileo to Newton had declared, is mathematical. But, for Condillac this has a special meaning. The world is not operationally mathematical; that is, it is not matter-in-motion to be measured and quantified. It is qualitatively mathematical, to be understood as a structure of logical identities, pure, harmonious, symmetrical—the perfect expression of a geometer's God. (Knight 1968, 299)

It was this geometric impulse which permeated the whole Enlightenment, a will to sort out everything (ideas, things, people) into clearly delimited categories (with secure, stable, repeatable meanings and identities attached to them) which could be precisely laid out—discretely set apart, not overlapping or mixed up together—both in the imaginative geographies of the human mind and in the material geographies of nations, cities, and institutions. Several different academic geographers have considered this geometric impulse, to an extent even tracing its current incarnation in and beyond academic geography to the installation of Enlightenment reason in the academies of Europe. They have argued that the invisible lines structuring conventional "cartographies of thought" are inevitably ones which map out (and then police) a range of exclusions (either/or relations) and hierarchies ('higher' and 'lower') in the imagination (and see Heffernan, Bravo, and Edney here for parallels), and then just as inevitably inform the practices of powerful societal decision makers in their programs of state building, urban planning, and institutional provision.[30]

This is also very much the project of Markus, and his texts thereby detail at length the connections between the ordering propensities of Enlightenment thought and those actions of public bodies which began to press upon the disorder of the masses (and which would increasingly do so in both nineteenth-century Europe and its overseas possessions). His prime focus is obviously architectural, but this does not prevent him speculating about the Enlightenment "meta-orders" or common "compositional, geometric, and stylistic principles" (Markus 1982a, 7) which underlay both the design of new carceral-welfare institutions and the planning of new urban areas (notably Edinburgh New Town). This chapter has sought to consider both institutions and public spaces, and to offer a 'geographical' gloss on Markus's architectural concerns by charting the diverse tracts and sites integral to the geographies of reason, unreason, and their encounter, noting both the broader picture and the specificities of the Edinburgh experience in this respect. Not all of the chapter fits into one overall argument—it suggests, for instance, that some urban spaces of unreason were left largely untouched by the Enlightenment—but the prevailing angle is nonetheless one following a loosely Foucauldian line on how reason began to produce geographies in its own 'geometric' image by shaping either gridded tracts of reason (such as Edinburgh New Town) or bounded sites of unreason (such as the Edinburgh City Bedlam where Fergusson spent his last days).

Notes

Particular thanks are due to David Livingstone, Charlie Withers, and two readers for their encouragement and invaluable advice. An earlier version of this paper was presented at the Geography and Enlightenment conference which preceded this volume, and other versions have since been presented at seminars in the Departments of Geography at University College London, the University of Dundee and the University of St. Andrews. Thanks to all of the people who commented generously on these occasions, and would that I had been able to incorporate all of your excellent suggestions. Thanks as well to Kay Anderson, Hester Parr, and David Sibley for the nice words.

1. See Daiches (1978, chaps. 8 and 11), who discusses Edinburgh as "the Athens of the North" and also uses Fergusson's poetry to frame a discussion of "social life" in eighteenth-century Edinburgh.

2. The evidence and claims relevant to Fergusson's condition are assessed in Beveridge (1990). See also Smith (1978, Vol. 1, 183–192).

3. The description quoted here is from Robert Louis Stevenson.

4. Known as "Lang Sandy Wood," he is reported to have visited his patients accompanied by a pet sheep and a raven (Beveridge 1990, 317).

5. The edition being used here is the English translation, initially published in 1965, of the original French text (Foucault 1961). Much debate surrounds what has been 'lost' between the French original and the English translation: see, e.g., Gordon (1992).

6. There is a whole industry of commentary on *Madness and Civilization*

CHRIS PHILO

as a contribution to psychiatric history, much of it critical, a taste of which can be obtained from Still and Velody (1992). For very different kinds of assessment, compare relevant passages of Sedgewick (1982) with Jones and Fowles (1984, chap. 2).

7. In the course of an extended critique of *Histoire de la Folie,* Derrida attacks Foucault's recourse to 'metaphysical' notions, and in so doing objects to Foucault's positing of two great oppositional entities (Reason and Unreason or Madness) seemingly fully present in struggle throughout history: for a translation of this essay, see Derrida (1981), and see also the translated response in Foucault (1974). In a later text, Foucault criticized his earlier work for "admitting an anonymous and general subject of history" (Foucault 1972, 16), by which he meant the 'referent' of Unreason or Madness, and he suggested instead that "we are not trying to reconstitute what madness itself might be, in the form in which it first presented itself to some primitive, fundamental, deaf, scarcely articulated experience" (Foucault 1972, 47). Nonetheless, he accepted—in a footnote on the same page— that just such an effort of reconstitution was implied by the original Preface, which presumably is why he suppressed it in the second French edition.

8. In this chapter I will not capitalize first letters of such terms, except when directly quoting from or paraphrasing Foucault, and I thereby indicate my wish to skirt around the grander metaphysical and phenomenological trappings of *Madness and Civilization.* The possibility that my approach continues to tremble on the brink of accepting such trappings is, however, something of which I am acutely aware.

9. Much has been made of Foucault's remarks about the policy of setting idle people to work as "one of the answers [that] the seventeenth century gave to an economic crisis that affected the entire Western world: reduction of wages, unemployment, scarcity of coin" (Foucault 1967, 49). One reading of these remarks has enabled some writers, including geographers, to propose that Foucault basically sets his whole inquiry into the origins of the asylum within the framework of Marxian political economy: for critical comments, see Philo (1997b, 80).

10. A standard criticism is simply that Europe never experienced the mass incarceration of mentally disordered people during the seventeenth and eighteenth centuries which Foucault apparently describes in *Madness and Civilization,* but it must be remembered that Foucault recognizes that mad people per se were only incidentally swept up in the broader impulse to shut away all manifestations of unreason (as it was then being defined, identified, and captured). He is not envisaging any systematic and all-encompassing confinement of people suffering mental health problems at this time, even if from my own research and secondary reading (notably of Hay 1979) I reckon that there *was* more sociospatial segregation of mad people in nonspecialist institutions than a scholar such as Porter (1987, chaps. 1 and 3) allows. It is also important to emphasize that—as is typical of Foucault—his claims about "the great confinement" are at least as much about the *idea* of such confinement as a preferred society-wide response to unreason, as they are about the actual practice of such incarceration (which Foucault fully expects to lag unevenly behind the widespread circulation of the idea).

11. In *Madness and Civilization* Foucault tends to use the capitalized first-letter terms of Unreason and Madness interchangeably, although in the detail of his text more subtle discriminations are clearly introduced, and it is evident that he regards the Age of Enlightenment as when European society first solidified a conceptual elision of the two terms. This point is nicely explored by Radden, whose reading of Foucault for guidance on the history and philosophy of the "insanity defence" leads her to consider this elision at some length. She concludes that "[t]he Age of Reason, during which madness was unreason, was one in which the concept of reason dominated, explained and shaped all understanding" (Radden 1985, 46), meaning that for contemporaries "[m]adness was a failure of reason within the person" (Radden 1985, 48).

12. This stressing of how the call for specialist asylums to house the mad per se grew out of the depths of a nonspecialist confinement, one unable to cope with true madness in its midst, is a key—but too rarely acknowledged—component of the overall narrative in *Madness and Civilization* (1967, especially chap. VIII). Again, from my own research and secondary reading I feel that Foucault's claims here do indeed have considerable empirical veracity.

13. For general appreciations of Foucault's spatialization of history, see Philo (1992a) and also Soja (1989, chap. 1). Much of Philo (1992b) is concerned with exploring further the 'spatial history' present, if sometimes underplayed, in *Madness and Civilization*.

14. To borrow Said's famous phrase in yet another context; see Said (1978, e.g., 54–55 and 71–72).

15. Foucault's own discussion and marshalling of evidence in this connection appears in Foucault (1967, chap. II). There is a tangled mass of scholarship relevant to the question of just how extensive was the provision of such carceral institutions: think of Sellin's (1944) research on the Amsterdam houses of correction and their Europe-wide "imitators" dating from the sixteenth and seventeenth centuries; the Webbs' (1927) research on the geographical coverage in England of the county house of correction or bridewell system; or Adams's (1990) measured assessment of Foucault's claims in the course of demonstrating the widespread establishing of *dépôts de mendicité* in Enlightenment France. See also Mellosi and Pavarini (1981). See also note 10.

16. These are all terms used by Foucault when discussing houses of confinement.

17. When considering the geography of the Scottish Enlightenment, mention must be made of the obvious orientation toward the central Lowland belt and of the relative lack of involvement from the Highlands (even though some agricultural improvements did diffuse into, and certainly plans for improvements did encompass, the Highland regions). It is worth noting one historian's observation that "[o]utside of Edinburgh geography was in no way responsible for the intimacy of the central, lowland area in which the Scottish Enlightenment thrived" (Chitnis 1976, 36). Here Chitnis is contrasting the spatial propinquity of like-minded people made possible by the compact 'geography' of Edinburgh with the aspatial 'common interests' which united like-minded members of institutions and societies.

CHRIS PHILO

However, and as Chitnis's own remarks also indicate, matters of space, context, and physical copresencing of individuals were not irrelevant to the workings of institutions and societies: and so geography *was* still a decisive influence on the course of the Scottish Enlightenment beyond the simple fact of crowding people together in Edinburgh.

18. Underlying my observations here are several different lines of thinking. First, there are the persuasive claims of Livingstone (1994, 1995; see also Thrift 1985) about the irreducible situatedness of debates, learning, experiments, field studies, and the like in very specific time-space contexts, and about how the material and intellectual resources made available by these contexts serve to shape the knowledges, attitudes, and practices generated from within their bounds. A flavor of what might be said in this respect by Enlightenment scholars is suggested by the entries on subjects such as universities, salons, coffeehouses and clubs in Yolton et al. (1991). Second, there is the suggestion prompted by Habermas that the Enlightenment witnessed the flowering of a "new public world" or "oppositional bourgeois sphere" "created by new systems of communication in eighteenth-century cities" (Cohen 1991, 92), which then broadens what needs to be said about the social inclusions and exclusions (perhaps along axes of class and gender) that marked sites such as universities, salons, and so on (and which arguably entered into the grain of Enlightenment thinking in ways which still resonate today).

19. In such sites the 'high' and the 'low' seemingly mixed together, and one writer suggests that—while it would be wrong to suppose that there were not certain social distinctions between the likes of Fergusson and the main figures of 'improving' Edinburgh—"[i]t is true that social lines were not clearly drawn in eighteenth-century Edinburgh when it came to matters of conviviality in taverns, and some very high-placed people could be found enjoying plebian food and drink in some fairly low places of public resort" (Daiches 1978, 140). Interestingly, Daiches speaks of the 'Jekyll-and-Hyde character' of Enlightenment Edinburgh, observing that it "was a city of contrast: New Town and Old Town; elegance and filth; humanity and cruelty" (Daiches 1978, 170); and this is a peculiarly appropriate observation given that Robert Louis Stevenson, the author of the Jekyll and Hyde story, set his tale in a 'London' which in many ways was modeled on the polarised character of his home city, Edinburgh. Thanks to Jacquie Burgess for pointing out this connection.

20. Much of the text of this document is reproduced in Youngson (1966, 3–12), and see also Daiches (1978, chap. 7) and Youngson (1967, 16–18).

21. Adam Smith was a key figure of the Scottish Enlightenment, being associated with the universities in both Glasgow and Edinburgh. Watson also remarks on how David Hume, perhaps *the* key figure in this respect, "spoke of geometry as beauty" and rejected "the fuzzy, the irregular, the wayward, the mystical and the idiosyncratic" (Watson 1989, 101).

22. Details about Craig's plans, its influences, the municipal regulations, the activities of the builders and the like are recounted in Youngson (1966, especially chap. IV). It is important not to overstress the geometric purity of the New Town, nor to forget that people from the Old Town might end up working (as servants, porters, and so on) in the New Town and maybe even

living in poor-quality accommodation behind the impressive facades of the New Town houses. Thanks to Elspeth Graham for these qualifications.

23. More generally, it is supposed that, while an "urban design language" of street and square predated the eighteenth century, under the sign of the Enlightenment there arose "a new awareness of dynamic possibilities for juxtaposing sequences of spaces" (Samuels and Samuels 1991, 534), which simultaneously urged uniformity (similar building heights, consistent frontage lines, and the like). One historian argues that in many Enlightenment cities, "town space is laid out as essentially administrable, hinting at the perfect geometry of Utopia" (Meakin 1994, 6, discussing Andries 1994), while Andries (1994) also stresses how in such cities "rational control" was exercised through the categorizing procedures beloved of the *philosophes*. In addition, I have found persuasive Porter's (1994) claims about the new 'urbanity' associated with both the public spaces and the new town houses of Enlightenment London.

24. The use of the term 'Bedlam' to describe a building in which mentally disordered people were confined derived from Bedlam as the corruption of Bethlem or Bethlehem Hospital, the infamous charitable lunatic asylum in London which had become known for performing this particular function as early as the fifteenth and sixteenth centuries.

25. This proposal is considered at length in Markus (1982b, 65–66), who reckons it to have been remarkable for looking to enforce clear classification, separation, and "centrical" surveillance of inmates in a manner anticipating the specifications of Jeremy Bentham's "Panopticon." Markus's overall argument is that the late-Enlightenment gave rise to a whole new architectural order in the "briefed spaces" of diverse institutions, particularly carceral and welfare ones, and that this entailed the propelling of a geometric precision through these spaces so as to bring out "the essential ingredients of order: visibility, light, classification, unity and clear, stable boundaries" (Markus 1982b, 26–27).

26. This structure was designed by Robert Adam and constructed after he died in 1792 by his brother, James, and it betrays the influence of Bentham, with whom Robert was in correspondence. This being said, "[i]t was a serious disappointment to Bentham that the only British prison design which came anywhere near a strict interpretation of the Panopticon was, in its execution, a departure from it" (Markus 1982b, 82).

27. By 1795 it appears as if the managers had discontinued this practice, however, and had seemingly decided that mentally disordered patients would be better treated elsewhere (Risse 1986, 107). This being said, Risse's extensive analysis of the hospital's General Register of Patients, 1770–1800, suggests that a not insignificant number of patients suffered from 'neurological and mental diseases' (Risse 1986, 153–157).

28. As Markus also notes, "[t]his combination of practical, clinical function, religious worship and cosmic science was a peculiarly Scottish Enlightenment response by Adam to the established European practice of locating a church, chapel or altar at the centre" (Markus 1982b, 34).

29. There are various meanings attached to the term 'spatiality' by human geographers, but all of them "refer to the human and social implications of

space," and thereby "are united in their opposition to the conventional separations between 'space' and 'society'" (Gregory 1984, 582–584).

30. These themes recur throughout the work of Olsson; see Olsson (1980, 1991; see also Philo 1994; Reichert 1992). Related themes figure in the work of Sibley, although he is more concerned with how 'boundaries' drawn in the minds of the powerful translate into 'boundaries' constructed on the ground which exclude the powerless (all manner of marginal, minority, and outsider peoples): see Sibley (1981a, 1995; see also Philo 1986). Interestingly, following Sibley's publication of a short piece on the ordering obsessions of spatial science (Sibley 1981b), Olsson wrote to Sibley, sending copies of his own papers on boundaries and "suggest[ing] that boundaries were a taboo subject in geography" (Sibley 1986, n.p.). Arguing through a feminist critique of the "masculinism" endemic to much academic knowledge production, Rose offers a further account of the conventional geometries which organize how geographers and others appropriate the world—confining these appropriations to an objectivist, rigid sense of how people interact with space, thus neglecting the roles of imagination, emotion, desire, fear, and the like in these interactions—and which pivot around specific a priori 'maps' or 'geographies' of how and what knowledge is to be sought: see Rose (1993, 1995a, 1995b).

References

Adams, T. M. 1990. *Bureaucrats and Beggars: French Social Policy in the Age of Enlightenment.* Oxford: Oxford University Press.

Andries, L. 1994. "Paris et l'imaginaire de la Ville dans les Almanaches Francais du Xviiie Siècle," in *The Secular City: Studies in the Enlightenment,* eds. T. D. Hemming et al. Exeter: University of Exeter Press, pp. 12–26.

Anon. n.d. *The Other Georgian Edinburgh: Working Life in the City.* Edinburgh: Scottish Records Office.

Beveridge, A. W. 1990. "Edinburgh's Poet Laureate: Robert Fergusson's Illness Reconsidered." *History of Psychiatry* 1: 309–329.

Campbell, R. H., and A. S. Skinner, eds. 1982. *The Origins and Nature of the Scottish Enlightenment.* Edinburgh: John Donald.

Carter, J. J., and J. H. Pittock, eds. 1987. *Aberdeen and the Enlightenment: Proceedings of a Conference.* Aberdeen: Aberdeen University Press.

Chisick, H. 1989. "David Hume and the Common People," in *The 'Science of Man' in the Scottish Enlightenment: Hume, Reid and their Contemporaries,* ed. P. Jones. Edinburgh: Edinburgh University Press, pp. 5–32.

Chitnis, A. C. 1976. *The Scottish Enlightenment: A Social History.* London: Croom Helm.

Cohen, E. 1991. "Clubs," in *The Blackwell Companion to the Enlightenment,* ed. J. W. Yolton et al. Oxford: Basil Blackwell, pp. 92–93.

Daiches, D. 1978. *Edinburgh.* London: Hamish Hamilton.

———. 1986. *The Scottish Enlightenment: An Introduction.* Edinburgh: Saltire Pamphlets.

Derrida, J. 1981. "Cogito and the History of Madness," in *Writing and Difference.* London: Routledge, pp. 31–63.

Foucault, M. 1961. *Folie et Deraison: Histoire de la Folie a l'Age Classique.* Paris: Plon.

———. 1967. *Madness and Civilization: A History of Insanity in the Age of Reason.* London: Tavistock.

———. 1972. *The Archaeology of Knowledge.* London: Tavistock.

———. 1974. "My Body, this Paper, this Fire." *Oxford Literary Review* 4: 9–28.

Gilhooley, J. 1988. *A Directory of Edinburgh in 1752.* Edinburgh: Edinburgh University Press.

Gordon, C. 1992. "*Histoire de la Folie:* An Unknown Book by Michel Foucault," in *Rewriting the History of Madness: Studies in Foucault's Histoire de la Folie,* eds. A. Still and I. Velody. London: Routledge, pp. 19–42.

Gregory, D. 1994. "Spatiality," in *The Dictionary of Human Geography* (3d ed.), eds. R. J. Johnston et al. Blackwell: Oxford, pp. 582–585.

Hay, M. G. 1979. *Understanding Madness: Some Approaches to Mental Illness, circa 1650–1800.* Unpublished Ph.D. thesis, University of York, Department of Sociology.

Hook, A., and R. B. Sher, eds. 1995. *The Glasgow Enlightenment.* East Linton, East Lothian: Tuckwell Press.

Jones, K., and A. J. Fowles. 1984. *Ideas on Institutions: Analysing the Literature on Long-Term Care.* London: Routledge & Kegan Paul.

Knight, I. 1968. *The Geometric Spirit: The Abbé de Condillac and the French Enlightenment.* New Haven: Yale University Press.

Livingstone, D. N. 1994. "Science and Religion: Foreword to the Historical Geography of an Encounter." *Journal of Historical Geography* 20: 367–383.

———. 1995. "The Spaces of Knowledge: Contributions towards an Historical Geography of Science." *Environment and Planning D: Society and Space* 13: 5–34.

Markus, T. A. 1982a. "Introduction," in *Order in Space and Society: Architectural Form and Its Context in the Scottish Enlightenment,* ed. T. A. Markus. Edinburgh: Mainstream, pp. 1–23.

———. 1982b. "Buildings for the Sad, the Bad and the Mad in Urban Scotland, 1780–1830," in *Order in Space and Society: Architectural Form and Its Context in the Scottish Enlightenment,* ed. T. A. Markus. Edinburgh: Mainstream, pp. 25–114.

———. 1988. "Buildings and the Ordering of Minds and Bodies," in *Philosophy and Science in the Scottish Enlightenment,* ed. P. Jones. Edinburgh: John Donald, pp. 169–224.

———. 1989. "Class and Classification in the Buildings of the Late-Scottish Enlightenment," in *Improvement and Enlightenment: Proceedings of the Scottish Historical Studies Seminar,* ed. T. M. Devine. Edinburgh: John Donald, pp. 78–107.

McGirr, E. M. 1990. *Cullen in Context.* Stevenson Lectures in Citizenship, University of Glasgow.

Meakin, D. 1994. "Topographies of the Secular City," in *The Secular City: Studies in the Enlightenment,* eds. T. D. Hemming et al. Exeter: University of Exeter Press, pp. 3–11.

Mellosi, D., and M. Pavarini, 1981. *The Prison and the Factory: Origins of the Penitentiary System.* London: Macmillan.

CHRIS PHILO

Nimmo, I. 1991. *Edinburgh the New Town*. Edinburgh: John Donald.

Olsson, G. 1980. *Birds in Egg/Eggs in Bird*. London: Pion, London.

———. 1991. *Lines of Power/Limits of Language*. Minneapolis: University of Minnesota Press.

Philo, C. 1986. *'The Same and the Other': On Geographies, Madness and Outsiders*. Loughborough University of Technology, Department of Geography, Occasional Paper No. 11.

———. 1992a. "Foucault's Geography." *Environment and Planning D: Society and Space* 11: 137–161.

———. 1992b. *The Space Reserved for Insanity: Studies in the Historical Geography of the 'Mad-Business' in England and Wales*. Unpublished Ph.D. thesis, University of Cambridge, Department of Geography.

———. 1994. "Escaping Flatland: A Book Review Essay Inspired by Gunnar Olsson's *Lines of Power/Limits of Language*." *Environment and Planning D: Society and Space* 12: 229–252.

———. 1997a. "The 'Chaotic Spaces' of Medieval Madness: Thoughts on the English and Welsh Experience," in *Nature and Society in Historical Context*, ed. M. Teich et al. Cambridge: Cambridge University Press, pp. 51–90.

———. 1997b. "Across the Water: Reviewing Geographical Studies of Asylums and other Mental Health Facilities," *Health and Place* 3: 73–89.

Porter, R. 1987. *Mind-Forg'd Manacles: A History of Madness in England from the Restoration to the Regency*. London: The Athlone Press.

———. 1990. *The Enlightenment*. London: Macmillan, London.

———. 1994. "Enlightenment London and Urbanity," in *The Secular City: Studies in the Enlightenment*, eds. T. D. Hemming et al. Exeter: University of Exeter Press, pp. 27–41.

Radden, J. 1985. *Madness and Reason*. London: George Allen & Unwin.

Reed, P. 1982. "Form and Content: A Study of Georgian Edinburgh," in *Order in Space and Society: Architectural Form and Its Context in the Scottish Enlightenment*, ed. T. A. Markus. Edinburgh: Mainstream, pp. 115–145.

Reichert, D. 1992. "On Boundaries." *Environment and Planning D: Society and Space* 10: 87–98.

Rendall, J. 1978. *The Origins of the Scottish Enlightenment*. London: Macmillan.

Risse, G. B. 1986. *Hospital Life in Enlightenment Scotland: Care and Teaching at the Royal Infirmary of Edinburgh*. Cambridge: Cambridge University Press.

Rose, G. 1993. *Feminism and Geography: The Limits of Geographical Knowledge*. Cambridge: Polity Press.

———. 1995a. "Distance, Surface, Elsewhere: A Feminist Critique of the Space of Phallocentric Self/Knowledge." *Environment and Planning D: Society and Space* 13: 761–781.

———. 1995b. "Tradition and Paternity: Same Difference?" *Transactions of the Institute of British Geographers* 20(ns): 414–416.

Sack, R. D. 1980. *Conceptions of Space in Social Thought*. London: Macmillan.

Said, E. 1978. *Orientalism*. London: Routledge & Kegan Paul.

Samuels, I., and O. V. Samuels. 1991. "Urban Planning," in *The Blackwell Companion to the Enlightenment*. ed. J. W. Yolton et al. Oxford: Basil Blackwell, pp. 534–535.

Sedgewick, P. 1982. *Psycho Politics*. London: Pluto.

Sellin, T. 1944. *Pioneering in Penology: The Amsterdam Houses of Correction in the Sixteenth and Seventeenth Centuries.* Philadelphia: University of Philadelphia Press.

Sibley, D. 1981a. *Outsiders in Urban Societies.* Oxford: Basil Blackwell.

———. 1981b. "The Notion of Order in Spatial Analysis." *Professional Geographer* 33: 1–5.

———. 1986. Personal communication, 15 May 1986.

———. 1995. *Geographies of Exclusion: Society and Difference in the West.* London: Routledge.

Smith, C. J. 1978. *Historic South Edinburgh* (2 vols.). Edinburgh: Charles Steilton.

Soja, E. W. 1989. *Postmodern Geographies: The Reassertion of Space in Critical Social Theory.* London: Verso.

Still, A., and I. Velody, eds. 1992. *Rewriting the History of Madness: Studies in Foucault's Histoire de la Folie.* London: Routledge.

Thrift, N. 1985. "Flies and Germs: A Geography of Knowledge, in D. Gregory and J. Urry, *Social Relations and Spatial Structures.* London: Macmillan, pp. 366–403.

Watson, J. W. 1989. "People, Prejudice and Place," in *The Behavioural Environment: Essays in Reflection, Application and Re-evaluation,* ed. F. W. Boal and D. N. Livingstone. London: Routledge, pp. 93–110.

Webb, S., and B. Webb. 1927. *English Local Government: English Poor Law History, Part 1: The Old Poor Law.* London: Longmans, Green & Co.

Withers, C. W. J. 1988. "Improvement and Enlightenment: Agriculture and Natural History in the Work of the Rev. Dr. John Walker (1731–1803)," in *Philosophy and Science in the Scottish Enlightenment,* ed. P. Jones. Edinburgh: John Donald, pp. 102–116.

———. 1995a. "Geography, Natural History and the Eighteenth-Century Enlightenment: Putting the World in Place." *History Workshop Journal* 39: 136–163.

———. 1995b. "How Scotland Came to Know Itself: Geography, National Identity and the Making of a Nation, 1680–1790." *Journal of Historical Geography* 21: 371–397.

Wood, P. 1988. "Science and the Aberdeen Enlightenment," in *Philosophy and Science in the Scottish Enlightenment,* ed. P. Jones. Edinburgh: John Donald, pp. 39–68.

Yolton, J. W., et al, eds. 1991. *The Blackwell Companion to the Enlightenment.* Oxford: Basil Blackwell.

Youngson, A. J. 1966. *The Making of Classical Edinburgh, 1750–1840.* Edinburgh: Edinburgh University Press.

———. 1967. "The City of Reason and Nature," in *Edinburgh in the Age of Reason: A Commemoration,* ed. D. Young. Edinburgh: Edinburgh University Press, pp. 12–20.

Chapter Fourteen

Lisbon 1755: Enlightenment, Catastrophe, and Communication[1]

PETER GOULD

14.1. The Event

At some time between 9:30 and 9:40 in the morning local Lisbon time—and recall that we are in a world knowing only 'local time'—on All Saint's Day, November 1, 1755, the western edge of a thinning European plate released a surge of pent-up energy (Hoernle, Zhang, and Graham 1995), probably in an area about 36°N, 10°W, reported to this day as a zone of almost continual seismic activity lying about 300 kilometers southwest of Cape St. Vincent (Udias, Lopez Arroyo, and Mezcua 1976). It was one of the largest earthquakes in recorded human history (Gutdeutsch, Grünthal, and Musson 1992; Reid 1914), estimated by Richter in 1954, on his scale, by then carefully modified by hundreds of empirical observations, as 8.75 to 9.00 (Gutenberg and Richter 1954; Richter 1935). It is thought, for reasons of material failure, that shocks over 9.5 are unlikely, while magnitudes of 10 or more would be felt sensibly over the entire earth. Of course, there was no Richter, or the more recent seismic scales around in those days, but it is worth noting that even these numerical estimates on a logarithmic scale are still translated today into levels of sensible movement and degrees of building destruction to make them humanly meaningful (Andrews 1963; Verney 1979). Although taking into account various construction methods, these modern mappings from numbers to words follow essentially a classification used in a massive study commanded by the King of Portugal, who required carefully assessed reports from all parts of his kingdom after the event (Pereira de Sousa 1919–32). A similar massive and detailed report was compiled by the Royal Academy of History of Spain (Real Academia de la Historia 1756).

The effects of the earthquake were observed all over Europe, although one might approach such accounts cautiously because of the temptation to apocryphal recall. But there are so many accounts, so well-documented days and weeks before the news of Lisbon's destruction could have arrived, that they can be discarded only with great caution. A letter from Cork, Ireland, almost the same longitude as Lisbon, records "at thirty six minutes after nine o'clock, a very sensible shock" (*Western Flying Post,* November 17, 1755), while a similar letter from Kinsale relates that "sudden and surprising fluxes and refluxes of the sea continued from three o'clock in the afternoon until ten at night," with tidal waves strong enough to tear a 60 ton sloop from its moorings, while "fishing boats were likewise whirled about like so many corks" (*Western Flying Post,* December 1, 1755). Similar accounts of tidal surges come from Swansea, St. Ives, and Hayle; terribly agitated waters were recorded at Leyden, Amsterdam, and other Dutch provincial towns; while Edinburgh received "advice from the west that Loch Lomond was agitated in a very surprising manner" (*Western Flying Post,* December 1, 1755). Two women rowing across a lake in Sweden observed that the driftwood on the bottom was suddenly swirling around (Kjellén 1909), and the shock itself was felt at "Gluckstadt and other parts of Holstein" (*Western Flying Post,* December 15, 1755). But there are literally scores of accounts from all over Europe, the most common being reports of water in wells suddenly rising and falling (Mucciarelli and Albarello 1991). Allowing for longitudinal, and so time differences, all reports record almost identical times, indicating that the shallow seismic waves traveled very rapidly.

The enormous energy released initially, perhaps 2×10^{27} ergs[2], and that resulting from a series of almost equally severe aftershocks, devastated many of the great buildings of Lisbon, a sparkling jewel of Europe, which, with its commercial, colonial, and dynastic ties was not then in its present peripheral location in either geographic or other human spaces. I can do no better than quote excerpts from a letter written by a survivor, Lawrence Fowke, to his brother in London, on November 17, 1755 (Fowke 1755):[3]

> . . . I ran, they followed; houses and streets seemingly dancing under us . . . but it pleased God to carry me and Joze Alvez safe to the arch; but poor Francisco, being last, perished on the way . . . 'tis surprizeing how I was tossed about under the arch for the space of two minutes, when the motion ceased; and a little after, the dust settling, we saw a little light . . . we missed your son Harry . . . [but] calling out, he answered from the ruins, in which he was buried to the head . . . we took him out, wounded in the legs, his body and face bruised, but all his bones hole, thank God . . . I was then on Mr. Morrogh's house, there being no more streets—houses and

PETER GOULD

streets being all leveled; We all embraced, congratulated, &c., which was followed by the lamentation of Mr. Morrogh, for his wife and younge child . . . in the midst of which came on another—the second terrible earthquake. Our fear was then that our house would fall on us, which we saw bending like the mast of a ship in a storm . . . I immediately called to all our family to follow me . . . to St. Nicholas's church; but the horror there threw them into confusion, and, I believe, was the occasion of their scattering, numbers expiring, and other shocking spectacles, with the clergy running about over the ruins to confess and absolve them who were yet alive—all shouting to God for mercy. We went . . . through Rua das Arcos, to Rosia, and saved some poor people in our way; but when we gote there, scenes of horror were doubled, and I can compare it to nothing but the idea I had formed in my youth of miserable sinners at the last day crying to God for mercy, to which you must add the numberless objects around us expireing with groans and misery . . . we then proceeded to the fields, and on the hills behind the Puppet showhouse, we had a third violent earthquake, in which we were oblidged to lie down or kneel, not being able to keep our feet . . . In the afternoon I sent [Neb] and two boys to see if they could save anything; but all the passages were stopped by the fire, which began to rage, and at night we saw the whole city in flames . . . The King's Palace, Custom House, churches, are all consumed. The Court, the populace of the city, nuns, fryers, all lie out in the fields; and God knows what will become of us, as we still have one or two earthquakes every twenty-four hours . . . 'tis not to be express'd the misery the populace suffer by cold, hunger, and rain in the fields . . . the king seems determined to build Lisbon again in the same place; and the militia are now clearing the streets, but this must be a work of time.

Lawrence Fowke notes three enormous jolts that first and terrible day, with large aftershocks continuing for more than two weeks. What he does not record from the hills of Saceavem overlooking Lisbon, where he and the remains of his household had taken refuge, is that associated with each huge shock was a *tsunami,* whose force was probably focused by the shape of the coast framing the mouth of the Tagus estuary, one of the finest harbors in Europe. About an hour later, most of the people who had taken refuge along the waterfront with its open wharves and squares were drowned in the swirling mass of ships torn from their moorings or crushed by the piles of goods on quays or in warehouses.[4] These waves would be recorded about five or six hours later in England, Wales, and Ireland, and in the West Indies about nine hours later. In the ruined city, as Fowke records, fires had broken out everywhere, and many trapped in rubble were burned to death. The best estimates we have are that 10,000–15,000 people perished in a city of roughly a quarter of a million.[5]

The clergy displayed enormous courage in moving through the carnage to give absolution to the dying. As Kendrick (1956, 36) notes, "of all the horrors of the Lisbon earthquake, no horror was worse than the supreme horror of dying unconfessed and unforgiven."

Recall that this was one of the richest cities of Europe, not only in merchandise and bullion, but in the hundreds of priceless works of art, and tens of thousands of books, manuscripts, and maps, nearly all of which were destroyed. But when a number of major works have described the destruction and rebuilding of Lisbon (de Matos Sequeira 1916–34; 1939–41; Vieira da Silva, 1940–41), no summary here can serve, and we must leave the event itself and turn to the manner in which the news traveled.

14.2. The News of the Event

Simply tracing the news of the event, when it arrived, and the immediate evidence of its impact, has proved much more difficult than anticipated. Many potential sources remain untapped, not the least the surviving records of the great banking houses of Europe, some of which had their own, and highly efficient, courier services (Heuberger 1994). Standard sources are remarkable for the meagerness of their information. It is a source of great puzzlement that this devastating environmental event, right at the center of the century of Enlightenment, so often leaves hardly a trace in many standard works by biographers and historians. In Peter Gay's (1977) marvelous two-volume work on the Enlightenment, his long and detailed bibliographical essay (an essay that has guided my footsteps so often) mentions the earthquake only in connection with Voltaire's turning away from optimism. The same is true for his collection of essays (Gay 1971); Furbank (1992) does not mention it; while Broc (1972) mentions it only in passing, again in connection with Voltaire's *Candide*. If Voltaire (1756) had not written his celebrated *Poème sur le désastre de Lisbonne,* one wonders whether the event might have disappeared from human memory altogether. To paraphrase Pierre Nora (1989, 7–8), the terror of Lisbon's earthquake has been lost to the dialectic of remembering and forgetting, just a part of the slippage into a historical past that is gone for good.

What we have, in the news of this catastrophic environmental event, is a 'tracer,' analogous to a short-lived radioactive injection in the human body that discloses paths and blockages of circulation. In another, mathematical language, it is a traffic transmitted on a backcloth, whose transmission might tell us something about those mid-eighteenth-century structures of communication (Atkin 1974). Using tracers to disclose the impact of technological structures is not a new idea: isochronic maps, and map sequences indicating the course and intensification of epidemics, are well known (Gould 1993; Pyle 1969; Pyle and Patterson 1983). But here my focus is not upon the physiological impact of disease, but the intellectual impact upon people standing squarely in the century of Enlightenment.[6]

It goes without saying that the arrival of the news was greeted with consternation everywhere. How it was interpreted we shall come to shortly. Obviously, in referring to the "*Lisbon* earthquake," the city was not the only casualty, for coastal towns of Morocco (Levret 1991) and Spanish towns on the Atlantic, like Cadiz and Huelva (Solares, Lopez Arroyo, and Mezcua 1979), sustained severe damage from both shocks and waves. Oporto also felt the shocks, and large merchantmen lying outside the bar were carried clear over it and into the river by the waves, "to the Astonishment of the poor Sailors and Pilots, who thought of nothing but immediate Death" (*Dublin Gazette* 549, 1755, 1). But such were communications that it would take days, weeks, and even months, for such items of news to be exchanged. Several ships sailing near the presumed epicenter felt shocks so violent in a calm sea that boats and a compass were unshipped, and the glass of cabins shattered, but upon landfall, for example in Cork, people had no inkling of the Lisbon tragedy, and these things were simply reported as strange marine incidents. Ships sailing from other ports, for example from Gibraltar on or about November 17, arriving in Falmouth seven days later, "did not know of the dismal Affair at Lisbon which is strong Proof that the Shipping there must have suffered greatly: and, besides, we don't hear of any Ship being sailed from Lisbon to any Place since the Misfortune" (*Dublin Gazette* 550, 1755, 1). In fair and favorable weather, we would expect news to travel faster by sea than on land, and, indeed, at Malta, Captain Hervey of the Royal Navy received the news of Lisbon's destruction on November 9 (Eskine 1953), so one or two ships must have escaped the general destruction suffered by the enormous number at anchor or tied up along the wharves, including the newly arrived Dutch convoy on its annual voyage to the East Indies, and the Brazil fleet about to be loaded (*Dublin Gazette* 550, 1755, 1). But the weather was not kind everywhere: a boat carrying the first news to "the Downs" off the east coast of Kent was delayed by gales and had to take refuge at Guernsey, so delaying the arrival of news to London directly by sea.

Meanwhile, the King of Portugal had sent an express messenger with a letter to the King of Spain, leaving Lisbon on November the 4th and arriving at Madrid on the 8th. To anyone who has driven this distance on a modern highway, this is an astonishing time.[7] Equally astonishing is the speed with which a letter, sent also on November 4 by the French ambassador at Lisbon, reached the court at Versailles on the 18th (Besterman 1957, 156). Six days later, November 24, letters from Geneva are already 'rebroadcasting' the terrible news, not the least by Voltaire to his friend Jean Robert Tronchin in Lyon (ibid. 157). From Paris, clearly a major 'information node,' the news is dispatched to London, presumably before the publication of the weekly *Gazette de France* on November 22. The first published announcement appears in the *London Gazette* of November 25, but the news must have arrived a day or two before and seems to be the result of several sources arriving almost simultaneously: letters and dispatches from Bordeaux, sent on November 4; from Malaga, sent November 8; from the British ambassador in Madrid, November 10 (Kendrick 1956, 142), and a curious note from Escurial, Spain, dated November 3, which notes that "the 1 ft Instant, about Ten o'Clock in the Morning, we felt here an Earthquake, that lasted near four Minutes, but without doing any Damage" (*Dublin Gazette* 549, 1755, 1). From London, the *Gazette* carries the news to Dublin, where its *Gazette* in turn announces the catastrophe on November 29, to be picked up by the *Cork Journal* on December 18. Long before this, the news has reached Hamburg by letter from Paris by November 28 (Tiderman 1991), presumably by boat, since Utrecht records it only on December 2, one assumes by messenger over land. From Hamburg it arrives at Stockholm to be published in the *Post-Tidningar* of December 8, and the same day it is carried to Uppsala "by a rider on a foaming horse" (Frängsmyr 1994). This happens to be the same date it was published in the *Boston Gazette* and *Boston Evening Post;* while the *New York Mercury* reported the news eleven days later on December 19. The *South Carolina Gazette* only published the news on February 5, 1756 (Carozzi 1983), yet another indication of the importance of coastal ties in North America, particularly in the winter season.

To speak of lacunae in any isochronic map at this point would be laughable: what we have at the moment are little more than a few 'spot heights' to direct our thinking towards any 'communication space' in the mid-eighteenth century. But even this spatially spotty record makes us realize that any isochronic surface, interpolated from many more 'temporal anchors,' would only represent an imperfect snapshot of a highly dynamic surface whose swirling isolines

PETER GOULD

would be controlled not simply by such physical factors as wind, rain, terrain, and season, but such human aspects as road conditions, public safety, political conflicts, and, not least, the wealth and social station of those corresponding. To talk of temporal variance around some mean isochronic surface may appear pretentious, and yet we have indications that for long periods of time such surfaces and structured spaces exhibited great overall stability, judged by the repetitive course of innovative traffic moving over these communication backcloths (Hägerstrand 1966). One certainly has a sense of major, arterial connections—Lisbon, Madrid, Paris, London, and so on—and an equal sense that once off these major alignments, news traveled very much more slowly. It seems worthwhile to continue the search for other temporal spot heights, for this crucial aspect of eighteenth-century human geography appears to have been hardly explored. Like Rupke's much fuller account of Humboldt's work (chap. 11), it should be possible to conceive of this as a geography of knowledge reception in the Enlightenment. But what of the impact of the news, apart from the initial and humane reactions of horror, disbelief, and pity? How was the news greeted?

14.3. Reaction

It is when we turn to the reaction to the news of Lisbon's virtual destruction that all the superficial clichés about Enlightenment appear risible. The fact is that Europe in the century of Enlightenment (and clearly in the centuries both before and after) is so complex, so 'contradictory,' that ideologues of almost any persuasion can find what they are looking for to support their current cause. Science versus Religion, Faith versus Reason . . . all the favorite opposites are there and at the same time, paradoxically, not there; not there as oppositions, but as compatible combinations, existing amicably side by side, as well as in conflict. If you approach Enlightenment from a particular perspective, from a 'situatedness' two and a half centuries later, but always one of your own choosing, you find what you want to find, you read what you want to read. But for me, Enlightenment is a multidimensional space, neither linear nor orthogonal, so that differences can be close together, and similarities far apart, depending upon where you want to locate yourself in the hyperspace of a later modernity containing it and how you are prepared to squint down those twisting dimensions. We need animated, spatiotemporal maps to limn the geographies of Enlightenment: with considerable license, one can imagine, over a dark gray Europe, a rosy glow appearing in Edinburgh, spreading, all too slowly, to the south, jumping the channel to the Low Countries, haven for so many exiles and home to publishers of so many books still forbidden in the dark gray areas, although France is definitely getting brighter. I am, of course, describing, very roughly, the spread of free masonry, so brilliantly delineated by Margaret Jacob (1991) in her study of early freemasonry in Europe. Like the Portuguese Monseigneur, who noted that a distance of ten leagues from Lisbon took you back one hundred years in time, or the French philosopher Duclos (Gay 1977, Vol. 2, 4), who wrote "Those who live a hundred miles from the capital [Paris?] are a century away from it in their modes of thinking and acting" (Duclos 1750), it is tempting to say that Scotland in 1755 is Spain in 1855. But this would be a gross, arrogant, and unjust comparison, because if you look closely enough there are already points of light all over Spain which, in time, will coalesce to clear the darkness away.

We are in a European world where the god giving meaning is being displaced from the center and being replaced by the human as the measure of all things, positing increasingly the way the world *will* be, positing even the god being displaced. But this is no new and sudden thing, and some of the roots go back to Aquinas and the intensive discussions of beauty in the Paris of the twelfth and thirteenth centuries (Eco 1986). Once you allow that allegorical truth emerges from custom, not God's word, once you raise the question that educated sensibility judges what is beautiful, you have arrived at "beauty lies in the eye of the beholder," a relative, not an absolute, and therefore not a god-given, truth. Five hundred years later, after a Renaissance, after the appearance and translation of Greek and Arabic texts into readily accessible Latin, Enlightenment becomes a distinct brightening, but a faint glow was already there.

It was, of course, a field day for clergy and theologians of almost all persuasions, an occasion to point to the Sodom and Gomorrah that Lisbon had been, and how God in his wrath had brought punishment upon the wicked. A pamphlet by John Wesley (1755), a tract positively reveling in horror, sold like hotcakes, but Anglican churchmen like the Bishop of Exeter noted that, while some national improvement would be highly desirable, hysterical panic and "wild stories about Christ's immediate advent" (Kendrick 1956, 159) were not helpful. An aging Archbishop of Canterbury would agree that "A presumptuous forewardness in pronouncing on extraordinary events we leave to raving designing monks, methodists and ignorant enthusiasts" (Kendrick 1956, 161).[8] There is a large genre of

PETER GOULD

"earthquake sermons," from Europe and North America, itself an indication of the impact of the event, but we must remember that many of the clergy were themselves among the most literate and best-educated men of their day. Charles Darwin was still a century away, and few found any intellectual conflict with their belief in the Christian God, and their interest in, and sometimes a passion for, physical science, including that of the earth itself.

Parallel to, and sometimes intertwined with the sermons preached, was a large body of scientific writing on the causes of earthquakes, indicative of a thriving "natural philosophy" all over educated Europe. The news of the Lisbon earthquake certainly catalyzed scientific interest in these extraordinary natural events, but five years before the *Philosophical Transactions of the Royal Society, London* (1750, 601–750) included a 150-page appendix containing 57 communications on earthquakes, some of which related Benjamin Franklin's (1751) lectures of 1749 on electricity and lightening to what seemed to be the similar, and virtually instantaneous, propagation of earthquake shocks.

For many, this was already a seriously considered, not to say popular, theme, for the phenomenon of 'electricity' was all the rage in scientific circles, an interest that would have a sustained impact for early investigations into mountain building (Carozzi and Newman 1995). Indeed, Horace-Bénédict de Saussure, a pioneer in the geological sciences, noted the virtual simultaneity of Lisbon's earthquake and the time reported from Copenhagen in an account about how a sudden shock threw a ship's captain and crew to the deck of their vessel as they were loading it for the West Indies that same Sunday morning (Saussure 1784). Surely only electricity could travel that fast over such a distance. Certainly it was a phenomenon considered seriously by the 80-year-old Benedictine, Benito Jerónimo Feyzóo y Montenegro, perhaps Spain's most distinguished man of science, well known for his courageous and outspoken declamations against superstition and the neglect of modern scientific knowledge. We must recall that this was a time, nearly seven decades after Newton's *Principia*, when those responsible for "natural philosophy" at theological colleges in Spain were still refuting "the presumption of Copernicus, who had tried to halt the sun and move the earth" (Kendrick 1956, 67). Feyzóo also speculated that earthquakes were due to a contraction and cracking of the earth, which was not so far off in modern plate-tectonic terms. In contrast, Kant favored explosions in underground caverns, also a popular hypothesis, and one well argued, to account for the near simultaneity of these huge events. For we are in a world where wave propagation through solids was not yet a condition of possibility for thinking, so explosions are really all that are left to explain such things in natural terms (Oeser 1992, 27–30).

It is in this quasi-secular world, withdrawing from, but still attached to, even imbedded in, the reluctant theological matrix, that the natural explanations are most heated and needed. No one exemplifies such need better than Pombal, at the center of Lisbon's aftermath and reconstruction, a thoroughly practical despot who contained all the contradictions of Enlightenment (Maxwell 1995). His rough reply to the young King's despairing question of what to do, "to bury the dead and feed the living," may be apocryphal, but it catches completely the pragmatic voice of a man of action and science, one who could 'request' a Cardinal to order his priests to tone down their lamentations and dire warnings. What was needed was not a terrified and paralyzed population but people understanding that this was a purely natural event so they could bestir themselves to doing rather than trembling.

But disturbing questions are not easily dismissed: how could a loving God do this to His people, especially after His Son had taken all of the sins of the world upon Himself? Indeed, was there a god after all? The great natural philosopher and Swiss pastor, Elie Bertrand (1766), obviously did not consider the latter question, and acknowledged that we cannot answer the former, before spending 150 pages carefully considering the possible natural explanations. It is fair to say that this is a common Protestant response (Mitchell 1761; Stukeley 1756), and one with a considerable precedent (Van de Wetering 1982), but, by rationalizing natural cause and effect, you are slowly chipping away at the mystery of the god's presence, and, therefore, the need of the god as a source of mystery. Rousseau, responding to Voltaire's *Poème sur le désastre de Lisbonne,* points out, not terribly helpfully from his rustic retreat at L'Hermitage, that if people insist on living in cities, and building houses six or seven stories high, they have only themselves to blame, not God. "If the population had been dispersed over a wider area, and housed in different sorts of buildings, there would have been less damage, and perhaps none at all" (Cranston 1991, 30). But this I-told-you-so defense of the Christian Deity is precisely the catalyst for Voltaire's *Candide,* after which the whole Panglossian *tout est bien* school of Leibniz, Pope, and others will be laughed out of court across literary, educated Europe (Gay 1971, 125; Glacken 1967, 522).

In this respect, and carefully qualified, the earthquake and *Candide* become a turning point from optimism to pessimism in eighteenth-century thought. On both sides of the Atlantic, verses of lamentation were written within days of receipt of the news,[9] and it also became a source of catastrophic simile, even to our own times. One doctor of medicine, Walsh (1793, pp. 19–28) of Dublin, felt compelled, in a hyperbolic form now mercifully out of fashion, to note how "Huge TAGUS heav'd his panting bosom bare/Above his waves,"[10] and by 1858, Oliver Wendell Holmes (1908) was recalling "At half past nine by the meetin' house clock,/Just the hour of the Earthquake shock." But the earthquake and its effects are not over. Here, and closer to our own times, is the Swedish poet Hjalmar Gullberg (1959, 138):[11]

> Vad gjorde detta år hans prästerskap för miner
> när de I mässan drack hans vin och åt hans bröd?
> Nyheten spreds. Voltaire skrev på alexandriner
> Ett ode och en värld förstod att Gud var död.
>
> [What looks did his priests have on their faces that year
> when at mass they drank his wine and ate his bread?
> The news spread. Voltaire wrote an ode in alexandrines,
> And a world understood that God was dead.]

14.4. Epilogue

For all those who would use the title "student" properly as an honorific, let me say simply this: in the same way that there are many histories of Enlightenment, so there are many geographies of Enlightenment waiting "out there" to be written. For geographers, *The Past [may be] a Foreign Country* (Lowenthal 1985), but the "spatial domain" still seems to be a place where the thinking of historians, except for an occasional Braudel, feels a sense of tentativeness, even trespass. I feel we have to reach out, and perhaps one day join hands to write, possibly along the lines of Paul Carter's (1987, 1996) "spatial histories," more complete "spatiotemporal histogeographies," perhaps as a small token of thanks to all those historians who have enlightened us.

Notes

1. My interest in the Lisbon earthquake was stimulated many years ago by Thomas Kendrick (1956). His book, *The Lisbon Earthquake,* has been a constant guide and starting point for numerous lines of related inquiry, many

of which remain to be followed. Too often, and sometimes with a sense of petty chagrin, one finds one of Kendrick's "intellectual cairns" along a line of investigation one thought might have a touch of originality. But his are invigorating footsteps to follow.

2. I have used here Gutenberg's and Richter's equation for shallow shocks, $\text{Log } E^* = 19.4 + 0.9M$ (Gutenberg and Richter 1954, 19).

3. The account is listed by Kendrick (1956, 166) in his "Bibliographical note to Chapter II," but my copy was found for me by Mr. Richard Harrison of Bantry, County Cork, Ireland, who noted it was "received from a friend of mine and have no further detail" (letter dated 8.xii.94).

4. Not to mention huge stones from the strongly constructed buildings and warehouses destroyed by the *tsunami*. Reports from Cadiz, which also experienced terrible damage from the tidal waves, noted that although the force of the waves had been somewhat attenuated by the rocks of the shore, the force on the walls was sufficient to carry large building stones of 8 to 10 tons up to 40 or 50 yards from their locations (*Dublin Gazette* 550, 1755).

5. Initial reports of deaths were highly exaggerated, some recording up to 100,000 people. In Lisbon, corrections were forthcoming on a parish-by-parish basis by responsible clergy as early as the first few weeks. Kendrick (1956, 34), after weighing up various sources of evidence, says "What seems to be the best and most careful estimates agree that probably between 10,000 and 15,000 people lost their lives in Lisbon. . . ."

6. Upon careful reflection, I have decided to adopt Furbank's (1992, 450–451) 'complaint' regarding the use of the definite article, *The* Enlightenment, and the way it reifies 'enlightenment' into some *thing,* some sort of event with the implication of a definite beginning and end. In English, it appears to be a phrase first appearing in twentieth-century writing, although whether from a mistranslation of *Die Aufklärung* or not, as Furbank postulates, is difficult to say. By refusing the definite article, I feel that the whole spirit of 'Enlightenment' is kept open, and so becomes a continuing condition of possibility for our time. I am sure this is how Jacques Derrida (1995, 400, 427–428) feels about his frequent uses of *lumières* in the context of present and future oriented discussions.

7. To cover the modern 642 kilometers in 96 hours means traveling at an average of nearly 7 kilometers per hour. Allowing for rough roads and November conditions, let alone the necessity for some sleep, this is a remarkable testament to the fortitude of both rider and, presumably, horses.

8. This is attributed by Kendrick to an anonymous tract, *The Power of God over the Constitution of Nature.*

9. Poems, or perhaps better, verses, appeared almost immediately in English, French, Swedish, and Portuguese, and possibly other European languages. Few are worth repeating except for their role as historical documents.

10. As an example of late eighteenth-century "verbal orgy," I am grateful to Andrew Carpenter, University College Dublin, for sending me a copy of this verse from a surviving, but rare, copy of this doctor turned "poet."

11. I am grateful to Torsten Hägerstrand and Olof Wärneryd, Lund University, for sending me copies. My translation is a very slight modification of theirs.

PETER GOULD

References

Andrews, Allen. 1963. *Earthquake.* London: Angus and Robertson.

Atkin, Ronald. 1974. *Mathematical Structure in Human Affairs.* London: Heinemann Educational Books.

Bertrand, Elie. 1766. "Mémoires historiques et physiques sur les tremblements de terres," in *Recueil de divers traités sur l'histoire naturelle de la terre et des fossiles.* Avignon: Louis Chambeau: 227–379.

Besterman, Theodore. 1957. *Voltaire's Correspondence. Volume XXVIII, September–December 1755, Letters 5796–5996.* Genève: Institut et Musée Voltaire, Lettre 5932 Jean Louis Du Pan to Suzanne Catherine Freudenreich, and Lettre 5933 Voltaire to Jean Robert Tronchin.

Broc, Numa. 1972. *La Géographie des Philosophes: Géographes et Voyageurs Français au XVIIIᵉ Siècle.* Paris: Editions Ophrys.

Carozzi, Albert, and John Newman. 1995. "Saussure's Manuscript Oration on Earthquakes and Electricity (1784) Influenced by William Stukeley and Benjamin Franklin." *Archives des Sciences* 48: 209–237.

Carozzi, Marguerite. 1983. "Reaction of British Colonies in America to the 1755 Lisbon Earthquake: A Comparison to the European Response." *History of Ideology* 2: 17–27.

Carter, Paul. 1987. *The Road to Botany Bay.* Chicago: Chicago University Press.

———. 1996. *The Lie of the Land.* London: Faber and Faber.

Cranston, Maurice. 1991. *The Noble Savage: Jean-Jacques Rousseau 1754–1762.* Harmondsworth: Penguin Books.

de Matos Sequeira, Gustavo. 1916–1934. *Depois do Terremoto: Subsidios para a História dos Bairros Ocidentais de Lisboa, Vols. 1–4.* Lisbon: Academia das Ciências.

———. 1939–1941. *O Carmo e a Trindade.* Vols. 1–3. Lisbon: Academia das Ciências.

Derrida, Jacques. 1995. *Points . . . Interviews, 1974–1994.* Stanford: Stanford University Press.

Dublin Gazette, Extract of a letter from a Gentleman at Oporto to his Friend in London, dated November 2, 1755, 549, 1.

Dublin Gazette, 1755, 550, 1.

Duclos, Charles. 1750. *Considérations sur les Moeurs de ce Siècle.* Paris: Prault.

Eco, Umberto. 1986. *Art and Beauty in the Middle Ages.* New Haven: Yale University Press.

Erskine, David. 1953. *Augustus Hervey's Journal. Being and Intimate Account of the Life of a Captain in the Royal Navy Ashore and Afloat.* London: W. Kimber.

Fowke, Lawrence. 1755. Letter to his brother, November 17.

Frängsmyr, Thomas. 1994. Letter to Peter Gould, March 10.

Franklin, Benjamin. 1751. *Experiments and Observations on Electricity, Made at Philadelphia by Mr. Benjamin Franklin, and Communicated in Several Letters to Mr. P. Collinson of London.* London: E. Cave.

Furbank, Philip. 1992. *Diderot: A Critical Biography.* New York: Alfred Knopf.

Gay, Peter. 1971. *The Party of Humanity: Essays in the French Enlightenment.* New York: W. W. Norton.

———. 1977. *The Enlightenment: An Interpretation.* Vol. 1, *The Rise of Modern*

Paganism. Vol. 2, The Science of Freedom. New York: W. W. Norton.

Glacken, Clarence. 1967. *Traces on the Rhodian Shore: Nature and Culture in Western Thought from Ancient Times to the End of the Eighteenth Century*. Berkeley: University of California Press.

Gould, Peter. 1993. *The Slow Plague: A Geography of the AIDS Pandemic*. Oxford: Blackwell.

Gullberg, Hjalmar. 1959a. "Jordbävningen i Lissabon," *Ögon, Lappar: Dikter av Hjalmar Gullberg*. Stockholm: Nordstedts.

Gutdeutsch, Rudolph, Gottfried Grünthal, and Roger Musson, eds. 1992. *Historical Earthquakes in Central Europe*. Vienna: Abhandlungen der Geologischen Bundesanstalt.

Gutenberg, Beno, and Charles Richter. 1954. *Seismicity of the Earth and Associated Phenomena*. Princeton: Princeton University Press.

Hägerstrand, Torsten. 1966. "Aspects of the Spatial Structure of Social Communication and the Diffusion of Information." *Papers and Proceedings of the Regional Science Association* 16: 27–42.

Heuberger, Georg. 1994. *The Rothschilds: A European Family*. Rochester, NY: Boydell and Brewer.

Hoernle, Kaj, Yu-Sheng Zhang, and David Graham. 1995. "Seismic and Geochemical Evidence for Large-Scale Mantle Upwelling Beneath the Eastern Atlantic and Western and Central Europe." *Nature* 374: 34–39.

Holmes, Oliver. 1908. *The Complete Poetic Works of Oliver Wendell Holmes*. Boston: Houghton Mifflin.

Jacob, Margaret. 1991. *Living the Enlightenment: Freemasonry and Politics in Eighteenth Century Europe*. New York: Oxford University Press.

Kendrick, Thomas. 1956. *The Lisbon Earthquake*. London: Methuen.

Kjellén, Rudolph. 1910. "Sveriges Jordskalf: Försök till Seismisk Landsgeografi." *Göteborgs Högskolas Årsskrift* 15: 211.

Levret, A. 1991. "The Effects of the November 1, 1755, 'Lisbon' Earthquake in Morrocco." *Tectonophysics* 193: 83–94.

Lowenthal, David. 1985. *The Past is a Foreign Country*. Cambridge: Cambridge University Press.

Maxwell, Kenneth. 1995. *Pombal: Paradox of the Enlightenment*. Cambridge: Cambridge University Press.

Mitchell, John. 1761. "Conjectures Concerning the Cause, and Observations Upon the Phenomena of Earthquakes." *Philosophical Transactions of the Royal Society* 51: 566–634.

Mucciarelli, Marco, and Dario Albarello. 1991. "The Use of Historical Data in Earthquake Prediction: An Example from Water Level Variations and Seismicity." *Tectonophysics* 193: 247–251.

Nora, Pierre. 1989. "Between Memory and History: Les Lieux de Mémoire." *Representations* 26: 7–24.

Noticia que de la Real Academia de la Historia del Terremonto de Noviembre de 1755. Madrid, 1756.

Oeser, Erhard. 1992. "Historical Earthquake Theories from Aristotle to Kant," in Gutdeusch, Grünthal, and Musson, 11–31.

Periera de Sousa, Francisco. 1919–1932. *O Terremoto do 1° de Novembro de 1755 em Portugal e um Estudo Demografico*. Vol. 1: *Faro, Bejae Evora [1919]*. Vol. 2: *Santarem e Portalegre [1919]*. Vol. 3: *Lisboa [1928]*. Vol. 4: *Leira*

Castelo Branco Coimbra Guarda Aveiro e Vizeu [1932]. Lisbon: Serviços Geologicos, Tipografia do Comercio.

Philosophical Transactions of the Royal Society, London. 1750. 46: 601–750.

Pyle, Gerald. 1969. "Duffusion of Cholera in the United States." *Geographical Analysis* 1: 59–75.

Pyle, Gerald, and Keith Patterson. 1984. "Influenza Diffusion in European History: Patterns and Paradigms." *Ecology of Diseases* 2: 173–184.

Reid, Harold. 1914. "The Lisbon Earthquake of November 1, 1755." *Bulletin of the Seismological Society of America* 4: 53–80.

Richter, Charles. 1935. "An Instrumental Earthquake Magnitude Scale." *Bulletin of the Seismological Society of America* 25: 1–32.

Saussure, Horace-Bénédict. 1784. *Oratio 1784. Discours aux Promotions,* designated as *Oration on Earthquakes and Electricity.* Geneva: Archives Saussure, *mss.* 59/13.

Solares, J. Martinez, A. Lopez Arroyo, and J. Mezcua. 1979. "Isoseismal Map of the 1755 Lisbon Earthquake Obtained from Spanish Data." *Tectonophysics* 53: 301–313.

Stukeley, William. 1756. *The Philosophy of Earthquakes. Natural and Religious, or, An Inquiry into Their Cause and Purpose* (3d ed., with Part III added). London: C. Corbet.

Tiderman, Karl. 1991. "Katastrofen, Tidningen, Människan, Tanken: Idéhistoriska Betraktelser över Stockholms Post-Tidningars Rapportering om Jordbävningen i Lissabon, 1755–56." Uppsala: Uppsala Universitet, Institutionen för Idé- och Lärdomshistoria (uppsats) 1–8.

Udias, A., A. Lopez Arroyo, and J. Mezcua. 1976. "Seismotectonic of the Azores-Alboran Region." *Tectonophysics* 31: 257–289.

Van de Wetering, Maxine. 1982. "Moralizing in Puritan Natural Science: Mysteriousness in Earthquake Sermons." *Journal of the History of Ideas* 43: 417–438.

Verney, Peter. 1979. *The Earthquake Handbook.* London: Paddington Press.

Vieira da Silva, Augusto. 1940–41. *As Muralhas da Ribeira de Lisboa.* Vols. 1–2. Lisbon: C.M.L.

Voltaire. 1756. "Poème sur le désastre de Lisbonne," in *Mélanges.* Paris: Bibliothèque de la Pléiade, 310–309.

Walsh, E. 1793. "The Shining Guinea," *Bagatelles or Political Sketches.* Dublin: N. Kelly.

Wesley, John. 1755. *Serious Thoughts Occasioned by the Late Earthquake at Lisbon.* London (pamphlet).

Western Flying Post, or The Sherborne and Yeovil Mercury, Extract of a letter from Kinsale, dated November 2, 17 November 1755.

Western Flying Post, or The Sherborne and Yeovil Mercury, Extract of a letter from Cork, dated November 3, 17 November 1755.

Western Flying Post, or The Sherborne and Yeovil Mercury, Extract of a letter from Edinburgh, 1 December 1755.

Western Flying Post, or The Sherborne and Yeovil Mercury, Extract of a letter from Hamburg, 15 December 1755.

Afterword

Roy Porter

Until quite recently, it used to be almost a reflex action for geographers to denigrate their Enlightenment forebears, dismissing them either as dry-as-dust regurgitators of facts, or as perpetuators of ancient fantasies—as, for example, the idea of *terra australis*. In the old authorized version of the subject's development, the eighteenth century was largely consigned to 'prehistory'; it is only with the nineteenth century—with Romanticism, with Alexander von Humboldt and Carl Ritter—that modern geography is generally reckoned to have found its true principles and place in the sun (Bowen 1981). Such a view was, perhaps ironically, given a new lease of life by Michel Foucault's postulation of a radical epistemological break around 1800, when the old grid of representations characteristic of the static classical knowledge order was suddenly and mysteriously superseded by new organic, temporalized discourses in such fields as philology, biology, and political economy—and, by inference and extension, in geography as well (Foucault 1970, 1989; see discussion in Gutting 1989).

This standard vision of geography's past, which has been dominant since the middle of the twentieth century, awarded high priority to the new paradigms formulated by Alexander von Humboldt. It was also preoccupied with the emergence in the early nineteenth century of the new research-oriented university—in large measure, in Germany at least, the achievement of his brother, Wilhelm. A revitalized academe provided a cradle for the growth of the infant discipline of geography. This mode of writing geography's history was given further authority by impressive work in the sociology of the professions, and it became colored from the 1960s by the high profile accorded to the flourishing discipline of the history of science.

A salient feature of the present book, by contrast, is that it transcends such discipline- and profession-oriented views of geography's development and destiny. Repudiating that particular teleology, the contributors examine the geographies of the Enlightenment era with a fresh gaze, challenging the perspectives and preconceptions of the last century and a half (Cosgrove and Daniels 1988; Driver and Rose 1992; Ogborn 1998; Pile and Thrift 1995). To say this is not, of course, to imply that the approaches here in evidence are somehow value-free, magically cleansed of all presentist concerns. Far from it, for the rediscovery of Enlightenment geography being conducted in this volume clearly mirrors and boosts the geographical renaissance taking place nowadays in Britain and North America—a geography fascinated by space as such, by geographies of the mind, and by attempts to reestablish the subject within the humanities. Today's is a geography more intrigued by the philosophical and the discursive than by the evolution of proper academic structures and institutions. In that light, the geographies expressed up in the writings of such *gens de lettres* as Montesquieu or the Abbé Volney matter at least as much to the contributors to this book as do the heroes of the earlier teleological and disciplinary canon.

All these are fascinating matters, deeply relevant to the identity of geography today, but they are ones which I must respectfully, though regretfully, leave to others. In my guise not as geographer but as historian, my aim in this Afterword is neither to unveil the rhetorics of modern geographers nor to appraise the contribution of eighteenth-century travelers and cartographers to the edifice of geographical truth or the forging of geographical concepts. Rather my intention will primarily be to suggest some ways in which the new geographical initiatives and imperatives of the eighteenth century shaped, reinforced, and reflected the views and values of the Enlightenment itself.

Here, it goes without saying, the concept of the Enlightenment is itself problematic. It can no longer be treated as some homogeneous entity. For one thing, it needs to be broken down geographically: there were as many Enlightenments as there were local networks of *savants* engaged in intellectual endeavor, as, indeed, this volume makes so clear. There was, historians are increasingly agreed, no one, single, uniform movement, but rather distinctive Enlightenment identities in France, England, and Scotland, as well as in the German states, Italy, and Scandinavia—and beyond Europe too, for instance in the Americas. As well as such geographical contrasts, there were conceptual differences too. Major controversies flared among enlightened intellectuals themselves; *philosophes* rarely saw eye to eye (Outram 1995).

It is aptly emphasized by many of the contributors to this volume that to enlightened minds *geography* could be a way of representing *history*, while *history* in its turn could be a back-projection of *geography:* the 'there and now' mapped onto the 'here and then' (and vice versa), while the 'beyond' could be another way of expressing the 'before.' This 'double vision' was embraced partly because Enlightened minds strategically sought to grasp and broadcast the wider picture, to command those grander Archimedean vantage points which seemed necessary for mounting critiques of the 'here and now'—that is, the political and mental structures of *ancien régime* Europe. When Edmund Burke conjured up the idea of the "Great Map of Mankind" being "unrolled before us," he was, one might say, favorably contrasting the new Enlightenment grasp of space— the entire clustering of peoples and environments throughout the globe—against the partial and opaque knowledge available to the historian. Therein lay a characteristically Burkeian judgment for, when it came to the past, the conservative statesman was always more concerned to sustain reverence for tradition than to pretend (unlike many *philosophes*) to be able to pry into and determine the beginnings of things: he was a man ever suspicious of those claiming to hold the key to historical origins.

For a mind like Burke's, geography might thus offer the prospect of a reliable display of human nature. For other thinkers, however, history and geography, time and place, doubled as the coordinates of an understanding simultaneously of the varieties of humankind and its settings but also of the unity of human nature. Take, for example, James Cook. On his three great Pacific voyages, Cook was a stickler for factual accuracy, pleased to puncture the mythic balloons inflated by French explorers like Bougainville, with all their 'fantasies' about the exotic mores of the Tahitians and noble savages in an original state of nature. With characteristic Yorkshire dourness, Cook was convinced that close observation would establish the true character of the South Sea Islanders, the authentic, if astonishing, facts about their customs and sexual habits. He was no less certain that sober experience would confirm the common denominators of human nature at large—for instance, the tendency for societies to hold property, uphold a social order, and obey defined sexual codes—rather than being the carefree children of nature envisioned by armchair fantasists (Porter 1989; Rennie 1995; Smith 1960; Stafford 1984).

As mention of Cook might hint, many of the papers in this volume attest a genuine respect among eighteenth-century travelers and writers for the surprising verities about peoples and places uncovered by the eighteenth-century voyages of discovery. But they equally show how readings of such new findings became inscribed in dialogues about Enlightenment norms. Such values included human diversity—an excited glorying in human difference and the respectful tolerance it commanded—but they also comprised profound commitments to models of the unity of human nature. All this, in turn, was underpinned by the growing appreciation that, at long last, the system of the terraqueous globe itself had now been elucidated. All its coastlines and continents were becoming familiar—no longer did there need to be those embarrassing gaps on the maps which had for so long been the breeding grounds of foolish fantasies. As Lorraine Daston and Katharine Park have recently emphasized, the Enlightenment was the era when the time-honored reverence for 'wonder' was yielding to imperatives of organized curiosity under the impetus of the new science (Daston and Park 1998; Porter 1980).

By the close of the eighteenth century it was possible, therefore, for *philosophes* to propose integrated images of geocosmic reality and the physical and moral laws governing the earth and its peoples. Perhaps this was what enabled Gibbon playfully to imagine, in the *Decline and Fall of the Roman Empire,* a future New Zealander voyaging North and witnessing the ruins of Europe itself—Gibbon's antipodean is one of those novel Enlightenment geographic touches which attest the new perception of an integrated planetary economy (Gibbon 1776–88). For its part, Volney's saga of the ruins of empire bodied forth a mighty vision of the ceaseless interplay between the inexorable forces of Nature and the drives of civilization, constructive and destructive alike. Volney was teaching mankind lessons, warnings against hubris. A similar unifying philosophy—one, however, much more optimistic in tone—may be found in James Hutton's *Theory of the Earth* (1795), in which humanity's physical presence is reduced to the status of a tiny yet privileged element in the stupendous economy of a quasi-eternal Nature which reveals no vestige of a beginning and no prospect of an end. By possessing understanding, however—by the very act of being able to comprehend this system—man was able to transcend his own physical puniness. Something similar could be said, for a later generation, of Humboldt's *Cosmos*—a transitional work which, as several contributors to this volume observe, resists being pigeon-holed as either purely Enlightenment or Romantic. Humboldt too envisaged a stupendous system in which mankind was enrolled within a wider natural economy (Glacken 1967; Humboldt 1846; Hutton 1795).

To highlight in this way the Enlightenment idea of the unity and sovereignty of Nature must not lead us to underestimate how far Enlightenment aspirations involved a developing and upbeat ideology of human superiority and domination. Many of the foregoing essays rightly emphasize the contemporary quest for control over Nature, which was to be achieved thanks to scientific knowledge and technical skills, be it through agriculture, engineering, exploration, or military conquest itself. The eighteenth century, after all, was the age which brought the final and much-acclaimed triumph of the Moderns over the Ancients, the inspirational idea of progress, the notion—itself crucially dependent, as has been shown here, upon a new geography—of a new world supplanting the old, one which literally came into political being in 1776 in the wake of the American Declaration of Independence. The hoary idea, still put around by Buffon, that the new world was enfeebled and degenerate gave way to a fresh geographical balance of power in which Europe's role would be to serve as a counterpoise between an East whose ancient empires were circumscribed and familiar and a West whose destiny was not only unexplored but barely yet imagined (Bermingham 1986; Daniels 1993; Said 1978).

It is, of course, one of the historical ironies of the Enlightenment that Europeans were becoming overawed by Nature's grandeur, sensing—partly thanks to disasters like the Lisbon earthquake—their own puniness, tossed on the great tidal waves of Nature, at the very same time as they were hatching the most grandiose ambitions to subdue the entire terraqueous globe and all its denizens. This irony must partly be explained in terms of a continuing Christian, or at least theistic, input into ideologies of the natural world. While the book contains a number of studies pertaining to matters of religious belief and practice in a variety of Enlightenment situations, the religious input might, nevertheless, have been more prominently represented. For that reason, I shall now try to indicate the enduring, though changing, Christian contribution to Enlightenment geographies.

Certain Christian geographies, of course, become undermined or deeply problematic in the Enlightenment, not least the traditional mappings of heaven and hell—future topographies of the soul—which lost much of their purchase in an age in which Biblical literalism was marginalized among the intelligentsia. Horizons, one might say, were spectacularly shrinking, for when visualizing the universe, our *philosophe,* unlike his Reformation or Baroque precursors, had probably excluded from his sights heaven, hell, and all the satanic squadrons of demons and witches that suffused eschatologies from Augustine to Milton (Almond 1994; Gijswijt-Hofstra, Levack, and Porter 1999, Vol. 5). "The truth is," pontificated Thomas Carlyle in 1829:

> Men have lost their belief in the Invisible, and believe and hope and work only in the Visible . . . Only the material, the immediate practical, not the divine and spiritual, is important to us (*Edinburgh Review* 1829, 59, 439–59).

But though we here detect anticipations of the Weberian disenchantment of the world, the planet had not yet—in the eyes of the great majority of Enlightenment thinkers at least—been reduced to the meaningless mass of congealing magma which horrified Tennyson and other Victorian honest doubters. Guided by Alexander Pope, the spokesmen of the English Enlightenment read Nature as the great masterwork of Divine artistry—one should look "from Nature up to Nature's God." Filing out of church of a Sunday, the devout gazed up in awe:

> The Spacious Firmament on high
> With all the blue Etherial sky
> And spangled Heav'ns, a Shining Frame
> Their great Original proclaim
> (Psalm 111, rhymed and regularized by Joseph Addison
> [Pope 1965, 546; Willey 1962, 51])

In that confident, Christian, or deistic worldview, there could be no such thing as mere *Nature* pure and simple; Nature was rather *Creation,* and Creation remained a perfectly designed amphitheater with designated roles, costumes, and scripts for all creatures great and small, from stones, herbs, and beasts, up (through the Chain of Being), to Addison's great Original. It is no accident that Haydn's *Creation* was perhaps the Enlightenment's greatest oratorio. Perceptions of the terrestrial economy as a drama or, equally, as a well-managed estate, matched the daily material realities of the symbiosis of the human with the natural world, which has been so superbly delineated by Keith Thomas (Schama 1995; Thomas 1983). After all, most people in *ancien régime* Europe still lived on the land—in 1700 only thirteen percent of England's population resided in towns of over 5000; and there was an overwhelming proximity, physical, mental, and emotional—be it friendly, be it frightening—between farmers, flocks, and fields. The sense that everything had its rank and station in Creation chimed with a popular mentality whose folk-tales mingled children, wolves, giants, and monsters; with an elite culture exemplified by Gilbert White's *Natural History of Selborne,* where swallows and hedgehogs were humanized into honorary parishioners (Warner 1994; White 1977); and, not least, with a temper which remained—despite Voltaire's *Candide* and its reflections upon Lisbon—breathtakingly anthropocentric. Unlike many other world religions, Christian theology affirmed, and the mechanical philosophy in both its Cartesian and its Newtonian versions seemed to confirm, that all had been divinely adapted for mankind, because, as Descartes philosophized, humans alone had immortal souls and so could be saved. Genesis had granted man "dominion over the fish of the sea, and over the fowl of the air, and over the cattle, and over all the earth and over every creeping thing that creepeth upon the earth." And even after the Fall and Flood, had not the Lord reissued His mandate? "Be fruitful and multiply and replenish the earth and subdue it" (Glacken 1967; Passmore [1968] 1972, 1980).

Nature, in other words, was no longer perceived by eighteenth-century minds as some disputed alien territory occupied and lorded over by Satan; nor was it intrinsically enchanted—the Christian Churches had always battled to quash paganism or any hint of pantheism. Rather it was a resource, "principally designed," asserted Richard Bentley—Cambridge don, Anglican divine, and Newtonian popularizer—"for the being and service and contemplation of man" (Bentley 1838, Vol. 3, 175). 'We can, if need be, ransack the whole globe," maintained his fellow physico-theologian, the Anglican Rev. William Derham, "penetrate into the bowels of the earth, descend to the bottom of the deep, travel to the farthest regions of this world, to acquire wealth, to increase our knowledge, or even only to please our eye and fancy." And so providently benevolent was the Creator that, no matter how acquisitive man might be, "still the Creation would not be exhausted, still nothing would be wanting for food, nothing for physic, nothing for building and habitation, nothing for cleanliness and refreshment, yea even for recreation and pleasure" (Derham 1713, 112, 54–55). Even at the dawn of the nineteenth century, the Quaker geologist William Phillips could still reassure his readers that *"everything* [was] *intended for the advantage of Man,"* by the "Lord of Creation," a sentiment mirrored in Paley's *Natural Theology* (1802) and in the *Bridgewater Treatises* of the 1830s (Brooke 1991; Gillispie 1951; Phillips 1815, 193, 191).

This representation of Nature as an ideal habitat arose in part because Addison's generation, that "class of 1688" which gloried in the Glorious Revolution, had inherited a profound environmental crisis which it had then zealously combatted and confuted. "The opinion of the World's Decay is so generally received," George Hakewill had observed back in 1630, "not onely among the Vulgar, but of the Learned, both Divines and others" (Hakewill 1630). Reformation commentators had affirmed the old Classical tropes and Biblical prophecies: our Earth, this vale of tears, was a wreck, old and decrepit; the end of the environment was nigh (Davies 1969; Tuan 1968).

Everywhere, subscribers to the idea of *mundus senescens* had declared, global cooling was taking place, the soil was growing exhausted, and pestilences were multiplying. Originally, explained Thomas Burnet in his *Sacred Theory of the Earth* (1684), the face of the Earth had been eggshell smooth; but the very existence of mountains, and, furthermore, their perpetual denudation, showed that all was cracking up, becoming reduced to a pile of "Ruines and Rubbish"; nowadays the habitat was a *"little dirty Planet,"* a superannuated sphere, punishment for original sin (Burnet 1684–1690, quoted in Glacken 1967, 411).

If Burnet's prose smacked of the Baroque pulpit, rhetorically reflecting on the motif of mutability, others could point to environmental decay of a wholly tangible kind: collapsing cliffs, landslips, earthquakes, volcanic eruptions, silting estuaries, and the like. At home John Evelyn deplored smoke pollution and deforestation while abroad, as Richard Grove has brilliantly shown in his seminal study *Green Imperialism,* observers on Barbados and other new colonies were alarmed at how rapidly slash-and-burn deforestation and plantation monocultures like sugarcane brought on droughts, flash floods, and devastating soil erosion, turning once fertile terrains arid almost overnight (Evelyn 1776; Grove 1995). Original sin and modern greed together explained what many diagnosed as the symptoms of a planet terminally sick.

But such theological eco-pessimism was to meet resolute and triumphant challenge from Enlightenment optimists. The Glorious Revolution of 1688/9 enthroned a new regime which professed to stand for freedom, order, prosperity, and progress; and its apologists, including the Boyle Lecturers, provided environmental visions that vindicated the new governmental order by naturalizing it. Like William III's political settlement, and all the more so the Hanoverian succession of 1714, the natural order now became eulogized for

ROY PORTER

its stability: the "grand design of Providence," deemed the Newtonian geologist and physician, John Woodward, was thus the "Conservation of the Globe" in a "just aequilibrium" (Jacob 1976; Woodward 1695, 35).

In his *An Essay Towards a Natural History of the Earth* (1695), Woodward frankly admitted that events like the Deluge *prima facie* suggested "nothing but tumult and disorder":

> Yet if we draw somewhat nearer, and take a closer prospect of it . . . we may there trace out a steady hand producing . . . the most consummate order and beauty out of confusion and deformity . . . and directing all the several steps and periods to an end, and that a most noble and excellent one, no less than the happiness of the whole race of mankind. (Woodward 1695, 61)

As with English politics, the Earth's turbulent revolutionary career was over; all was now equilibrium, the body terrestrial was healthily balanced; and the final global revolution—the Deluge—had been constructive not punitive, a "Reformation" introducing a new "constitution" "into the Government of the Natural World." Through that revolution the Lord in His consideration had transformed mankind "from the most deplorable Misery and Slavery, to a Capacity of being Happy," by rendering the postdiluvial Earth niggardly, thereby forcing man to labor by the sweat of his brow, and compelling industriousness (*ibid.* 94).

Eighteenth-century theorists of the terraqueous globe further insisted that the laws of Nature governing the Earth were "immutable" and "progressive," and familiar phenomena were reinterpreted in the light of a rational theology of design (Porter 1979). Decomposing mountains had formerly been taken as dysfunctional and as betokening environmental degradation; now their positive uses were stressed—"the plains become richer, in proportion as the mountains decay," explained Oliver Goldsmith (1774, I, 163). No mountains, no rainfall, no fertility—thus reasoned a new generation of geographers, discrediting the ecological doomsters. The Scottish physician and philosopher, James Hutton, showed in his *Theory of the Earth* (1795) how decomposition of mountains produced the detritus which, flowing down the rivers to form the seabed, would, millions of years hence, become the basis of new strata, whose ultimate decay would once again form rich soil, and so on, in endless cycles. Likewise with volcanoes and earthquakes, as this book shows, a sore topic after Lisbon: all such apparently destructive processes were actually integral to Nature's operation. Hutton insisted that the globe was self-sustaining and self-repairing, forming an enduring habitat, perfect for man. Praising Hutton, a reviewer observed the switch from ego-gloom to eco-glory: "the dreary and dismal view of waste and universal ruin is removed, and the mind is presented with the pleasing prospect of a wise and lasting provision for the economy of nature" (Hutton 1794; Jones 1985).

The Enlightenment's new environmental thinking married Newton to Locke. Along with this law-governed Earth-machine went a Lockean possessive individualism which rationalized the Divine donation of dominion through a labor theory of property and value: man had the right, divine and natural, to appropriate the Earth and its fruits (MacPherson [1964] 1983; Pagden 1995). The Biblical mandate to man to master the Earth and multiply was thereby given rational sanction. The age of Donne and Shakespeare had experienced mutability—"tis all in pieces"—while in the Civil War era the saints and sectaries had anticipated the apocalyptic overthrow of Anti-Christ in fire and floods; but from the 1690s the environment was philosophically stabilized (Harris 1966; Hill 1971; Williamson 1961). Both pious Christians like John Ray, and later Deists like Hutton, portrayed a steady-state terrestrial economy, rather as Hutton's crony, Adam Smith, would praise the free-market economy. Illustrating these views, Oliver Goldsmith depicted the Earth as a godsent "habitation," a mansion for the Lord's tenant to enjoy—on condition he toiled to improve his estate, for:

> While many of his wants are thus kindly furnished, on the one hand, there are numberless inconveniences to excite his industry on the other. This habitation, though provided with all the conveniences of air, pasturage, and water, is but a desert place, without human cultivation.
>
> A world thus furnished with advantages on the one side and inconveniences on the other, is the proper abode of reason, is the fittest to exercise the industry of a free and a thinking creature. (Goldsmith 1774, 1, 400)

So: the Earth was not, after all, in crisis; it was a self-adjusting system, governed by universal laws, and made for man. In England, Latitudinarian Anglicanism backed such thinking: God was Benevolent, the Devil was *de facto* discounted (there might be a *ghost*, but there certainly were no *gremlins*, in the machine). And this environmental philosophy propped up the politics of the Hanoverians: God was the architect of natural order rather as Walpole was the manager of political stability. On the Continent, the *philosophes* re-iterated and recast such views (Plumb 1967).

And more than stability, all this betokened *improvement*. As was long ago maintained by Weber and Tawney, Protestant theology accentuated the individual's duty of self-realization; cultivating Nature promised spiritual nourishment no less than daily bread. Authors had few qualms about man's right—his *duty* even—to harness Nature, "bringing all the headlong tribes of nature into subjection to his will," according to Goldsmith, "and producing . . . order and uniformity upon earth" (Goldsmith 1774, 1: 401; Tawney 1926; Weber 1930). Had not the noble Lord Verulam proclaimed "Knowledge itself is power" and that "the end of our foundation is . . . the enlarging of the bounds of human empire, to the effecting of all things possible"? Through natural philosophy, maintained Joseph Glanvill, reiterating Bacon, "nature being known . . . may be mastered, managed, and used in the services of humane life" (Bacon 1857–1874, 7: 253; 3:156; Glanvill 1668, 87).

Such views, of course, rationalized and underwrote what Europeans had anyway been doing to the environment for centuries—clearing forests, embanking, ploughing, planting, mining. Draining and deforestation had been praised for freeing the land from dankness and disease, and thus turning waste into wealth. Radical and feminist historians have, however, recently reproved the aggressive, macho element in Baconian thinking, for replacing notions of Mother Earth with a new model of Nature exploited, raped, and forced to yield up her fruits. Man's dominion must not be hindered by sentiment or superstition: "Know that by nature," Descartes wrote, "I do not understand some goddess or some sort of imaginary power. I employ this word to signify matter itself." "The veneration wherewith men are imbued for what they call nature," grumbled Robert Boyle in a similar antisuperstitious vein, "has been a discouraging impediment to the empire of man over the inferior creatures of God: for many have not only looked upon it, as an impossible thing to compass, but as something of impious to attempt." An end to such scruples! (Boyle 1774, 4:363; Descartes 1973, Vol. 1, 349 quoted in Easlea 1981, 72; Kolodny 1975; Merchant 1980).

It is right to take note of these undertones of environmental violence, but they must be kept in perspective. For the key Enlightenment paradigm of man's relation to the environment was not *conflictual* but *cooperative,* indeed often positively georgic. "I have now placed thee in a spacious and well-furnish'd World," the botanist and natural theologian John Ray imagined God informing mankind:

> I have provided thee with Materials whereon to exercise and em-
> ploy thy Art and Strength . . . I have distinguished the Earth into
> Hills and Vallies, and Plains, and Meadows, and Woods; all these
> Parts, capable of Culture and Improvement by Plowing, and Car-
> rying, and Drawing, and Travel, the laborious Ox, the patient Ass,
> and the strong and serviceable Horse. . . . (Ray 1691, 113–114)

Once the Deity had explained to man his place in the divine scheme
of things, Ray reflected upon God's assessment of what might be
called the Divine Assessment Exercise:

> I persuade myself that the bountiful and gracious Author of Man's
> Being . . . is well pleased with the Industry of Man, in adorning the
> Earth with beautiful Cities and Castles; with pleasant Villages and
> Country-Houses; with regular Gardens and Orchards, and Planta-
> tions of all Sorts of Shrubs and Herbs, and Fruits, for Meat, Medi-
> cine, or Moderate Delight . . . and whatever differenceth a civil and
> well-cultivated Region, from a barren and desolate Wilderness.
> (*ibid.* 484)

The model typically defining the proper relations between man
and Nature was thus the farm. According to Ray's contemporary, Sir
Matthew Hale, God was the great freeholder, the world his estate,
and man his tenant. "The end of man's creation," Chief Justice Hale
explained in legal terminology, "was to be God's steward, *villicus*,
bailiff or farmer of this goodly farm of the lower world." For this reason man had been "invested with power, authority, right, dominion,
trust and care, to correct and abridge the excesses and cruelties of the
fiercer animals, to give protection and defence to the mansuete and
useful"—in short, "to preserve the face of the earth in beauty, usefulness and fruitfulness" (Hale 1677, sec. 4, 8: 370). Everyone would
have understood Hale's paternalistic metaphor of the good steward,
be he in the Bible or Bedfordshire, in Burgundy, or Belgium. Nature
would yield, and yield well, but only if the principles of good husbandry were upheld: matching stock and crops to soils, adopting
sound rotations, planning for long-term sustainability—quite literally, to use the Weberian metaphor, ploughing back the profits
(Mingay 1990).

Such images of stewardship—paternal rather than plundering—
sanctioned action and ordained environmental ethics and aesthetics. Pioneering in this respect was the work of John Evelyn, whose
*Silva, A Discourse of Forest Trees and the Propagation of Timber in His
Majesty's Dominions* (1662) condemned wasteful land practices, exposing how, in the provision of charcoal and pasture, "prodigious

havoc" had been wreaked through the tendency "to extirpate, de-molish, and raze . . . all those many goodly woods and forests, which our more prudent ancestors left standing" (Evelyn 1776, 1). Evelyn's conviction that economic growth depended on sound conservation practices set the tone for the new managerial approach to Nature advocated in the eighteenth century.

I have been suggesting that liberal and optimistic *religious* ideologies upheld, in the mind of Enlightenment Europe, a conviction of the orderliness of Nature and of man's legitimate domination over it. There was also a continuing Christian, and more broadly theistic, contribution to thinking about mankind, working powerfully to uphold belief in the unity of the human race. With certain notable exceptions—Lord Kames, for instance—monogenism retained its hold upon the minds of Enlightenment theorists, thanks in part to rather Stoic notions of a common human nature, and in part to the Lockeian belief in infinite human plasticity and malleability (Passmore 1972, 159f).

Such models of mankind assuredly embodied profound ironies and ambiguities. The twists and turns of such commitments are poignantly exemplified in the writings of one of Humboldt's English contemporaries and admirers, James Cowles Prichard, a man who consistently juxtaposed geography and anthropology. Educated as a doctor in late Enlightenment Edinburgh, Prichard simultaneously serves both as an exemplar and also as a critic of Enlightenment attitudes to man and his place in Nature. Born into a Quaker family but converting to an Evangelical Anglicanism, he made it his life's business to delineate the relationship between environment and human race, milieu and civilization.

It was essential, Prichard believed, to uphold the superiority of the human—that is, the spiritual and the eternal—over the merely material and secular, thereby denying the cruder sorts of Enlightenment environmentalism. It was the duty of enlightened Christians to awake the spark of religion in those savage peoples around the globe whose culture, though pagan, already bore witness to the unity of human consciousness—in that respect, Prichard was of another generation than Cook. The psychic unity of the human soul was also to be stressed. True civilization was desirable in so far as it realized the unity of the human race within a comprehensive notion of the order of Nature in harmony with the Christian dispensation. So-called civilization was, by contrast, a threat and a menace if it meant, as it did to Prichard, the unloosing of European greed throughout what we would today call the undeveloped world (Augstein 1996). Convinced of the value of geographical and ethnological knowledge, but strikingly ambivalent about the impact of Europeans throughout the globe, Prichard in many respects exemplifies the themes this book has been investigating.

In short, the new geographies of the Enlightenment unleashed great hopes but also awakened in Europeans doubts about their own place in the scheme of things. Uncertainties were triggered regarding the old Biblical geography, history, and destiny of the human race. The relationships between Europe, the Old World, and the New, were problematized. At the same time, through its myths of progress and the Baconian dictum that knowledge was power, they instilled the beliefs that it was Enlightenment man's right, duty, privilege— and burden—to be the lords of human kind (Kiernan 1969).

References

Almond, Philip C. 1994. *Heaven and Hell in Enlightenment England.* Cambridge: Cambridge University Press.

Augstein, Franziska A. 1996. "James C. Prichard's Views of Mankind: An Anthropologist Between the Enlightenment and the Victorian Age." University of London, Ph.D. thesis.

Bacon, Francis. 1857–1874. "New Atlantis," iii, 156, and "Of Heresies," vii, 253, in J. Spedding, R. L. Ellis, and D. D. Heath, eds., *The Works of Francis Bacon.* 14 vols. London: Longman, vii, 253.

Bentley, Richard. 1838. "Eight Sermons Preached at the Hon. Robert Boyle's Lecture in the Year MDCXCII," in A. Dyce, ed. *The Works of Richard Bentley.* London: Francis Macpherson, vol. 3, 175.

Bermingham, Ann. 1986. *Landscape and Ideology: The English Rustic Tradition 1740–1860.* Berkeley: University of California Press.

Bowen, Margarita. 1981. *Empiricism and Geographical Thought: From Francis Bacon to Alexander von Humboldt.* Cambridge: Cambridge University Press.

Boyle, Robert, 1774. "A Free Inquiry into the Vulgarly Received Notion of Nature," in *The Works of the Honourable Robert Boyle.* London: A. Millar, 4:363.

Brooke, John Hedley. 1991. *Science and Religion: Some Historical Perspectives.* Cambridge: Cambridge University Press.

Burnet, Thomas. 1684–1690. *Sacred Theory of the Earth,* translated from the 1681 Latin original. London: printed by R. Norton.

Cosgrove, Denis, and Stephen Daniels, eds. 1988. *The Iconography of Landscape: Essays on the Symbolic Representation, Design and Use of Past Environments.* Cambridge: Cambridge University Press.

Daniels, Stephen. 1993. *Fields of Vision: Landscape Imagery and National Identity in England and the United States.* Cambridge: Polity Press.

Daston, Lorraine, and Katharine Park. 1998. *Wonders and the Order of Nature 1150–1750.* New York: Zone Books.

Davies, Gordon. 1969. *The Earth in Decay.* London: MacDonald.

Derham, William. 1713. *Physico-Theology: or a Demonstration of the Being and Attributes of God, from His works of Creation*. London: Innys.

Descartes, Réné. 1973. *Le Monde* in F. Alquié, ed. *Oeuvres Philosophique de Descartes*. Vol. 1. Paris: Garnier, 349.

Driver, Felix, and Gillian Rose, eds. 1992. *Nature and Science: Essays in the History of Geographical Knowledge*. Cheltenham: Historical Geography Research Group of the Institute of British Geographers, No. 28, especially the "Introduction: Towards New Histories of Geographical Knowledge," 1–7.

Easlea, Brian. 1981. *Science and Sexual Oppression: Patriarchy's Confrontation with Woman and Nature*. London: Weidenfeld and Nicolson.

Edinburgh Review. 1829, 59, 439–59. "Sign of the Times" [T. Carlyle]. Pp. 452–453.

Evelyn, John. [1662] 1776. *Silva, or a Discourse of Forest Trees*. York: printed for J. Dodsley.

Foucault, M. [1970] 1989. *The Order of Things: An Archaeology of the Human Sciences*. London: Tavistock; London: Routledge.

Gerbi, Antonello. 1973. *The Dispute of the New World: The History of a Polemic*, trans. J. Moyle. Pittsburgh: University of Pittsburgh Press.

Gibbon, Edward. 1776–88. *The Decline and Fall of the Roman Empire*. 6 vols. London: printed for W. Strahan and T. Cadell.

Gijswijt-Hofstra, Marijke, Brian Levack, and Roy Porter. 1999. *History of Witchcraft and Magic*, Vol. 5. London: Athlone.

Gillispie, Charles C. 1951. *Genesis and Geology: A Study in the Relations of Scientific Thought, Natural Theology, and Social Opinion in Great Britain, 1790–1850*. Cambridge, MA: Harvard University Press.

Glacken, C. 1967. *Traces on the Rhodian Shore: Nature and Culture in Western Thought from Ancient Times to the End of the Eighteenth Century*. Berkeley: University of California Press.

Glanvill, Joseph. 1668. *Plus Ultra*. London. Printed for James Collins.

Goldsmith, Oliver. 1774. *A History of the Earth and Animated Nature*. London: Printed for J. Nourse.

Grove, Richard. 1995. *Green Imperialism. Colonial Expansion, Tropical Island Edens and the Origins of Environmentalism 1600–1800*. Cambridge: Cambridge University Press.

Gutting, Gary. 1989. *Michel Foucault's Archaeology of Scientific Reason*. Cambridge: Cambridge University Press.

Hakewill, George. [1627] 1630. *An Apologie*. 2d ed. Oxford: printed by William Turner; Preface.

Hale, Matthew. 1677. *The Primitive Origination of Mankind*. London: printed by William Godbid.

Harris, V. I. 1966. *All Coherence Gone*. London: Frank Cass & Co.

Hill, Christopher. 1971. *Antichrist in Seventeenth Century England*. London: Oxford University Press.

Humboldt, Alexander von. 1846. *Cosmos*. 4 vols. London: Longman, Brown, Green and Longmans.

Hutton, James. 1794. *An Investigation of the Principles of Knowledge, and of the Progress of Reason, from Sense to Science and Philosophy*. 3 vols. Edinburgh: printed for A. Strahan and T. Cadell.

———. 1795. *The Theory of the Earth*. Edinburgh: William Creech.

Jacob, M. C. 1976. *The Newtonians and the English Revolution, 1689–1720*. Ithaca, NY: Cornell University Press.

Jones, Jean. 1985. "James Hutton's Agricultural Research and his Life as a Farmer." *Annals of Science* 42: 573–601.

Kiernan, Victor. 1969. *The Lords of Human Kind: European Attitudes to the Outside World in the Imperial Age*. Harmondsworth: Penguin.

Kolodny, A. 1975. *The Lay of the Land: Metaphor as Experience and History in American Life and Letters*. Chapel Hill: University of North Carolina.

MacPherson, C. B. [1964] 1983. *The Political Theory of Possessive Individualism: Hobbes to Locke*. Oxford: Oxford University Press.

Merchant, C. 1980. *The Death of Nature: Women, Ecology and the Scientific Revolution*. San Francisco: Harper and Row; London: Wildwood House.

Mingay, G. E. 1990. *A Social History of the English Countryside*. London: Routledge.

Ogborn, Miles. 1998. *Spaces of Modernity: London's Geographies, 1680–1780*. New York: Guilford Press.

Outram, Dorinda. 1995. *The Enlightenment*. Cambridge: Cambridge University Press.

Pagden, Anthony. 1995. *Lords of all the World: Ideologies of Empire in Spain, Britain and France c. 1500–c. 1800*. New Haven: Yale University Press.

Passmore, John. [1968] 1972. *The Perfectibility of Man*. London: Duckworth.

———. 1980. *Man's Responsibility for Nature*. London: Duckworth.

Phillips, William. 1815. *An Outline of Minerology and Geology*. London: printed and sold by William Phillips.

Pile, Steven, and Nigel Thrift, eds. 1995. *Mapping the Subject: Geographies of Cultural Transformation*. London and New York: Routledge.

Plumb, J. H. 1967. *The Growth of Political Stability in England, 1675–1725*. London: Macmillan.

Pope, Alexander. 1965(?). "Essay on Man," in J. Butt, ed., *The Poems of Alexander Pope*. London: Methuen, 546.

Porter, Roy. 1979. "Creation and Credence: the Career of Theories of the Earth in Britain, 1660–1820," in B. Barnes and S. Shapin, eds. *Natural Order*. Beverly Hills, CA: Sage Publications, 97–123.

———. 1980. "The Terraqueous Globe," in G. S. Rousseau and R. Porter, eds. 1980. *The Ferment of Knowledge*. Cambridge: Cambridge University Press, pp. 285–324.

———. 1989. "The Exotic as Erotic: Captain Cook at Tahiti," in Rousseau and Porter 1989, pp. 117–144.

Ray, John. 1691. *The Wisdom Of God Manifested in the Works of the Creation*. London: Printed for Samuel Smith.

Rennie, Neil. 1995. *Far-fetched Facts: The Literature of Travel and the Idea of the South Seas*. Oxford: Clarendon Press.

Said, E. W. 1978. *Orientalism*. Harmondsworth: Penguin.

Schama, Simon. 1995. *Landscape and Memory*. London: Harper Collins.

Smith, B. 1960. *European Vision and the South Pacific, 1768–1850: A Study in the History of Art and Ideas*. Oxford: Clarendon Press.

Stafford, Barbara Maria. 1984. *Voyage Into Substance: Art, Science, Nature, and the Illustrated Travel Account 1760–1840*. Cambridge, MA: M.I.T. Press.

Tawney, Richard. 1926. *Religion and the Rise of Capitalism.* New York: Harcourt, Brace and Co.

Thomas, Keith. 1983. *Man and the Natural World.* London: Allen Lane.

Tuan, Yi-fu. 1968. *The Hydrologic Cycle and the Wisdom of God: A Theme in Geoteleology.* Toronto: University of Toronto Press.

Warner, Marina. 1994. *From the Beast to the Blonde: On Fairy Tales and Their Tellers.* London: Chatto & Windus.

Weber, Max. 1930. *The Protestant Ethic and the Spirit of Capitalism.* London: Allen and Unwin.

White, Gilbert. 1977. *The Natural History and Antiquities of Selborne,* ed. Richard Mabey. Penguin: Harmondsworth (first published 1789).

Willey, Basil. 1962. *The Eighteenth Century Background: Studies on the Idea of Nature in the Thought of the Period.* London: Penguin.

Williamson, G. 1961. "Mutability, Decay and Seventeenth Century Melancholy," in *Seventeenth Century Contexts.* London: Faber & Faber, 73–101.

Woodward, John. 1695. *An Essay Towards a Natural History of the Earth.* London: printed for R. Wilkin.

Contributors

Michael Bravo is University Research Fellow and Lecturer in Social Anthropology at Manchester University. Drawing on extensive fieldwork in the Arctic regions, he studies the history and philosophy of science, language, and culture from the eighteenth century to the present day. His forthcoming book examines the creation of the Inuit as an icon of "homo technologicus" and its allegorical context in European debates about land, society, and politics.

Paul Carter is a Senior Research Fellow at the University of Melbourne and Professor within the Faculty of Constructed Environment, Royal Melbourne Institute of Technology. His research into the mythopoeic foundations of imperial cultures frequently involves creative collaborations with artists and architects. His books include *The Road to Botany Bay* (1987), *Living in a New Country* (1992), *The Lie of the Land* (1996), and *Lost Subjects* (1999).

Denis Cosgrove is Professor of Human Geography at Royal Holloway College, University of London. His principal publications include *Social Formation and Symbolic Landscape* (1984), *The Palladian Landscape: Geographical Change and Its Cultural Representations in Sixteenth-Century Italy* (1993), and, as co-editor with Stephen Daniels, *The Iconography of Landscape* (1988).

Stephen Daniels is Professor of Cultural Geography at the University of Nottingham. He is the author of *Fields of Vision: Landscape Imagery and National Identity in England and the United States* (1993), *Joseph Wright: Art and Enlightenment* (1999), and *Humphry Repton: Landscape Gardening and the Geography of Georgian England* (1999), and, as co-editor *The Iconography of Landscape* (1988). He is currently working on geography, education, and citizenship in eighteenth-century England.

Matthew H. Edney is Associate Professor of Geography-Anthropology and of American and New England Studies at the University of Southern Maine, with scholarly oversight of the Osher Map Library and Smith Center for Cartographic Education. His *Mapping an Empire: The Geographic Construction of British India 1765–1843* was published by the University of Chicago Press in 1997.

Anne Godlewska is Associate Professor of Geography at Queen's University, Canada. She is the author of *The Napoleonic Survey of Egypt* (1988) and *Geography Unbound: French Geographic Science from Cassini to Humbolt* (University of Chicago Press, 1999). She also co-edited *Geography and Empire* (1994) with Neil Smith. She is currently working in a number of overlapping areas including the role of mapping in the evolution of the modern state; maps, globes and the rhetoric of "golbalization"; and the mapping impulse in Western imperialism, science, and art.

Peter Gould is the Evan Pugh Professor of Geography at Pennsylvania State University. He has a long-term interest in spatiotemporal processes and has recently concluded research on the geographic diffusion of the AIDS pandemic, *The Slow Plague*. He is the author/editor of sixteen other books, some of which have appeared in French, Italian, and Japanese.

Michael Heffernan is Professor of Historical Geography at the University of Nottingham. His research focuses on European geographical thought since the eighteenth century, the political and cultural geography of France since the eighteenth century, and the cultural and intellectual implications of French imperial expansion in the nineteenth and twentieth centuries. Recent publications include a co-edited volume on *Geography and Imperialism, 1820–1940* (1995) and *The Meaning of Europe: Geography and Geopolitics* (1998).

David Livingstone is Professor of Geography and Intellectual History at the Queen's University of Belfast and a Fellow of the British Academy. He is the author of *Nathaniel Southgate Shaler and the Culture of American Science* (1987), *Darwin's Forgotten Defenders* (1987), *The Preadamite Theory* (1992), and *The Geographical Tradition* (1992). He is currently working on a historical geography of science.

Dorinda Outram is the Clark Professor of History at the University of Rochester, New York. She has also been a Fellow at the Dibner Institute for Science and Technology at MIT, a Guest at the Max-Planck-Institut für Wissenschaftsgeschichte in Berlin, and visiting Associate Professor at Harvard. She is the author of *The Body and the French Revolution* (1989) and *The Enlightenment* (1995).

Chris Philo is Professor of Geography at the University of Glasgow. He has research interests in the historical geography of the "mad business" in Britain, and in geography and social theory. He is the co-author of *Approaching Human Geography: An Introduction to Contemporary Theoretical Debates* (1991), and co-editor of *Selling Places: The City as Cultural Capital, Past and Present* (1993).

Roy Porter is Professor of the Social History of Medicine at the Wellcome Institute for the History of Medicine, London. Recent books include *Doctor of Society: Thomas Beddoes and the Sick Trade in Late Enlightenment England* (1991), *London: A Social History* (1994), and *"The Greatest Benefit to Mankind": A Medical History of Humanity* (1997). He is currently working on a general history of the Enlightenment in Britain. He is interested in eighteenth-century medicine, the history of psychiatry, and the history of quackery.

Nicolaas Rupke is Professor of the History of Science at Göttingen University and Director of the Institut für Wissenschaftsgeschichte. His areas of expertise are the late-modern earth and life sciences. In studying these, he has favored the approach of scientific biography and has written on such Victorian men of science as William Buckland in *The Great Chain of History* (1983) and *Richard Owen* (1994). Currently, he is working on Alexander von Humboldt.

Susanne Seymour is Lecturer in Geography at the University of Nottingham. Her main research interests are in rural cultural and historical geography. She is a contributor to *An Historical Geography of England and Wales* (1990), and co-author of *Moralizing the Environment: Countryside Change, Farming and Pollution* (1997). She has recently completed several articles (with Stephen Daniels and Charles Watkins) on the theme of picturesque landscapes and landed estates in the eighteenth century.

Charles Watkins is Reader in Rural Geography at the University of Nottingham. His books include *Woodland Management and Conservation* (1990), *Rights of Way: Policy, Culture, Management* (1996), and *European Woods and Forests: Studies in Cultural History* (1998). He has published several other works on landscape history and rural geography.

Charles Withers is Professor of Geography at the University of Edinburgh. He is the author of *Gaelic in Scotland 1698–1981* (1984),

Gaelic Scotland: The Transformation of a Culture Region (1988), and *Urban Highlanders: Highland-Lowland Migration and Urban Gaelic Culture, 1700–1900* (1998). Recent work in the history of geography includes discussion of encyclopedism and geographical knowledge, and geography and the public sphere. He is currently writing a book on geography, science, and national identity in Scotland since 1550.

Index

Boldface page numbers refer to figures in the text.

Chinese, discussed by Lapérouse, 214, 220, 221, 224, 226, 228
chorography, 41, 172
Choiseul-Goufier, Comte de, 179–81
Christendom, idea of, 1, 38, 132–33, 153n. 6
Christian geographies, 419–22
Christianity
  authority of, 30
  contribution of, to Enlightenment geographies, 419–25, 427
  place of paradise in, 30, 67–78, 85–86
  views of, on origin of humans,100–101, 129–30, 132
  world view of, 34, 49, **143,** 419
  See also authority, scriptural
Christian republic, ideas of a, 94, 107, 109–10
chronometer, 167
chronopolitics, idea of, 15
civilization, ideas about, 14–15, 19, 133–34
classification
  as an element in emergence of 'modern' geography, 1
  in natural history, 242–43
  See also taxonomy
Classicism, 36
Claval, Paul, 247
climate, 103–5
climatic determinism, 103–4
Cockburn, William, 143
coffee houses, as sites of knowledge, 4
Colombus, Christopher, 70, 202
colonialism
  and mercantilism in the France of Louis XIV, 36
  place of in the Enlightenment, 12
Colbert, Jean-Baptiste, 45, 62
commensurability
  ethnic, 122, 218–22, 224, 230–33
  linguistic, 122, 218–22, 224–26, 230–33, 267
  See also translation
commerce, age of, 131. See also progress, idea of
Common Sense, philosophy of, 3, 105
Commons, House of, 149
Comte, Auguste, 147, 153n. 5
Condamine, Charles-Marie de La, 282
Condillac, Etiénne Bonnot, Abbé de, 80, 126, 284, 388–89
Condorcet, Marie-Jean de, 9, 95, 134, 151
Connecticut, 111

Cook, James
  discoveries of as an element in emergence of 'modern' geography, 1
  on instrumentation, 287
  similarities with Lapérouse, 209
  views of, concerning paradise, 84
  voyages of, 1, 83, 84, 200, 204, 209, 277, 282, 287, 292, 298, 417, 427
Copernicanism, 34, 39, 407
Copernicus, 407
Corbin, Alain, 296
Cork, 400, 403, 404
Cork Journal, 404
Coronelli, Vincenzo
  and the *Atlante Veneto,* 35, 39–41, **52,** 60
  and the *Biblioteca Universale,* 38
  and the *Corso Geografico Universale,* 38, 40
  and the *Epitome Cosmografica,* 38
  cosmographic conceptions in the work of, 29, 33, 39–42, **43,** 44, 52, 63nn. 1, 2, 172
  encyclopedism in the work of, 38–39
Corso Geografico Universale, 38, 40. *See also* Coronelli
cosmography
  'crisis of,' 41–42, 49
  idea of, 33, 39–63
  'mathematical,' 165
Cosmos, 263, 278, 320, 322–23, 325, 334–36, 418
cosmos
  Alexander von Humboldt on, 265–66
  ideas about the, 33, 238
Cote d'Ivoire, 169
Counter-Reformation, 133
courts, as sites of knowledge, 4, 17, 48
Craig, James, 382
credibility, ideas of, 15, 283–84
Cree people, 169
Critical Review, 199, 232
Crocker, Leslie, 6
Crome, Auguste, 252, 272n. 25
Crosse, John, 328
Cullen, William, 380
culture, idea of, 133
Cuvier, Georges
  discusses knowledge formation, 248–49, 278, 285–87, 290, 293n. 4
  reviews Alexander von Humboldt, 286

Dahur people, 207
d'Alembert, Jean, 5, 72, 175, 186

Danish Chronicle, 98
d'Anville, Jean Baptiste Bourguinon
  and 'positive geography,' 186
  place of in Enlightenment, 179–80,
    186, 205–6, 251, 256
Darwin, Charles, 126, 407
Darwin, Erasmus, 347, 350, 351
Daston, Lorraine, 293n., 418
Davis, Gulf of, 96
Davy, Humphrey, 306, 311, 350, 355,
  364
Decandolle, Augustin-Pyramus
  on plant geography, 242, 243–44,
    247–49
  on the natural region, 248
deductivism, 75–76, 278, 295–97
d'Eichtal, Gustave, 150
Defoe, Daniel, 177–78
Degerando, Baron, 136, 137
De Langle Bay, 203, 213, 214, 217, **220,**
  224, 226, **227**
Delisle, Guillaume, 169, 206
Delphi, **174**
Deluge, discussion of the Biblical, 69,
  130–31
Denmark, 96
Derbyshire, 341, 346
Derham, William, 421
Descartes, Rene, 301–3, 305, 307, 425
description
  as a geographical practice, 5, 167–70,
    222–24
  realism in, as an element in emer-
    gence of 'modern' geography, 1
D'Estaing, Bay, 214, 217, 229
d'Estrées, Cardinal Cesar, 37, 44, 61
d'Halloys, Omalius, 248
determinism, climatic, 103–5, 121
Dettelbach, Michael, 238–39, 267n. 2,
  293n. 4
Dickens, Charles, 140
Diderot, Denis
  and the *Encyclopédie,* 5, 72, 175, 186
  influenced by Rousseau, 135
  views of, concerning Tahiti, 83
  views of, on paradise, 84
Dirks, Nicholas, 12
displaying as a geographical practice, 13
discovery as a geographical practice, 1, 14
  idea of, 14
  voyages of, 1
  See also Lapérouse
Dolomieu, Dieudonné (called Déodat),
  de Gratet de, 249

Dorset, 348
Downton, 347, 357–59, 364–65
Dublin, 403, 409
Dublin Gazette, 403, 404, 410n. 4
Dunbar, James, 83
Duncan, Dr. Andrew, 373, 380
Dupain-Triel, J. L., 252
Dutrochet, Henri, 355, 361, 365
Duveyrier, Charles, 150
Dwight, Timothy, 109–10

Eagleton, Terry, 137
Earth, debates on the age of the, 30,
  418, 422–24
earthquakes, 399. *See also* Lisbon
Ebeling, Christoph, 171
École Normale, 144
École Polytechnique, 147, 148, 150
Ecuador, 168, 176
ecumene, ideas on the classical, 40
Eden
  as a garden, 67
  discussed by Kircher, 60
  See also paradise, terrestrial
Edinburgh
  City Bedlam in, 380
  City Bridewell of, 386
  as the 'City of Reason,' 378–82, **385**
  Enlightenment in, 342, 352, 372–98,
    399nn. 19, 20, 406, 427
  Musical Society of, 379
  Philosophical Society of, 379
  Royal Asylum of, 373, 386
  Royal Infirmary of, **385, 387**
  Royal Society of, 379
  Select Society of, 379
  sites of madness in, 372–74, 382–87
Edinburgh Chronicle, 359
Edinburgh Review, 322, 328, 334
Egypt, 136, 138, 171
  Carte Topographique de l', 171
  Description de l', 171
  See also Volney, and *Voyage en Egypt et
    en Syrie*
elephants, imaginary, 167
Elton, 347–57
emblematic geography, 29, 37–43
empiricism
  in Enlightenment mapping, 168–70
  geography as a form of, 2, 49, 238–39
  in Humboldt's work, 241–44
  in Linnaeus's work, 82
  role of language in promoting, 79
encyclopedias, medieval, 39

Encyclopedie
  geography in the, 3
  as a symbol of the Enlightenment, 6–7
  as a 'world map,' 186
encyclopedism
  and eighteenth-century intellectual
    culture, 78, 122, 133
  and geographical knowledge as an
    archive of, 165, 170–71, 175, 190–
    93
  and mapping, 186–90
  See also Diderot
Endeavour, 346
England, 7, 176, 205, 401, 424
  Enlightenment in, 7, 10–12, 345–46,
    416
  horticulture in, 347–49
  improvement in Georgian, 345–65
Enfantin, Prosper, 150
Enlightenment
  Age of, 2, 18, 113
  in America, 11, 107–9, 113, 416
  anthropology and, 10
  attitudes to the environment in the,
    18, 84, 87
  in Austria, 11
  in Bohemia, 11
  definitions of, 3–7, 113, 410n. 6,
    416–17
  in Edinburgh, 342, 372–98, 427
  in England, 7, 10–12, 345–46, 416
  eroticism and, 10, 18
  exoticism in the, 10, 18
  exploration as a geographical practice
    in the, 4, 168, 281–93
  faith and, 6
  in France, 11, 113, 125–52, 238, 416
  geography and, 3–4, 5, 13–21, 111–
    13, 128, 230–33, 416–17
  geography in, 2–13, 128
  geography of, 3–9, 10–21, 93–113,
    128, 230–33, 279, 341–44, 415–17,
    427–28
  in Germany, 11, 416
  in Greece, 11, 178, 281
  historiography of the, 3–21
  in Hungary, 11
  in Italy, 11, 12, 38, 416
  and the idea of global illumination,
    33–34, 37–40, 44–50, **57**
  literary histories of, 10
  medical knowledge and the, 10, 18
  mentalité of, 4, 112

national contexts of, 10–12
  in the Netherlands, 11
  origins of the, 4–9, 29–31
  placing of, 4
  in Poland, 11
  political authority and the, 6
  provincial culture in the, 10
  resistance to, 11
  science in the, 10
  sexual underworlds in the, 10
  sites of, 341–44
  in Sweden, 11, 416
  in Switzerland, 11
  visual representation in the, **174**
  See also graphic representation;
    mapping
environment, attitudes to, 18, 84, 87.
  See also Nature
environment, degradation of, 87
epistemic shifts
  Foucault on, 236–37, 375–76
Epitome Cosmografica, 38. *See also*
  Coronelli
Escurial, 404
*Essai Politique sur le Royaume de la
  Nouvelle-Espagne,* 326, 327–44. *See
  also* Humboldt, Alexander von
Essai sur la Géographie de Plantes, 261,
  263, **264**. *See also* Humboldt,
  Alexander von
Ethiopia, 69
ethnogeography, Latour's account of,
  226–30
ethnographic navigation
  accuracy in, 221–22
  as questions of commensurability
    and incommensurability, 212–14,
    221–22
  use of the body and mapping and,
    221–22
  in the work of Lapérouse, 122, 175,
    202–15, 217–26
ethnography, 83, 122, 217–26, 231–33
ethnology, 82–83, 103
Etna, 51, 55
Euphrates, River, 70, 75
European attitudes to the rest of the
  world, 166. *See also* Mankind, Great
  Map of; New World
Europe, idea of, 12, 14, 132–33, 291
Evans, Lewis, 189
Evelyn, John, 353, 422, 426
Exeter, Bishop of, 406

geography (*continued*)
  and polite sociability, 2
  'positive,' 186
  in the public sphere, 2, 4
  rational, 3
  of reason, 374–78, 387–90
  teaching of, in Enlightenment, 2,
    114n. 9
  in textbooks, 1, 189
  of time, 404–5, 406, 409
  understood as regional description, 1
  of unreason, 372, 374–78, 382–387
  See also Enlightenment
geology, 236, 309, 312–13
'geometric spirit,' 387–90
geopolitics, idea of, 15
George III (of Great Britain), 362
Georgia, 176
Germany, Enlightenment in, 11
Gibbon, Edward, 3, 146, 418
  and rational geography, 3
Gibbs, Thomas, 350
Gibraltar, 403
gift, geographical, 203–30
Gihon. *See* Ganges, River
Giornale Enciclopedico di Napoli, 334
Glacken, Clarence, 146
Glasgow, 135, 378, 399n. 21
globes
  celestial, 35
  as emblematic devices, 29, 33, 34–35,
    43, 44–50
  as instruments of Enlightenment, 44–
    40
  in Kircher's work, 56
  as symbols of geographical discovery,
    36
  terrestrial, 35
  in Volney's work, **142**
Gluckstadt, 400
Gods, Age of, 132. *See also* progress
Godwin, William, 134
Golden Age, idea of a, 125
Goldsmith, Oliver, 299, 423, 424
Gomorrah, 406
Goodenough, Samuel, 351
Gordon, General George, 68, 84
Gorgon, 281
Gorman, Michael John, 56
Göttingische Gelehrte-Anzeigen, 322, 328,
  334
governmentality, 133
Grafton, Anthony, 101, 103
Grammar, Port Royal, 78

grammars, geographical, 2, 20, 78
grammars, universal, 78
Grand Tour, 138
graphic representation
  by Barthelémy, 179–80, 183–85
  Foucault on, 239–40
  graticules in, 41, 48, 172, 193n. 1
  Alexander von Humboldt's use of,
    239–43
graticules, 41, 172, 193n. 1
Great Map of Mankind, 5, 14, 136, 417
Greece
  ancient, in Barthelémy's *Voyage du
    Jeune Anacharsis,* 178, 180–81, 182–
    84
  Enlightenment in, 11, 178, 281
Green, John, 187
Greenland, 95–98, **97**
Gregory, Derek, 14, 113n.
Grotius, Hugo, 98, 113n. 3
Grove, Richard, 422
Guernsey, 403
Guiana, 71

Habermas, Jurgen, 8–9, 393n. 18
Haeckel, Ernst, 307
Hakewill, George, 422
Hakluyt, Richard, 70
Hakluyt Society, 96
Hale, Sir Matthew, 426
Halevy, Leon, 150
Halle, Jean-Noel, 95
Hamburg, 404
Hampson, Norman, 6
Hare, Henry, 75
Hariot, Thomas, 70
Harraway, Donna, 16
Harrison, John, 167
Harvard, 108, 293n
Harvey, David, 13
Hawaii, 209, 299
Hazard, Paul, 6, 103
Hebrew
  images of creation in, 53, 83
  Tetragrammaton, 53–54, 56
Hegel, Georg Wilhelm Friedrich, 14
Heidelbergische Jahrbüecher der Literatur,
  322, 328
heliocentrism, 34–35
Hellenism, 166
Hellespont, 180
Hercules, 45
Herder, Johann Gottfried de, 79, 80
Herefordshire, 347, 353, 356–61

Herefordshire Agricultural Society, 358
Hermes Trismegistus, 49
hermeticism, 49–50
Herodotus, 137, 179
Heroes, Age of, 132. *See also* progress
Heyleyn, Peter, 74
hierglyphs, 55
Hindustan, 189
history, universal, 70, 86, 131. *See also*
    progress; stadial theory
Hokkaido, 206
Holkham Hall, 349
Holland, Henry, 328, 349, 351
Holstein, 45, 400
hope, geography of, 9
Hopkinson, John, 72
Hôpital General (Paris), 377
Horn, George, 98, 113n. 4
horticulture
    geographies of, 347–49
    Thomas Andrew Knight on, 355–61
    political significance of, 347–49
    practices of, 347–49
    See also agriculture
Horkheimer, Max, 8–9
Hudson Bay, 96
Huelva, 403
Huet, Bishop, 75, 77, 79
Hulme, Peter, 8, 12, 13–14, 33
human nature, character of, 417–19
    language of, 5
    universal theories on, 109
human origins, ideas of, 30, 71, 95–97
Humboldt, Alexander von
    and the *Ansichten der natur,* 324, 327–
        34, 335
    and the *Atlas Géographique et Physique
        du Royaume de la Nouvelle-Espagne,*
        251, **254, 257, 258, 259, 260,
        262**
    on the body as an instrument, 288–
        89
    and Aime Bonpland, 246, 285, 323
    cartographic methodology in the
        work of, 123, 239–65
    contemporary reception of the work
        of, 286, 321–27, 328–36
    discussion of, by Cuvier, 286
    discussion of, by Dettelbach, 238–39
    empiricism in the work of, 240–65,
        285
    *Essai Politique sur le Royaume de la
        Nouvelle-Espagne,* 326, 327–34,
        335–36

Essai sur la Géographie de Plantes, 261,
    263, **264**
as 'founding father' of modern geog-
    raphy, 1, 236–37, 319–20, 405,
    415
geognosy, 241, 244, **260,** 261, 269n.
    10, 270n. 15
graphic representation in the work of,
    123, 238–39, 252–65
historiographical interpretations of,
    319–21
ideas about the cosmos in the work
    of, 265–66
on instrumentation, 245–46
on interior structure and function,
    248–49
on isolines, 245
and *Kosmos,* 263, 278, 320, 322–23,
    325, 334–36, 418
on map of New Spain, 245
on mapping, 240, 250–66
on the moral history of man, 262
on movement, change and distribu-
    tion, 242–44
on paradise question, 72
plant geography, 241, 243–44, 247–
    49, 261, **264**
on precious metals, 244, **254,** 255,
    269n. 9
on the problem of commensurability,
    267
on rattlesnakes, 269n. 11
*Relation Historique du Voyage aux Re-
    gions Equinoxiales du Nouveau Conti-
    nent,* 323, 326, 328–29, 334, 335
on scale, landscape, and the natural
    region, 246–48
on theory, cause and explanation,
    240–242
on travel accounts, 270n. 12
travels of, in South America, 246,
    282, 285, 288–89
use of statistics by, 237, 245
visual thinking in work of, 122, 239,
    253–65
Humboldt, Wilhelm von, 285, 291,
    415
Hume, David, 106, 108, 283–84, 300,
    305, 383, 393n. 21
Hungary, Enlightenment in, 11
Hunters, Age of, 131. *See also* progress
Husbandmen, Age of, 131. *See also*
    progress
Hutcheson, Frances, 108, 109

Hutton, James
and gradualism, 307
the methodology of, 309–12
and *Theory of the Earth,* 305, 307,
309–12, 418, 423
hydrography, 240

Iceland, 95, 98
Icelandic Chronicle, 96
iconography, 35, 47
iconosophy, 61
ideas, geography of
'social theory of,' 9
See also geography, of ideas
ideology, history of, 137
ignegno, idea of, 295, 301–3
illumination, idea of, 56
imaginative geography. *See* geography,
imaginative
improvement
in Georgian England, 347–65
ideas of, 18, 345–47
practices of, 345–47
Independence, Declaration of, 419
inductivism, 278, 295–309, 310, 312
Industrialism, Age of, 148–49. *See also*
progress
Institut National des Sciences et Arts, 137,
144, 146
instrumentation
the body used as a form of, 42, 288–
89
in work of Cook, 287
in work of Humboldt, 245–46
in work of Sennebier, 288–89
Invege, Agostino, 72–73
Ireland, 176, 400, 401
isochronic mapping. *See* mapping,
isochronic
isolines, 244
Humboldt on, 245–46
isothermal mapping. *See* mapping,
isothermal
Italy, Enlightenment in, 11, 12, 38, 416
Iter Extaticum Celeste, 54. *See also*
Kircher

Jacobinism, 110
Jacob, Christian, 47, 64n. 4
Jacob, Margaret, 13, 406
Jamaica, 348
Japan, 70, 205
Jefferson, Thomas, 11, 245
Jesso, 205, 215, 217

Jesuits, 34
Johnson, Nuala, 113n
Johnson, Samuel, 3, 172
Jomard, Edme, 237
Jordanova, Ludmilla, 8, 12, 13–14, 33
Journet, Noël, 101
Jupiter, Temple of, 185

Kafka, Franz, 295
Kamchatka, 208, 215, 217
Kames, Lord, 106–7, 114n. 8, 427
Kamschatdale, 211, 226
Kant, Immanuel, 2, 5, 9, 407
Karafuto, 205, 232
Kastri Bay, 215, **223,** 224, **227**
Kempe, Andreas, 79
Kensington, 362
Kent, 403
Kenya, 69
Kepler, Johannes, 39
Kew, 346, 348–49, 365
Kinsale, 400
Kircher, Athanasius
as an advocate of 'new' science, 54
and the *Arca Nöe,* 60–61
and the *Ars Magnae Lucis et Umbrae,*
56, **57, 58,** 59
and the *Ars Magna Sciendi,* 54
as a chorographer, 50–63
as a cosmographer, 29, 33, 50–63,
64n. 7
and hermetic philosophy, 54–55
and the *Iter Extaticum Celeste,* 54
and the *Mundus Subterraneus,* 51, **52,**
55
and the museum as a site of global
display, 53
on paradise, 75
Knight, Richard Payne, 355
Knight, Thomas Andrew
and horticultural science, 341, 355–
61
reputation of, in London, 361–65
knowledge
body as a site of, 19, 221, 277–78,
283–86, 288–93
courts as a site of, 4, 17, 48
formation of, through graphic representation, 184–85
gaps in, 295–316
geography of, 14–15, 16–17, 21, 173,
199–211, 321, 341–44
in the field, 277–78
laboratory as a site of, 4, 17

madness
  Foucault on, 374–76, 390nn. 10, 11,
    392nn. 12, 16
  idea of, 18, 342, 374–78
  sites of, 342
  See also reason; unreason
Madrid, 147, 404, 405
Malaspina, Alejandro, 204
Malta, 297
Malte-Brun, Conrad, 85, 237, 328,
  334
Malthus, Rev. Thomas, 135
Manchuria, 207, 214
Mankind
  Great Map of, 5, 14, 136, 417
  history of, 295
  moral history of, in writings of
    Alexander von Humboldt, 263
Man, Science of
  geography as an element in the, 95
  place of language in the, 18, 68, 80
Mantcheoux, 225
mappaemundi, 70
mapping
  as emblematic of encyclopedic
    knowledge, 122, 165
  ethnic, 14
  as a geographical practice in the En-
    lightenment, 7, 12, 14, 17, 20–21,
    80, 121–23, 165–85
  idea of 'native,' 14, 225, 297
  ideas of, 121–23, 186–95, 244
  isochronic, 404–5, 406, 409
  isothermal, 253
  memoirs and, 187–90
  as a metaphor in the Enlightenment,
    7, 14, 20–21, 121, 186
  thematic, 250–53
  topographic, 250–53
  See also cartography; ethnographic
    navigation
Markus, Thomas, 382–84, 389–90,
  394n. 28
Marly, 36, 46, 64n. 4
Marsden, William, 63
Marshall, William, 351
Marx, Karl, 149
Mason, William, 351
mathematics, 51, 61, 62, 237,
  246–47. *See also* cosmography,
    mathematical
Mather, Cotton, 186
Mauritius, 84, 85
Mauss, Marcel, 204

Mayer, Tobias, 186, 256
Mayhew, Robert, 87n
measurement, place of in Enlighten-
  ment knowledge, 19, 221, 244
medicine, 10
Medusa, 281
memoirs, mapping and the production
  of, 187–90
memoria, idea of, 301–3
Men, Age of, 132. *See also* progress
Mendelssohn, Moses, 5
Mesopotamia, 60, 73, 74, 81
Messina, Straits of, 51, **52**
metals, trade in, 244, **254,** 255, 269n. 9
Mexico, 51, 147, 247, 251, 259, **262,**
  332
Middlesex, 353
Milan, 38
military campaigning, geography's use
  in, 2
Millar, John, 135
Milton, John, 35, 420
Minerva, 128, 183, 185
mining, geography's use in, 3
Minorites, 36
missionaries, 176
Mississippi, River, 36, 48
Mitchell, John, 188
moderatism, 31
modernity
  ideas of, 190–92
  map of, 190–93
monarchy, authority of, 33
*Monatliche Correspondenz zur Befoer-
    derung der Erd- und Himmelskunde,*
  322, 328
Moniteur Universel, 324, 328, 334
monogenism, 98–105
Montenegro, Benito Jeronimo Feyzóo y,
  407
Montesquieu, Charles-Louis de Secondat, Baron de la Brede et de, 95,
  105, 107, 112, 131, 300, 416
  and Turgot, 129–31
Monthly Review, 322, 351
Montrose, 373
moral integrity, 9
moral philosophy
  in America, 108–9
  as a matter of Enlightenment, 9, 30,
    93–96, 341–42
  in work of Samuel Stanhope Smith,
    104–9
Morea, 40

la Peyrère on, 99, 100–103
Lord Kames on, 106–7
Pontey, William, 349
Pope, Alexander, 408, 420
Popkin, Richard, 99, 101–3
population, as a category of analysis, 133, 135
Porter, Roy
  contribution of, to thinking on Enlightenment, 10, 345, 375
  on madness, 375–76
Portland, Duke of, 351
Portugal, King of, 399, 403–4
Portuguese, role of in voyages of discovery, 70, 83
'positive geography,' 186
postcolonialism, 296–97
poststructuralism, 296
Pownall, Thomas, 189
Pozzo, Andrea, 59
Praeadamitae, 100–103. *See also* Peyrère
PreAdamites, 101–4
Prichard, James Cowles, 427
Priestley, Joseph, 134, 144, 347
primitives. *See* primitivism
primitivism
  'hard,' 209–10
  ideas about, 19, 20, 209–10
  la Peyrère on, 'soft,' 209. *See also* barbarism
Princeton, 105, 110, 113
'principle of attachment,' 202, 213
progress, idea of
  as a matter of theoretical history, 80, 103, 125–52
  human, and the Enlightenment in relation to, 3, 9, 14, 121, 125–52, 153n. 1
Ptolemaic theories, 39, 50
Ptolemy (Claudius Ptolemaeus), 33, 64n. 5
public house, as a site of knowledge, 17
public sphere
  geography and the, 2, 13
  place of, in the Enlightenment, 9, 13, 17, 345, 393n. 18
Pufendorf, Samuel, 112
Purchas, Samuel, 70
Pyrrhonism, 99–101
Pyrrho of Elis, 99

Quakers, 94, 421, 427
Quarterly Review, 322, 328, 334
Quatrefages, Armand de, 328

race, ideas on, 30, 98
racial geography, 30–31, 96–99
racial variation, 93–99
Raleigh, Sir Walter, 71, 73, 82
Ramusio, Gianbattista, 38, 41
Raspe, Rudolph Eric, 177
rational geography, 3
rationality
  ideas about, 3, 8–10, 16, 72, 341–42
  situated nature of, 16, 341. *See also* reason
rational religion, 30, 92–113
Ray, John, 424, 425–26
reason
  age of, 8–10, 13, 342
  Edinburgh as 'the city of,' 378–82
  as an element of geographical narrative, 177
  as an element in Kircher's work, 56
  'gaps in,' 295–316
  geography of, 15, 387–90
  human, 9
  place of, in Enlightenment, 8–10, 13, 15, 29, 296–309
  sites of, 15
  triumph of, 387–90
Reconnaissance
  Age of, 168
  in the geographical work of Abbé Barthelémy, 178–90
  idea of, 168, 168–69, 172, 175–78, 193n. 2
  mapping as a practice of, 165, 168–69, 175–78
Reformation, 133
region, natural, in work of Alexander von Humboldt, 247–48
Reid, Thomas, 2–3, 108, 114n. 9
Relation du Groenland, **97,** 99–100. *See also* Peyrere
Relation Historique du Voyage aux Regions Equinoxiales du Nouveau Continent, 323, 326, 328–29, 334, 335. *See also* Humboldt, Alexander von
religion
  place of, in historiography of Enlightenment, 7, 93–96.
  and reasonable Christianity, 101
  See also Anglicanism; authority, sacred; authority, scriptural; Catholicism; Christianity; Evangelicalism; moderatism
Renaissance, 85, 132
Rennell, James, 169, 187

Sennebier, Jean, 287–88
Seward, Anna, 350
Seychelles, 68, 84
Shapin, Steven, 15
shepherds, age of, 131. *See also* progress
Siberia, 207, 290
Siccar Point, 312
Siena, 44
Sinclair, George, 350, 352
Sitwell, O. Francis George, 189
Smith, Adam, 108, 135, 157n. 7, 399n. 21
Smith, Bernard, 238
Smith, James Edward, 350, 351
Smith, Preserved, 6
Smith, Samuel Stanhope, 31, 103–13
sociability, geography and the promotion of, 341
Société de Geographie de Paris, 151
Society Isles, 83
Sodom, 406
Soulavie, Gerard de, 247
South Carolina Gazette, 404
South Seas. *See* Pacific
Spain, Enlightenment in, 147, 406
    King of, 404
    Royal Academy of History of, 399
stadial theory
    and the 'Great Map of Mankind,' 3, 5, 14, 103
    in the writings of Saint-Simon, 146–51
    in the writings of Turgot, 128–36
    in the writings of Volney, 136–46
statistics
    and graphic representation, 176, 237, 244
    the language of, in the Enlightenment, 133, 154n. 7, 176, 237, 244, 266
Steno, Nicolaus, 247
stewardship, idea of, 426
Stewart, Dugald, 108
Stillingfleet, Benjamin, 356
Stockholm, 365, 404
Stockholm Agricultural Society, 365
Stoddart, David, 1
Suffren Bay, 214
Sulemann II, 84n. 5
Superior, Lake, 169
survey, topographical, 172–73, **174,** 191, 193n. 2
Sweden, 11, 176, 365
Swedenborg, Emanuel, 81

Swift, Jonathan, 167, 177
Switzerland, Enlightenment in, 11
Syria, 136, 138. *See also* Volney, *Voyage en Egypt et en Syrie*

Tagus, River, 399
Tahiti, 4, 82–83, 200, 287, 417
Tartary, Bay of
    descriptions of people near, 204, 208–9
    Lapérouse comments upon, 214–17
    navigation in, 214–17
taxonomy, 186. *See also* classification
Tchoka, 217, **222**
textbooks, 2
theatre, the world as a, 14, 298
theatrum belli, idea of, 230
theatrum mundi, idea of the Pacific as, 210
thematic mapping. *See* cartography; mapping, thematic
Theory of the Earth, 305, 307, 309–12, 418, 423. *See also* Hutton
Theseus, Temple of, 183
Thevet, Andre, 40, 41–42, 51
Thierry, Augustin, 147, 149
Thomas Aquinas, Saint, 406
Thomas, Keith, 420
Tigris, River, 70, 75
time, 'abyss of,' 309
topographic mapping. *See* cartography; mapping, topographic
topography, 172
Tracy, Antione Destutt de, 137
trade, role of in Enlightenment, 4
translation, discursive practices of, 211–14, 218, 230–33. *See also* commensurability
travelling, as a geographical practice, 3, 13, 17, 19, 21, 80, 103, 202–3, 247, 277–79, 366, 418. *See also* ethnographic navigation
travel literature, 177, 184
triangulation, 191
trust
    idea of, 5, 282
    role of in Enlightenment knowledge, 5, 16, 19, 282–84
truth
    in development of geographical knowledge, 200, 282–84
    geography of, 16, 200
    ideas of, 9, 19
    and scriptural authority, 30